Purposeful Program Theory

Effective Use of Theories of Change and Logic Models

SUE C. FUNNELL AND

PATRICIA J. ROGERS

JOSSEY-BASS
A Wiley Imprint
www.josseybass.com

Published by Jossey-Bass

A Wiley Imprint

One Montgomery, Ste. 1200, San Francisco, CA 94104 —www.josseybass.com

Jossey-Bass books and products are available through most bookstores. To contact Jossey-Bass directly call our Customer Care Department within the U.S. at 800-956-7739, outside the U.S. at 317-572-3986, or fax 317-572-4002.

Jossey-Bass also publishes its books in a variety of electronic formats. Some content that appears in print may not be available in electronic books.

Library of Congress Cataloging-in-Publication Data

Funnell, Sue C., date.
    Purposeful program theory : effective use of theories of change and logic models / Sue C. Funnell and Patricia J. Rogers.
        p.   cm.
    Includes bibliographical references and index.
    ISBN 978-0-470-47857-8 (pbk.); ISBN 978-0-470-93988-8 (ebk);
    ISBN 978-0-470-93989-5 (ebk); ISBN 978-1-118-00433-3 (ebk)
        1. Evaluation research (Social action programs)   I. Rogers, Patricia J.   II. Title.
    H62.F855   2011
    001.4'2—dc22

                                                                            2010034870

Printed in the United States of America

FIRST EDITION

PB Printing   10 9 8 7 6 5 4

CONTENTS

PART ONE

Key Ideas in Program Theory

PART THREE

Developing and Representing Program Theory

FIGURES, TABLES, AND EXHIBITS

Figures

Tables

Exhibits

MANY PEOPLE HAVE helped us along the journey to this book. Our work builds on the contributions of the pioneers and innovators in program theory evaluation and those whose work underpins key concepts in program theory. The work of Dan Stufflebeam, Carol Weiss, Joseph Wholey, Claude Bennett, Len Bickman, Carol Fitz-Gibbon, and Lynn Morris laid the foundations for program theory. Benjamin Bloom, Thomas Hastings, and George Madaus's taxonomy of educational objectives, when linked with Bennett's Hierarchy, gave rise to the notions of outcome chains, and Tom Hastings's important article, "Curriculum Evaluation: The Why of the *Outcomes*," expanded thinking beyond a focus on outcomes. Bryan Lenne of the New South Wales Program Evaluation Unit contributed significantly to the development of the matrix approach to program theory and the classification of program archetypes. More recently, we have learned much from the work of Michael Patton, Rick Davies, and Boru Douthwaite exploring the use of network theory and systems approaches; the development of realist evaluation and realist synthesis by Ray Pawson and Nick Tilley; and outcome mapping by Barry Kibel, Terry Smutlyo, Fred Carden, and Sarah Earl in terms of using program theory for complicated and complex situations and interventions.

Over more than twenty-five years, we have had the good fortune to work with colleagues and clients who have stretched our thinking about program theory. Sue particularly acknowledges those whose work and insights have contributed to the development of the matrix approach: Carolyn Wells for development of the concept of attributes of success as part of program theory matrices; Steve Baxter for his contribution to thinking about managing external risk factors that affect outcomes and the implications of that for developing a program theory; and Larraine Larri for her assistance with portraying and explaining program theory matrices. Sue also acknowledges the contributions of those who helped identify program theories for particular program archetypes: Bryan Lenne for his contribution to the development of the generic theory for advisory programs drawing on the work of Claude Bennett; Ross Homel for his work on regulatory programs and contribution to the development of the sticks archetypal program theory; Harry Hatry's work on performance incentive programs that contributed to the development of the carrots archetype; Alison Matthews for her case management approach that drew attention to the need to have models that could accommodate different outcomes for different people using different paths, types of service, and activities; Greg Masters for his work on service delivery programs; and Barry Smith for his work on community capacity-building programs. She acknowledges as well the valuable work of Ellen Taylor-Powell, Larry Jones, and Ellen Henert of the University of Wisconsin Extension Center who, through their online course, Enhancing Program Performance with Logic Models, drew her attention to the potential usefulness of research-based theories of change and prompted her in-depth exploration of those theories.

Patricia's thinking about program theory owes a particular debt to her mentor, Carol Weiss, and colleagues Tim Hacsi, Tracey Huebner, and Anthony Petrosino at the Harvard Project on Schooling and Children, where she completed a postdoctoral fellowship supported by the Spencer Foundation. She is also grateful for the insights about program theory gained from discussions and evaluation projects with Brad Astbury, Fred Carden, Margaret Cargo, Jane Davidson, Rick Davies, Julie Elliott, Gerald Elsworth, Delwyn Goodrick, Irene Guijt, Ernest House, Bron McDonald, Michael Patton, Ray Pawson, Kaye Stevens, Gill Westhorp, Bob Williams, and Jerome Winston. Thanks also to Susie Elliott, Caitlin Nash, and Russell Stanbrough for their help with editing and manuscript preparation.

The book has benefited considerably from advice and encouragement from Andy Pasternack and Seth Schwartz at Jossey-Bass and from the constructive feedback on earlier drafts of the book from Fred Carden, Jane Davidson, Ernest House, Steve Montague, and especially Michael Patton. We bear responsibility for all remaining deficiencies.

Finally, we thank our families for their support and encouragement throughout the writing of this book and the staff and other stakeholders of the many programs from which we drew the experience on which this book is based.

To Peter, my partner and soulmate,
for his encouraging, good-humored, and practical support
throughout the writing of this book.
—SUE

To Paul, my dear man,
for all he does and has done to make this work possible.
—PATRICIA

SUE C. FUNNELL is a director of Performance Improvement, a company she established in 1992. She has more than thirty-five years of experience in program design, evaluation, and performance measurement. Since the 1980s, she has been one of the key contributors to the development, dissemination, and use of program theory in Australia. She has supported local, state, national, international, and global government and nongovernment organizations in developed and developing countries, and she has successfully used program theory for evaluation, monitoring, planning, and organizational learning. Her program theory work has been with the Joint United Nations Programme on HIV/AIDS, United Nations Center for Human Settlements, World Bank, Australian National Audit Office, Australian Department of Finance, and government agencies and departments in Australia and New Zealand. She has addressed communication technology; education, training, and leadership development; employment; energy and water; environmental protection; evaluation policy; family, youth, and community services, including alcohol and other drugs, disability, housing, mental health, problem gambling, volunteering, and welfare assistance; industry development;

legal systems; natural resources management; occupational health and safety; primary industries; and roads. Funnell, a past president of the Australasian Evaluation Society (AES) and now an AES Fellow, has been awarded the AES Evaluation Training and Services Award for outstanding contributions to the profession of evaluation.

PATRICIA J. ROGERS is Professor of Public Sector Evaluation at the Royal Melbourne Institute of Technology, Melbourne, Australia. She has worked in public sector evaluation and research for more than twenty-five years, with government and nongovernment organizations (international, national, state, and local) across a wide range of program areas, including agriculture, community development, criminal justice, early childhood education, health promotion, Indigenous housing, international development, and legal aid. She has worked on projects with the United Nations Development Programme, World Bank Institute, Network of Networks on Impact Evaluation, U.S. Department of Energy, the Office of the Presidency (South Africa), the Public Service Commission (South Africa), and the Development Bank of Southern Africa. She has presented keynote addresses at conferences of the Australasian, Aotearoa/New Zealand, European, United Kingdom, South African, and Swedish evaluation societies and associations and is on the editorial boards of the journals *Evaluation* and *New Directions for Evaluation.* She has been awarded the American Evaluation Association's Myrdal Award for Evaluation Practice, the Australasian Evaluation Society's Evaluation Training and Services Award for outstanding contributions to the profession of evaluation, the AES Caulley-Tulloch Prize for Pioneering Literature in Evaluation, and (with Sue Funnell) the AES Best Evaluation Study Award.

THE 1920S ENTREPRENEUR Carl Weeks once wrote, "If you can dream it, you can build it." This is the key idea that underpins program theory. Having a vision of where we are going and some clarity about how we plan to get there can help us work together to achieve our goals, and learn from both success and failure.

WHAT PROGRAM THEORY IS

A program theory is an explicit theory or model of how an intervention, such as a project, a program, a strategy, an initiative, or a policy, contributes to a chain of intermediate results and finally to the intended or observed outcomes. A program theory ideally has two components: a theory of change and a theory of action. The theory of change is about the central processes or drivers by which change comes about for individuals, groups, or communities—for example, psychological processes, social processes, physical processes, and economic processes. The theory of change could derive from a formal, research-based theory or an unstated, tacit understanding about how things work. For example, the theory of change underpinning some health promotion programs is that changes in perceived social norms lead to behavior changes. The theory of action explains how programs or other interventions are constructed to activate these theories of change. For example, health promotion programs might use peer mentors, advertisements with survey results, or some other strategy to change perceptions of social norms.

Program theory, under all its various labels, including "theories of change," "logic modeling," and "intervention logic," has grown in popularity over the past twenty years or so. Many government and nongovernment organizations across the world now encourage or require its use for planning, monitoring, and evaluating.

When done well, program theory can produce many benefits. It can develop agreement among diverse stakeholders about what they are trying to do and how, or identify where there are legitimately different perspectives. It can help to improve plans by highlighting gaps and opportunities for collaboration with partners. It can help to set realistic objectives. It can support the development of meaningful performance indicators to track progress and report achievements. It can be used to identify where and why unsuccessful programs are failing or what makes successful programs work, and how they might be reproduced or adapted elsewhere. It can provide a framework to bring together information from many sites, many projects, or many evaluations so that it is possible to learn from the past to improve the future.

Program theory, however, is not always done well. And when it is done badly, it misrepresents what an intervention does and what it can achieve. It can lead to monitoring systems and evaluations that produce an incomplete or distorted picture of what is happening and mistaken judgments about what is effective or efficient. It can demotivate staff and deflect attention from what is important to only what can be easily measured. It can silence important voices or fail to touch those who can act on it. It can take up time without adding value.

The promise of good program theory and the risk of bad program theory have motivated us to write this book. Over more than twenty years, we have worked with small and large organizations in countries all over the world; with municipal, state, and federal government agencies, and nongovernment organizations; on tiny local projects, multimillion-dollar national programs, and whole-of-government strategies; with service deliverers, policymakers, and funders; and in many sectors, including health, education, agriculture, justice, infrastructure, natural resources, community services, community development, and emergency management. Over this time, we have seen diverse approaches to program theory.

What we have learned from this experience, and from the expanding library of empirical research on program theory, is that program theory should be developed, represented, and used not in a formulaic way, but thoughtfully and strategically, in ways that suit the particular situation. We call this *purposeful program theory.*

PURPOSEFUL PROGRAM THEORY

Greek legend tells of the fearsome hotelier Procrustes who would adjust his guests to match the length of his bed, stretching the short and trimming off the legs of the tall. Guides to program theory that are too prescriptive risk creating such a Procrustean bed. When the same approach to program theory is used for all types of interventions and all types of purposes, the risk is that the interventions will be distorted to fit into a preconceived format. Important aspects may be chopped off and ignored, and other aspects may be stretched to fit into preconceived boxes of a factory model, with inputs, processes, outcomes, and impacts.

Purposeful program theory requires thoughtful assessment of circumstances, asking in particular, "Who is going to use the program theory, and for what purposes?" and, "What is the nature of the intervention and the situation in which it is implemented?" It requires a wide repertoire, not a one-size-fits-all approach to program theory.

Purposeful program theory also requires attention to the limitations of any one program theory, which must necessarily be a simplification of reality, and a willingness to revise it as needed to address emerging issues. As the American evaluator Daniel Stufflebeam (2001) has pointed out, evaluators who continue to use an unsuitable program theory are similarly at risk of creating a Procrustean bed for the evaluation.

OVERVIEW OF THE BOOK

The book is designed to help you assess your particular circumstances and develop, represent, and use program theory in appropriate ways. It has options at every stage and examples to help you decide which options to use and how to adapt them to your circumstances. Throughout the book, we draw on examples from our own work and the work of others. ("Our work" refers to projects we have done together and individually.) Each chapter includes exercises to try out new ideas and techniques.

If you are new to program theory, it will be most useful to read the chapters in sequence. If you have some experience or are coming back to the book during an evaluation, you can select the particular chapter you need.

Key Ideas in Program Theory

Part One sets out the key ideas of program theory and how it has developed over time. We explain in Chapter One the essential features of program theory, using the broad policy objective of eating an apple a day to keep the doctor away as an example of how program theory can be used in different ways to learn from success, failure, and mixed results. Chapter Two describes how program theory has developed over time and sorts out the confusion about the different terms that have been used. And Chapter Three introduces seven widespread myths about program theory and seven common traps to avoid.

Assessing Your Circumstances

A key message of this book is the need to approach program theory in a way that suits your circumstances. Therefore, Part Two examines how to analyze the intended uses of program theory and the nature of the situation and intervention.

We explain in Chapter Four why it is important to be clear about who is going to use program theory and for what purposes. A program theory that is useful for developing internal monitoring systems for incremental correction, for example, could be inappropriate for developing performance measures for external accountability. A theory to guide the design of an impact evaluation might not be sufficient to guide a process evaluation that aims to document an unfolding innovation. Being clear about the intended uses of program theory, reviewing this as circumstances change, and considering this when making decisions is an essential part of purposeful program theory.

Chapter Five discusses how to identify simple, complicated, and complex aspects of the program or policy and the situation in which it is being implemented. Program theory can be used for interventions that are simple; that is, they have a single implementing agency and a well-understood causal process that works pretty much the same everywhere. But most interventions have important complicated or complex aspects that program theory needs to address in order not to misrepresent how it works. The implications of complicated and complex aspects of interventions for developing, representing, and using program theory are addressed throughout the book.

Developing and Representing Program Theory

The chapters in Part Three focus on ways of developing and representing program theory.

Chapter Six discusses how to combine three approaches to developing a program theory. A deductive approach focuses on stated policies and procedures and previous research. An inductive approach builds from observing the intervention in action, reviewing previous observations of it, or observing similar interventions. A mental model approach works with stakeholders to articulate their tacit understandings of how the intervention works.

Chapter Seven sets out three steps to develop a theory of change. Step 1 is to undertake situation analysis to identify problems and opportunities and understand the causes and consequences of problems. Step 2 is to decide the program scope: agreeing which aspects of the problem—its causes and consequences—the program will focus on directly and primarily and which will be beyond the direct focus. The more complex the program is, the more fluid the boundaries should be. Step 3 is to articulate an outcomes chain that shows the assumed or hypothesized cause and effect or contingency relationships between immediate and intermediate outcomes and ultimate outcomes or impacts (both short and long term). In this chapter, we address each of these tasks by applying them to an employment program for mature workers, and we provide examples of how these can be done in different ways to suit any situation.

In Chapter Eight, we introduce a structured approach to developing the second part of the program theory, the theory of action, which spells out how the intervention is intended to activate the theory of change. For example, if a program aims to change health behaviors through increasing knowledge of their consequences, will this knowledge be achieved through a public advertising campaign, personal consultations from health professionals, viral marketing from peers, or some other activities? We introduce the program theory matrix: a structured approach that explores systematically the outcomes chain developed in the theory of change. For each of the outcomes in the outcomes chain, the matrix identifies the nature and quantity of program activities that are intended to achieve this and other factors that will affect whether and how well the outcome is achieved. It also defines what success will look like for the outcome. We continue with the example of an employment program for mature workers, introduced in Chapter Seven, to demonstrate the various components of a theory of action.

We look at different types of logic models in Chapter Nine that can be used to represent program theory. Pipeline models show an intervention as a linear series of boxes labeled something like "inputs, processes, outcomes, and impacts." Outcome chains, which show a series of results leading to the final impacts of interest, have the advantage of being able to represent more complicated and complex interventions where the activities occur throughout the causal chain and are not all present at the beginning of the process. Realist matrices focus on showing how interventions work differently for different groups or in different situations. We discuss what makes a good logic model, do some logic model makeovers, and review some technology for producing these models.

Chapter Ten discusses how to assess the quality of the program theory in terms of its internal coherence and its validity with respect to external considerations. A program theory can be poorly expressed, incompletely expressed, or just plain wrong. It is important to review it systematically during development and periodically throughout its use.

Resources for Developing Program Theory

The chapters in Part Four provide resources to help with the processes of developing and representing program theory. It can be helpful to draw on previous research and planning when developing the outcomes chain.

Chapter Eleven provides information about a number of theories of how change occurs for individuals, organizations, and communities. The theory of reasoned action (Fishbein and Ajzen, 1975), the theory of planned behavior (Ajzen, 1988), and the stages of change theory (Prochaska and DiClemente, 1983) are theories about changing behaviors of individuals. Empowerment theory (Perkins and Zimmerman, 1995) may relate to individuals, groups, or communities. Diffusion theory (E. Rogers, 1995) is largely about changing community behaviors (and behaviors of individuals en masse). Socioecological theory (Bronfenbrenner, 1979) is about mechanisms for change for individuals, families, groups, and communities and the interplay among all of those actors.

Network theory (Granovetter, 1973) is about how the relationships, networks, and connections among entities, and not just the characteristics of the entities themselves, affect outcomes. The entities could be individuals, organizations, special issues groups, or even whole countries. There are many other research-based theories of change, and the chapter lists some other potentially relevant theories that could be used as the basis for an intervention's specific theory of change.

Chapter Twelve outlines some common program archetypes that can be selected, adapted, and combined for particular situations. These include advisory, information, and education programs that seek to change individual behavior by informing decisions; "sticks and carrots," which work through incentives and sanctions; case management; community capacity development; and direct service delivery.

Chapter Thirteen provides examples of variations on pipeline and outcomes chain logic models.

Using Program Theory for Monitoring and Evaluation

The final part of this book describes how to use program theory specifically for monitoring and evaluation.

Chapter Fourteen explains how to use program theory to identify what aspects of the intervention, the context, and results should be measured and how to use key evaluation questions to focus an evaluation in terms of data collection, analysis, and reporting. Program theory can help to structure a coherent narrative report and a focused analysis, whether reporting the results of a single evaluation or bringing together data from many studies. We provide some suggestions on ways to do this for small and large evaluations.

Even when there is credible evidence that outcomes have occurred, can we be confident that an intervention has caused them or at least contributed to them together with other factors? In recent years there has been a vigorous debate about the suitability of different methods and designs to address the issue of causal analysis. In Chapter Fifteen, we set out a three-part framework for causal analysis when using program theory that can bring to bear the full range of research designs and methods for causal analysis. The starting point is looking for congruence of results with those predicted by program theory. The second part is finding relevant comparisons that indicate the difference that the intervention has made. These can include creating a control group or a comparison group or making other relevant comparisons. The third part is checking out alternative explanations for the results and exceptions to the patterns.

Chapter Sixteen describes ways to bring together information across the different levels of a program theory, or across several interventions that use the same program theory, and how to report this coherently and effectively.

TAKING A STRATEGIC AND ADAPTIVE APPROACH

Program theory can be developed, represented, and used in many ways. Throughout this book, we invite you to take a purposeful approach to program theory, matching it to your situation, checking how it is going, and adapting it as needed to ensure that it contributes to improved interventions and the outcomes you seek.

## PART 1

Key Ideas in Program Theory

1

The Essence of Program Theory

A N APPLE A DAY KEEPS the doctor away—or does it? Thinking about how we would find out if this is true and how we might use those findings shows the value of program theory. In this chapter, we set out the key ideas in program theory and show how program theory can be used to learn from success, failure, and mixed results to improve planning, management, evaluation, and evidence-based policy.

EVALUATION WITHOUT PROGRAM THEORY

Let us imagine that we have implemented a program based on the broad policy objective of an apple a day in order to keep the doctor away. This program, which we dubbed An Apple a Day, involves distributing seven apples a week to each participant. A representation of this program without program theory would simply show the program followed by the intended outcome of improved health (Figure 1.1).

Figure 1.1   **An Evaluation of An Apple a Day Without Program Theory**

This is what is often referred to as a black box evaluation: one that describes an evaluation that analyzes what goes in and what comes out without information about how things are processed in between.

ORIGINS OF "BLACK BOX"

Different sources have been suggested for the term *black box*. The current Wikipedia entry for *black box* traces the term, when used for flight data recorders, to World War II Royal Air Force terminology, when prototypes of new electronic devices were installed in airplanes in metal boxes, painted black to avoid reflections and therefore referred to as black boxes.

Former electronics buff turned evaluator Bob Briggs, on the American Evaluation Association's discussion list EVALTALK (Briggs, 1998), reminisced how electronics manufacturers would often cover components with opaque material to prevent consumers from "opening the black box" to see how it worked (and assembling their own version more cheaply). The parallel with evidence-based practices is useful: program theory aims to help policymakers and practitioners "open up the box" of successful programs to understand how it works rather than having to buy the whole package and plug it in.

However, as the evaluator and author Michael Quinn Patton (1998) pointed out in the same EVALTALK thread, the term can be seen as inappropriate: "Most uses of 'black box' or 'black box design' carry a negative connotation. The association of 'black' with negativity is what can be experienced as offensive, or at least insensitive" (Patton, 1998). He suggested using instead terms such as *empty box, magic box,* or *mystery box designs* to describe evaluations without program theory.

It can be difficult to interpret results from an evaluation that has no program theory. For an intervention that involves a discrete product for individuals, an experimental or quasi-experimental design might be appropriate for the evaluation. We will assume that people have been assigned to either a treatment group, who received the program, or to a control group, who went onto a waiting list to receive the program later if the evaluation shows it is effective. "Keeping the doctor away" has been operationalized as "maintaining or achieving good physical health." Data collection has been carefully designed to avoid measurement failure of outcome variables, with adequate sample size, appropriate measures of health, and systems in place to avoid accidental or deliberate data corruption.

Despite careful evaluation, it can be impossible to interpret evaluation results correctly in the absence of program theory. If the program failed to achieve significant differences in health outcomes between the groups (apple versus no apple), it might seem that the policy does not work—but it might also be that it has not been implemented properly. Maybe the apples were delivered but not eaten, or maybe they were too small, or too unripe, or too overripe to work as expected. Although the evaluation might include some measures of the quality and extent of implementation, it can be hard to know what aspects should be included unless there is a program theory.

An evaluation using program theory would identify how we understand this program works and what intermediate outcomes need to be achieved for the program to work. This allows us to distinguish between implementation failure (not done right) and theory failure (done right but still did not work). Without program theory, it is impossible to know if we have measured the right aspects of implementation quality and quantity.

If the results showed that the program seemed to have succeeded, as the treatment group had significantly better outcomes than the no-treatment group, we might also have trouble using these results more broadly. If we do not know what elements of the policy are important, we can only copy it exactly for fear of missing something essential. It does not provide any guidance for adapting the policy for other settings.

Finally, if we had mixed results, where the policy worked on only some sites or for some people, we might not even notice them if we were looking only at the average effect. If we did see differential effects in different contexts (for example, for men compared to women, or in urban areas rather than rural areas), an evaluation without program theory leaves us in the position of having to do simple pattern matching (for example, using the policy for the groups or sites where it has been shown to work) but with little ability to generalize to other contexts.

EVALUATION WITH PROGRAM THEORY

If we used a program theory approach, we would try to understand the causal processes that occur between delivering apples and improved health. We might start by unpacking the box to show the important intermediate outcome that people actually eat the apples. The logic model diagrams in Figure 1.2 show this: one in the form of a pipeline model and one as an outcomes chain. The pipeline logic model represents the program in terms of inputs, processes, outputs, and outcomes. The outcomes chain model shows a series of results at different stages along a causal chain.

Although these look like many logic models that are used regularly in evaluation, they are not much of a theory; rather, they are more like a two-step

---

Figure 1.2   Simple Pipeline and Outcomes Chain Logic Models

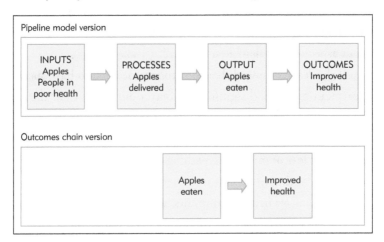

process, as Mark Lipsey and John Pollard (1989) called it, that identifies an intermediate variable without really explaining how it works. These models make it clear that eating the apples is understood to be part of the causal chain (rather than some other variable, such as social interaction with the apple deliverer or physical exercise from playing with the apples). But they do not explain how delivering apples leads to people eating apples or how eating apples improves health.

A plausible explanation would be that delivering apples increases the availability of fresh fruit, which leads to the apples being eaten, which increases the amount of vitamin C in the diet, which improves the physical health of participants. This is only one possible explanation, of course. Figure 1.3 shows this explanation as both a pipeline logic model and an outcomes chain.

The diagrams in Figure 1.3 represent a program theory that articulates the causal mechanisms involved in producing the two changes (changed behavior and changed health status). The first change relates to participants' willingness to act in the way the program intended and the second to the impacts of their actions. For many programs, it can be helpful to articulate both types of changes in the program theory.

Figure 1.3   **A Logic Model Showing a Simple Program Theory for An Apple a Day Based on Improved Vitamin Intake**

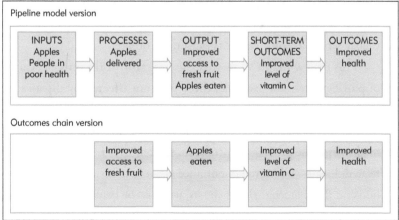

Learning from Failure

An evaluation based on this program theory would collect data about changes in access to fresh fruit, apple eating behavior, and nutritional status, as well as overall health. If the intended outcomes have not been achieved, we could work through the causal chain to see where it has broken down. If the apples were not even delivered, there is obvious implementation failure; if they were delivered but not eaten, then our theory of how to engage people in changing their behavior seems not to work. Similarly, if the expected health benefits had not been achieved, we would start by seeing if it was because the apples had not been eaten. If the apples had been delivered and had been eaten but without producing health improvements, then we have a problem with the theory of change that underpins the program. Based on these results, one option would be to reject the theory and look at other ways of improving health. Another would be to look at dosage: maybe vitamin C levels increased, but not enough to make a difference.

Learning from Partial Success

Developing a program theory also helps clarify differential effects, learning from those participants for whom the program was effective. The simple program theory is based on the assumption that the apples are both necessary and sufficient—that is, the apples will lead to good health in all circumstances and without contributions from other factors. Developing a more complicated logic model would focus on the differential effects we might expect for different types of participants, and we would collect and analyze data to examine these. Disaggregating the data would investigate whether the theory works in some contexts but not in others.

This review might show that the program works only for certain types of participants—for example, those who are affected by diseases related to inadequate nutrition. For people affected by infectious diseases, apples by themselves might not be enough to improve health. Based on these results, we might target the program to people most likely to benefit: those with nutrition-related diseases. Given the importance of the interaction between the intervention and the characteristics of clients, it would be helpful to revise the theory of change and its logic model to show this complicated causal path.

If the program works for some groups but not for others or at some sites but not others, it is important to try to understand why by identifying possible explanations and then checking these out empirically. For example, if the program worked for men but not for women, it might be because of differences in labor force patterns which affected access to fresh fruit or to differences in nutritional needs related to pregnancy. Finding exceptions to the pattern (the men who did not improve and the women who did) would provide more evidence to test these emerging program theories.

Learning from Success

Program theory has another benefit when an evaluation finds that something works: it helps in adapting the intervention to new situations. To be useful for evidence-based policy and practice, a program theory evaluation needs to identify the causal mechanism by which it works and determine whether this is different for different people and in different implementation contexts.

To explore this use, imagine that the evaluation has found that the program theory works: people are healthier when they eat an apple a day. Now the job is to implement a new program based on this evidence. In this case, the goal is not to understand failure but to understand success. Apples might produce these effects through quite different theories of change, which would lead us to quite different intervention theories and different program activities to suit the context. We would immediately have many questions about the statement. Does it work for everyone? Does it have to be a particular variety of apple (Granny Smiths? crab apples?), or does it apply to all varieties? What if apples are not available? Can we substitute other fruit, or apple juice, or vegetables? Would red onions work as well as red apples? An evaluation without program theory would reveal only that it works, with no guidance for how to translate the findings to a particular situation. Without this guidance, we can only blindly copy everything. With this guidance, we can understand how we might adapt it and still achieve the intended results.

We previously sketched out a program theory with a theory of change of providing a good source of vitamins in diets that are otherwise deficient. To test this out if we were implementing it would require data about people's nutritional status through either direct measures or relevant indicators so we could see if there was any change and also to identify the people we would expect to get the most benefit from the program. We would want to check that they actually ate the apples. And we would want to rule out alternative explanations by finding out if there had been other changes in their diets that might have contributed to changes in their nutrition. If this is the case, then other types of fruit are likely to be equally effective. In a country where apples are hard to obtain or expensive, distribution or subsidization of local fruit is likely to be an effective program, at least for people at risk of nutritional deficiency, if it is implemented well.

But maybe this is not how it works at all. Maybe it is not about the flesh or juice of the apples but their skin. The skin of apples contains a plant-based chemical called quercetin. Some research studies have suggested quercetin may help to prevent cancer, heart disease, and inflammation of the prostate. An evaluation would look at the intake of quercetin from various sources and outcomes in terms of these specific diseases, focusing on outcomes for people at risk of these diseases. If apples were not available, another source of quercetin could be used. Red onions, a rich source of quercetin, might be an effective substitute—an adaptation of the program that would not be immediately obvious if we were thinking only about fruit.

Another possible explanation focuses on apples as a substitute for high-calorie, low-nutrition snacks. Perhaps apples improve health by helping to reduce obesity as people stop eating potato chips and doughnuts and choose apples instead. An evaluation of this possibility would look at what people were eating in addition to apples and whether there had been a decline in their consumption of junk food. It also might measure short-term outcomes such as body mass index (BMI) and percentage fat, which have been linked to subsequent longer-term health outcomes. The evaluation would have to take into account criticisms that have been made of BMI as an indicator and predictor of health. Making other low-calorie snacks such as carrots and celery readily available might be equally effective. Figure 1.4 shows how these three different change theories might plausibly explain why the policy works.

Other possible explanations, involving different theories of change, would lead to different critical features in implementation that should be ensured. For example, if health improvements came about through increased

Figure 1.4   **Logic Models Showing Different Possible Causal Mechanisms Involved in Eating an Apple a Day**

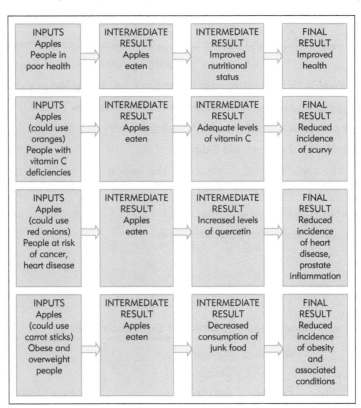

fiber consumption, eating the whole apple, not just drinking the juice, would be important. Once the plausible theories have been identified, they can be used to guide data collection and analysis of an evaluation. They can also be used to synthesize data from previous evaluations and research (we discuss this in Chapter Four).

Success in terms of achieving intended results might not mean success in terms of the theory. Another possible pattern of results is that the health outcomes have been achieved but not the intermediate results of changes in vitamin C. This would suggest that something other than the intervention had caused the health improvements or that a quite different theory of change was operating that did not involve vitamin C. Results like this would indicate theory failure despite success in terms of results.

Learning from "An Apple a Day"

Speculating on different possible causal mechanisms enables us to develop an evaluation that will collect and analyze data to be able to understand to what extent, for whom, and why an intervention does or does not work. (Chapter Fourteen describes how to use program theory to guide evaluation design.) Although a single evaluation is limited in its scope, program theory makes it easier to combine evidence from a number of studies. Table 1.1 summarizes how an evaluation informed by program theory can distinguish among different types of success and failure.

The apple a day example shows the importance of developing program theory that identifies the causal mechanism that is understood to be involved in producing the intended outcomes. This can help to produce more useful evaluations and better evidence for policy.

Table 1.1   Using Program Theory to Interpret Evaluation Findings

Apples Delivered	Apples Eaten	Vitamin C Levels Raised	Health Outcomes Improved	Interpretation
✗	✗	✗	✗	Implementation failure
✓	✗	✗	✗	Engagement or adherence failure (first causal link)
✓	✓	✗	✗	Theory failure (early causal link)
✓	✓	✓	✗	Theory failure (later causal link)
✓	✓	✓	✓	Consistent with theory
✓	✓	✓/✗	✓/✗	Partial theory failure (works in some contexts)
✓	✓	✗	✓	Theory failure (different causal path)

SUMMARY

This chapter has used a hypothetical example to explore how articulating a program theory—an explicit statement of how change will occur and how an intervention will produce these causal processes—can make evaluations more useful. Throughout the rest of the book, we use examples from actual evaluations to show how to develop, represent, and use program theory for evaluation and other purposes.

EXERCISES

1. If a social marketing campaign was used instead of direct delivery of apples for the Apple a Day program, what would implementation failure look like? What would theory failure look like? What would partial theory failure look like, where it works only in particular contexts?

2. Consider a policy that aims to increase student performance by increasing teachers' salaries. What might be some alternative causal mechanisms that would produce the intended outcomes?

2

Variations of Program Theory over Time

D ESPITE OCCASIONAL STATEMENTS that program theory is a new approach, its roots go back more than fifty years. Over time, there have been important variations and different emphases. This chapter reviews these developments, including the different terms that have been used inconsistently over time. We provide a guide for translating the different labels that have been used for concepts and the different ways these have been used.

A SHORT HISTORY OF PROGRAM THEORY

The history of program theory evaluation is not one of a steady increase in understanding. Instead, many of the key ideas have been well articulated and then ignored or forgotten in descriptions of the approach. It is not unusual to have statements that demonstrate a lack of knowledge of previous empirical and theoretical developments, such as a call for proposals from the Agency for Healthcare Research and Quality (2008) that claimed that "'theory-based evaluation' is a relatively new approach" (p. 14).

Philosophical Roots

The value of intentional action has long been recognized in philosophy. The Greek Stoic philosopher Epictetus wrote nearly two thousand years ago, "First say to yourself what you would be; and then do what you have to do." However, early planning and evaluation focused on specific objectives rather than on articulating the links between activities and these objectives.

Early Examples

Probably the first published use of what we would recognize as program theory was a series of four articles on evaluating training by Don Kirkpatrick (1959a, 1959b, 1960a, 1960b) published in the *Journal of the American Society for Training and Development*. Kirkpatrick looked at reactions of participants to training; learning in terms of new knowledge, skills, and attitudes; behavior in terms of applying the learning back in the work environment; and results in terms of targeted outcomes. Over time these four categories became known as Kirkpatrick's Four Levels of Learning Evaluation.

Kirkpatrick argued that this sequence of results should first be used for planning purposes, beginning with the ultimate targeted outcomes, working back to the behavior needed to produce these; the new knowledge, skills, and attitudes needed to be able to engage in the behavior; and the training experiences that would be needed to produce a positive reaction from participants. Then this sequence of results could be applied as a framework for planning evaluation, where evidence from each level could be used to build an argument about the contribution of the training.

This idea was more broadly applied by Edward Suchman, who argued in 1967 that program evaluation would benefit from explicitly examining the achievement of a "chain of objectives" (p. 55). He drew attention to the need to identify and examine an intervening process in between an activity and its objective. "The evaluation study," he wrote, "tests some hypothesis that activity A will attain objective B because it is able to influence process C which affects the occurrence of this objective. An understanding of all three factors—program, objective and intervening process—is essential to the conduct of evaluative research" (p. 177).

In the same year, Daniel Stufflebeam (1967), in a discussion of the limitations of experimental approaches to evaluating education programs, outlined a new evaluation model. It was in the form of a generic program theory, although it was not labeled as such. The CIPP model (Stufflebeam, 1967) set out an intervention in terms of four boxes—context, input, processes, and product—and asked a series of questions about each of them. This format put all the results together in one category, which avoids debates about whether a result is best classified as an output, an outcome, or an impact. It was one of the few models that incorporated context as an intrinsic part of the description of an intervention.

In 1969 a version of program theory, the logical framework approach, commonly referred to as the logframe, was developed for Practical Concepts Incorporated (1979). The title of the report outlining the approach, *The Logical Framework: A Manager's Guide to a Scientific Approach to Design and Evaluation,* shows that it was also intended to be used for both planning and evaluation. Subsequently the logframe was further developed for use by U.N. agencies by Gesellschaft für Technische Zusammenarbeit, the German international development agency. In the logframe, the causal chain was standardized into four components: activities, outputs, purpose (the rationale for producing the outputs), and goal (a higher-level objective to which this program and others contributed). For each component, four aspects were articulated: a narrative description, objectively verifiable indicators, means of verification, and assumptions (factors outside the control of the program on which the success of achieving that component depended). Because the logframe is still widely used in international development, we look at it in detail in Chapter Thirteen.

The idea of identifying and measuring intermediate variables was used in three evaluations related to safety: a speed control program, a motor vehicle inspection program, and an improved ambulance system (Hall and O'Day, 1971). They argued that this "causal chain" approach to evaluation provided more realistic indicators of the success of an intervention than evaluations that solely measured final results. In 1972, in one of the earliest books on evaluation, *Evaluation Research: Methods of Assessing Program Effectiveness,* Carol Weiss explained how an evaluation could identify several possible causal models of how a teacher home-visiting program might work and then investigate which of the versions was best, supported by evidence. Her representation of program theory emphasized the notion of chains of objectives, and in a single diagram that began with the intervention (home visits), she sketched out four causal chains: one relating to subsequent changes in teacher behavior, two relating to changes in parental behavior, and one involving referrals to secure special help for the child. Instead of the logframe format of just one level between activities and purpose, Weiss's program theory had four or five intermediate objectives between the activities and the ultimate objective of improved reading achievement. These two different approaches to representation—a standardized template of four or so boxes or a more flexible chain of results—continue to be used to this day and are discussed in more detail in Chapter Nine.

A few years later, Carol Fitz-Gibbon and Lynn Morris set out a detailed explanation of what was meant by theory-based evaluation and how it could add value to evaluation. They defined a theory-based evaluation as "one in which the selection of program features to evaluate is determined by an explicit conceptualization of the program in terms of a theory, a theory which attempts to explain how the program produces the desired effects" (Fitz-Gibbon and Morris, 1975, p. 1). In Chapter Fourteen, we explore how program theory can be used to guide evaluation design, and in Chapter Fifteen, we explore the use of program theory for causal inference.

The notion of evaluability assessment, described by Joseph Wholey (Wholey et al., 1975; Wholey, 1979); and Len Rutman (1977), incorporated many of the elements of program theory. Wholey suggested that before moving to an evaluation, it was important to review whether the program had specific goals and models, whether it was well defined and therefore could be evaluated, and whether stakeholders were likely to use the evaluation findings. Although this was similar to program theory in terms of articulating the expected causal chain, it was intended to be used only for programs that could be expected to become stable and well defined, not for those that were expected to continue to be changed and adapted. Recognition is increasing that program theory can be used for complex, emergent programs, as we discuss in Chapter Five.

An early generic program theory, built on Kirkpatrick's model for evaluating training, was developed by Claude Bennett (1975) for extension programs, where staff aim to influence the behavior of farmers, young people, and other rural community members through providing research-based information. The framework consisted of seven sequential components: inputs; activities; participation; reactions; changes in knowledge, aspirations, skills, and attitudes; behavior changes; and subsequent changes in social, economic, and environmental conditions. Bennett's hierarchy became known outside extension circles when Michael Patton included it in the first edition of *Utilization-Focused Evaluation* (1978) and applied it to evaluating evaluation itself.

The First Boom Period

During the 1980s, a number of key publications contributed to the development and popularization of program theory. Huey-Tsyh Chen and Peter Rossi published a series of influential articles and books on theory-driven evaluation (Chen and Rossi, 1980, 1983, 1987). Chen (1990a) distinguished six domain theories (treatment, implementation, environment, outcome, impact, intervening mechanisms, and generalization) that theory-driven evaluations could focus on either alone or in combination.

Three special issues of publications in quick succession focused on the approach. In 1987 Len Bickman edited "Using Program Theory for Evaluation" (Bickman, 1987a), a special issue of the American Evaluation Association's *New Directions for Program Evaluation* that explored the different uses for program theory (Bickman, 1987b) and different types of theories (Scheirer, 1987; Shadish, 1987; Conrad and Miller, 1987; McClintock, 1987). In 1989 Chen served as guest editor for an influential special issue of the journal *Evaluation and Program Planning* on the theory-driven approach to evaluation, which discussed key issues (Chen and Rossi, 1989), different types of models (Lipsey and Pollard, 1989), ways to address validity issues (Scott and Sechrest, 1989; Palumbo and Oliverio, 1989; Costner, 1989; Shapiro, 1989; Patton, 1989; Cordray, 1989), and barriers to use (Bickman, 1989). Shortly afterward, Len Bickman edited another special issue of

New Directions for Program Evaluation, this time focusing on advances in program theory (Bickman, 1990), describing how to develop program theory (Chen, 1990b) through path analysis (Smith, 1990), social science theory (Riggin, 1990), and pattern matching (Marquart, 1990); how to use program theory to understand program quality (Bickman and Peterson, 1990) and program types (Conrad and Buelow, 1990); and different ways of testing program theories (Mark, 1990; McClintock, 1990).

During the 1980s, in response to economic challenges and concerns about accountability, many countries introduced reforms within the public sector to focus on managing for results rather than only for compliance with processes, sparking the first boom in the use of program theory and logic models (U.S. General Accounting Office, 1995). In Australia, a system of program management and budgeting was introduced in 1983 requiring departments to articulate specific goals, explain how they planned to meet them, and report on performance. In Canada, the federal government implemented a new expenditure management system that incorporated a shift in focus from activities and outputs to impacts and results, and requiring departmental business plans to articulate priorities, set goals, and report on performance.

Some organizations adopting program theory focused on the Suchman notion of a chain of objectives. In the 1980s, Bryan Lenne led a team in the New South Wales Public Service in Australia that worked with departments across the state to use program theory in the form of outcomes hierarchies (Lenne and Cleland, 1987; Funnell and Lenne, 1990). Program activities were not necessarily at the front end of these chains, but they could contribute directly to later outcomes. External factors were also to be included in the program theory (in Chapters Seven and Eight, we discuss this approach in more detail). This approach was subsequently adopted by state and federal governments across Australia (Funnell, 1990; Milne, 1993). The Roundtable on Comprehensive Community Development, supported by the Aspen Institute, published an influential paper by Carol Weiss (1995) showing how theory-based evaluation, in the form of an outcomes chain, could be useful for programs where classic experimental and quasi-experimental approaches were not possible.

Other organizations focused on a pipeline approach to program theory. The United Way (1996), a nonprofit organization that works in a coalition of charitable organizations, developed a guide to developing and using logic models for outcome measurement that was widely accessed both through the Internet and in hard copy (United Way, 1996). It set out a four-box logic model of inputs (both resources and constraints), activities, outputs (direct services or products), and outcomes for participants. The W. K. Kellogg Foundation produced the *Logic Model Development Guide* (2004), which was also in the form of a linear template of five components: inputs, activities, outputs, outcomes, and impact. This template defined impacts as results for the broader community, organization, or system, beyond the individuals who participated in the intervention. NORAD, the Norwegian government international development agency, published a widely used guide to the logical framework approach that has now gone through four editions (NORAD, 1999).

More evaluation books and guides began to include discussion of program theory and logic models. For example, a widely used textbook, *Program Evaluation: A Systematic Approach,* added a chapter in its sixth edition (Rossi, Freeman, and Lipsey, 1999). Program theory was added to the repertoire in the revised edition of *Evaluation Models,* edited by Daniel Stufflebeam, George Madaus, and Tom Kellaghan (Rogers, 2000a). The British government included discussion of program theory in its *Magenta Book,* which provided a guide to planning and evaluation (Government Chief Social Researcher's Office, 2003).

A third issue of *New Directions for Evaluation* focusing on program theory was published (Rogers, Petrosino, Hacsi, and Huebner, 2000), discussing the promise of program theory for replication (Hacsi, 2000), meta-analysis (Petrosino, 2000), performance monitoring (Funnell, 2000), and developing a shared understanding (Hubener, 2000), and the challenges in terms of addressing causal inferences (Weiss, 2000; Davidson, 2000; Cook, 2000; Rogers, 2000b).

Innovations

A new conceptualization of program theory was introduced by Ray Pawson and Nick Tilley. In a paper in the *British Journal of Criminology* (Pawson and

Tilley, 1994) and their subsequent book, *Realistic Evaluation* (Pawson and Tilley, 1997), they set out a realist approach to evaluation, where program theory was understood as configurations of context-mechanism-outcome. Context was seen as a critically important part of program theory because causal mechanisms fire only in favorable contexts, that is, in particular implementation environments or for particular types of participants. We discuss the importance of differential program theories in Chapter Five and how to represent these in realist matrices in Chapter Nine.

Some different approaches to program theory were developed to address concerns that logic models make it seem that the entire causal process is under the control of program implementers. Sarah Earl, Fred Carden, and Terry Smutylo (2001), working with Barry Kibel, developed outcome mapping, an approach designed for interventions where implementers cannot control impacts but seek to influence these by affecting the behavior of boundary partners. Steve Montague, Gail Young, and Carolyn Montague (2003) described this in terms of circles of influence: the operational environment, the environment of direct influence, and the environment of indirect influence. We explore these ideas in more detail in Chapters Seven and Nine.

Variations on pipeline models were also developed, drawing on Bennett's hierarchy from 1975. In Canada, Steve Montague (1998) argued that logic models needed to include "reach," that is, articulating where and with whom particular results were intended. A generic logic model was developed at the University of Wisconsin that included articulating who was expected to participate, but also showing assumptions, external factors, needs, and priorities on the logic model (University of Wisconsin, 2003). We explore these variations in Chapter Thirteen.

The Current State

In more recent years, the use of program theory has become part of the mainstream of most approaches to evaluation. Many organizations in many countries are referring to program theory and often requiring its use in planning proposals and reporting. For example, the European Commission has included discussion of program theory in its guide to evaluability assessment (European Commission, 2009). More resources have been developed to support people in learning to develop and use program theory. A special topical interest group of the American Evaluation Association, Program Theory and Theory-Driven Evaluation, has been established. And a number of books focusing on program theory and logic models have been published (Frechtling, 2007; Donaldson, 2007; Knowlton and Phillips, 2008). However, the three issues of concern identified by Weiss in her review of twenty-five years (Weiss, 1997)—not having an articulated theory about how change comes about, or having a poor theory, or not really using it to guide the evaluation—are continuing features of many examples of program theory (Rogers, 2007; Astbury and Leeuw, 2010).

This explosion of activity has produced great diversity in what program theory is called, how it is represented, and how it is used (Coryn, Noakes, Westine, and Schröter, forthcoming). Sometimes organizations use a common template of components (such as a logframe or other pipeline model), and other times, there is scope for a more flexible representation, such as an outcomes chain. Sometimes the expectation is that a single program theory will be developed and used as a reference point throughout planning, implementation, and evaluation, and other times there is scope for developing several versions or revising the program theory throughout the process. Sometimes the program theory is developed by program staff, sometimes by an external evaluator, and sometimes this is done collaboratively. We discuss these options in Chapter Six.

This rich diversity of experience presents a wide range of options at each stage, which can be quite confusing. This book presents the diversity of options and also a way to select, combine, and adapt these to match the situation at hand.

TERMINOLOGY IN PROGRAM THEORY

Over the years, many different terms have been used to describe the approach to evaluation that is based on a "plausible and sensible model of how the program is supposed to work" (Bickman, 1987b):

- *Chains of reasoning* (Torvatn, 1999)
- *Causal chain* (Hall and O'Day, 1971)

- *Causal map* (Montibeller and Belton, 2006)

- *Impact pathway* (Douthwaite et al., 2003)

- *Intervention framework* (Ministry of Health, NZ 2002)

- *Intervention logic* (Nagarajan and Vanheukelen, 1997)

- *Intervention theory* (Argyris, 1970; Fishbein et al., 2001)

- *Logic model* (Rogers, 2004)

- *Logical framework (logframe)* (Practical Concepts, 1979)

- *Mental model* (Senge, 1990)

- *Outcomes hierarchy* (Lenne and Cleland, 1987; Funnell, 1990, 1997)

- *Outcomes line*

- *Performance framework* (Montague, 1998; McDonald and Teather, 1997)

- *Program logic* (Lenne and Cleland, 1987; Funnell, 1990, 1997)

- *Program theory* (Bickman, 1990)

- *Program theory-driven evaluation science* (Donaldson, 2005)

- *Reasoning map*

- *Results chain*

- *Theory of action* (Patton, 1997; Schorr, 1997)

- *Theory of change* (Weiss, 1998)

- *Theory-based evaluation* (Weiss, 1972; Fitz-Gibbon and Morris, 1975)

- *Theory-driven evaluation* (Chen and Rossi, 1983)

The labels and definitions used in this book are not the only ones in use. Given the wide variety of terms used, it is important to be aware of the other labels you might come across. These terms are not always used interchangeably; sometimes they have particular meanings, but these vary widely. In particular, although the terms *program theory* and *program logic* are often used interchangeably, they sometimes focus on particular distinctions. As we have worked on evaluations in different organizations, we have been confronted by very different uses of these terms. Table 2.1

shows some different distinctions that have been made between program theory and program logic. Sometimes they have been distinguished in terms of the nature of the theory, sometimes by the source of the theory, the level of detail, the detail of causal processes, or the focus, as shown in Table 2.1.

Because there is no consistency in the way these terms are used, if you find yourself in conversation with someone who is making a distinction between program theory and program logic, it is important to check exactly what the other person means. Because of the inconsistent way in which the terms are used, we do not use *program logic* again in this book.

Table 2.1 **Distinctions Sometimes Made Between Program Theory and Program Logic**

Distinction Made	Program Theory	Program Logic
Nature of the theory	A tightly articulated theory about how the program will achieve its intended outcomes, possibly including the causal mechanisms	A series of premises about the way a series of events might unfold
Source of the theory	Formal research-based theory	Informal set of causal assumptions held implicitly by practitioners or others
Detail of the theory	Diagram of the causal model	Detailed analysis of the assumptions underpinning the model, including standards of performance
Detail of causal processes	Detailed diagram showing causal processes linking program processes with outcomes	Simple set of boxes of inputs, processes, outputs, and outcomes
Focus	Entire results chain	Specific process involved in the key change in the intervention

Similarly, sometimes the terms *logic model* and *theory of change* have been distinguished in particular ways that are different to the ways we are using the terms here. For example Heléne Clark, from ActKnowledge, and Andrea Anderson, from the Aspen Institute Roundtable on Community Change, who employ program theory extensively, have used these terms to make the distinction between ways of representing program theory that we have labeled *pipeline logic models* and *outcomes chain logic models* (Clark and Anderson, 2004). Mhairi Mackenzie and Avril Blamey (2005) have used *theories of change* to refer specifically to the type of logic model advocated by the Aspen Institute (which we present in Chapter Fifteen).

The key message here is to define your terms carefully and ask others to do so as well. It cannot be assumed that you mean the same thing when you use the same term or that you mean something different when you use a different term.

Advantages and Disadvantages of Different Terminology

If you are introducing program theory into your organization, think carefully about which terms would be most appropriate. During our work with very different types of organizations and programs, we have learned that no label suits everyone. Moreover, labels that seem quite neutral sometimes have significant negative connotations.

Some people do not respond well to the "theory" part of the label "program theory," assuming it refers to academic analyses disconnected from practice or practical uses, or they get sidetracked in trying to relate it to a half-remembered grand theory. Some people confuse program theories—how the program works—and evaluation theories—how to go about focusing and managing evaluations, gathering and analyzing data, and supporting use. In these circumstances, the term *program logic* might be more suitable and focus on the overall logical sequence of the program. Of course, some people do not respond well to the "logic" part of *program logic,* feeling it makes their flexible, responsive program sound mechanistic.

The term *outcomes hierarchies* is useful for some people because it emphasizes the sequential and consequential nature of the chain of results. But it can lead to an emphasis on simple, linear results chains, which are not always appropriate, and the wording can suggest that earlier results are unimportant. For some people, any label with "program" in it will be problematic because "program" has a specific meaning, and they are managing or evaluating something other than a program (such as a project, strategy, initiative, or policy) and want a term that sounds as if it applies to their work.

Terms for Components in a Pipeline Model

Many pipeline models use the terms *input, processes, outputs, outcomes,* and *impacts.* But this is far from a universal practice. Outputs can be defined in very different ways: as completed activities, tangible products, or services, or as the first change in the causal chain. The distinction between outcomes and impacts is sometimes made in terms of sequence, but there is little consistency of approach (sometimes outcomes are before impacts, and sometimes vice versa) or scope (with outcomes defined as results for individuals and impacts as results for the broader community or organization). The following list shows the labels and definitions used by some major international organizations:

Australia: AusAID (Australian Agency for International Development) (Logframe)

Outputs: The tangible products or services that the activity will deliver

Component objectives or intermediate results: A level in the objectives or results hierarchy that can be used to provide a clear link between outputs and outcomes

Purpose or *outcome:* The medium-term results in terms of benefits to the target group that the activity aims to achieve

Goal or *impact:* The long-term development impact (policy goal) that the activity contributes to at a national or sectoral level (http://www .ausaid.gov.au/ausguide/pdf/ausguideline3.3.pdf)

Canada: Treasury

Activities: Key activities intended to contribute to the achievement of the outcomes

Outputs: Products or services generated by the activities

Immediate outcomes: Short-term outcomes that stem from the activities and outputs

Intermediate outcomes: Next links in the chain of outcomes that occur after immediate outcomes

Final outcomes: Final outcomes, or why these activities are being engaged in

(http://www.tbs-sct.gc.ca/cee/dpms-esmr/dpms-esmr05-eng.asp#_5.1_ Overview_of)

India: Comptroller and Auditor-General

Objectives

Aim—the broad mandate, legislative direction

Objectives—translation of the mandate into well-formed objectives

Targets—well-defined physical goals to achieve the objectives

Inputs: Resources that are to be transformed by program instruments into outputs

Processes: Planning, organization, and implementation of the program operations to produce the outputs

Outputs: Results achieved by the program operations that are within the control of the program manager

Outcome: The broad effects of the program outputs; expected to meet the program objectives and aim and influenced by external factors (http://www.cag.gov.in/publications/peraudrepcent/appendix%20E.pdf)

South Africa: Framework for Managing Programme Performance Information, National Treasury

Inputs: All the resources that contribute to the production and delivery of outputs: "What we use to do the work"

Activities: The processes or actions that use a range of inputs to produce the desired outputs and, ultimately, outcomes: "What we do"

Outputs: The final products, goods, and services produced for delivery: "What we produce or deliver"

Outcomes: The medium-term results for specific beneficiaries that are the consequence of achieving specific outputs: "What we wish to achieve"

Impacts: The results of achieving specific outcomes, such as reducing poverty and creating jobs (http://www.treasury.gov.za/publications/guidelines/FMPI.pdf)

United States: University of Wisconsin Extension Service

Inputs: What we invest

Outputs–Activities: What we do

Outputs–Participation: Who we reach

Short-term outcomes: Short-term results

Medium-term outcomes: Medium-term results

Impact: Ultimate impact

(http://www.uwex.edu/ces/pdande/evaluation/evallogicmodel.html)

United States: Innovation Network (InnoNET)

Resources: What you have to work with

Activities: What you will do with your resources in order to achieve program outcomes and, ultimately, your goal

Outputs: Tangible products of your activities

Outcomes: Changes expected to occur as a result of your work

Goal: Overall purpose of your program

(http://www.innonet.org/client_docs/File/logic_model_workbook.pdf)

United States: W. K. Kellogg Foundation

Resources/inputs: Resources needed to operate program

Program Activities: Processes, tools, events, technology, and action that are an intentional part of the program implementation

Outputs: Types, levels, and targets of service delivered

Outcomes: Specific changes in program participants' behavior, knowledge, skills, status, and level of functioning

Impact: Changes to organizations, communities, or systems as a result of program activities within seven to ten years

(http://www.wkkf.org/~/media/6E35F79692704AA0ADCC8C3017200208 .ashx)

United States: United Way

Inputs: What we invest to make the program happen

Activities: What we do in our program

Outputs: Products and participation—the "how manys"

Outcomes: Benefits for participants during and after program activities

(http://www.yourunitedway.org/media/Guide_for_Logic_Models_and_ Measurements.pdf)

KEY IDEAS IN PROGRAM THEORY

Program theory involves a particular use of the words *program* and *theory*. In this section, we address how we have used terms in this book. If you are working in an organization or with an organization that uses different terms for these concepts, it might be useful to pencil in the local equivalent.

Program

In *program theory*, the term *program* refers not only to something formally labeled as a program (for example, in a corporate management hierarchy of programs, subprograms, and components). It can refer to any intervention: a project, a strategy, a policy, a funding initiative, or an event. It includes interventions that are undertaken by a single organization, such as a direct service delivery project, and those that are undertaken by multiple organizations, such as a whole-of-government policy. It refers to both preplanned and tightly specified interventions, and broadly defined and emergent interventions.

Program Theory

A program theory is an explicit theory or model of how an intervention contributes to a set of specific outcomes through a series of intermediate results. The theory needs to include an explanation of how the program's activities contribute to the results, not simply a list of activities followed by the results, with no explanation of how these are linked, apart from a mysterious arrow. We find it helpful to think of a program theory as having two components: a theory of change theory and a theory of action. One of the benefits of articulating program theory is being able to systematically review the quality of the theory in terms of plausibility, consistency with evidence, and utility.

Theory of Change

This refers to the central mechanism by which change comes about for individuals, groups, and communities. For example, many health promotion programs are based on a theory of change that behavior changes in response to perceived social norms. It is possible to have more than one theory of change. There might be different theories of change at different stages of the program (for example, one about participants becoming engaged with the program and one about their changing their behavior) or for different groups of people (for example, some people might change their behavior in response to new information about risks and benefits, while others might change only in response to tangible incentives).

Programs are usually, but not always, about change. Sometimes a program aims to stop or reduce change or prevent something from happening—for example, maintaining the mobility of a person with a disability who might otherwise develop restricted movement, or maintaining biodiversity despite pressures from agriculture and industry. In these cases, the theory of change explains how pressure to change will be resisted or deflected. Another way to think of this is that the program is changing a situation from what it otherwise would have been.

Theory of Action

This explains how programs or other interventions are constructed to activate their theory of change—for example, what the program does to change social norms or make them more evident. The theory of action explains the activities that will be undertaken and what level of success will be needed for each outcome to produce the final intended results.

Logic Model

A program theory is usually displayed in a diagram called a logic model. The form that the logic model takes can also affect the way we think about a program theory and can shape it. It is therefore important to understand the strengths and weaknesses of different types of logic models for different purposes and to achieve an appropriate match between the type of logic model and the type of program and purpose.

Although logic models can be drawn in numerous way, they can be grouped into four broad types: pipelines, outcome chains, realist matrices, and narratives. Pipeline logic models look like production line flowcharts or a series of building blocks, and they are often light on theory. They show the inputs used, the activities undertaken, and short-term and longer-term results, all without necessarily explaining how the activities bring about the results. We refer to these as pipeline models because they represent the program as a linear process where inputs go in one end and outputs and outcomes come out the other end. Pipeline logic models are most useful when the activities are all up front, and then the rest of the results simply happen like a row of dominos (or at least with the help of other processes which are not the focus of the program). They can sometimes be a good starting point for organizations not accustomed to thinking about and reporting in terms of outcomes and impacts, but the risk is that they will entrench an oversimplified and unhelpful view of the program. We have found that logic models in the form of outcome chains are usually more effective: these represent the intervention and its consequences as a series of results. Logic models in the form of realist matrices focus on identifying the conditions under which the theory will apply. Both outcome chains and realist matrices emphasize the importance of understanding and focusing on relationships and linkages as much as on the building blocks. Narrative representations of program theory tell a story about how the intervention works, or the experience of participants as they pass through. They can be simple narratives that follow a pipeline causal process or more detailed narratives that explain how change occurs.

Evaluation

Many forms of evaluative activity explicitly combine facts and values. Evaluation includes up-front needs analysis that explores the nature and extent of a problem, existing strengths and capacities, and options for meeting needs. It includes process evaluation of different types: documenting an innovation, checking compliance with specifications, and identifying issues and problems that need to be resolved in ongoing, formative evaluation. It includes impact evaluation, analyzing the contribution of the intervention to particular results, and cost-benefit evaluation, where the costs of implementing an intervention (including resources consumed and any negative effects) are weighed against the benefits produced.

These definitions are summarized in Table 2.2, which includes space to pencil in any different terms that are used in your organization or in organizations with which you need to communicate.

DEFINING PROGRAM THEORY

Voltaire famously described the Holy Roman Empire as "not Holy, nor Roman, nor an Empire." Program theory could be described similarly as being "not program or theory." If a narrow definition is used, *program theory* can be used for policies, projects, strategies, funding initiatives, and practices—in other words, not only for interventions labeled as programs. The "theory" in *program theory* can be an articulation of practice wisdom or of tacit assumptions—that is, not only a formal, research-based theory.

Table 2.2   **Definitions Used in This Book**

Term	Definition	Your Organization's Label (if applicable)
Program theory	An explicit theory of how an intervention is understood to contribute to its intended or observed outcomes; ideally includes a theory of change and a theory of action	
Theory of change	The central processes or drivers by which change comes about for individuals, groups, or communities. It can be derived from a research-based theory of change or drawn from other sources.	
Theory of action	The ways in which programs or other interventions are constructed to activate these theories of change.	
Theory of Change	A research-based theory of change	
Logic model	A representation of a program theory, usually in the form of a diagram	

SUMMARY

The long and rich history of program theory has produced many different ways of developing, representing, and using it. It has also produced many different terms and definitions. Being aware of this can help you to make the best choices to suit your situation, and communicate effectively with other people who are using different terminology.

EXERCISES

To answer these questions, consider the organization you work in or one in your area of work.

1. What terms are commonly used in the organization to refer to program theory?

2. Does the organization prescribe or recommend a particular form of logic model? If so, what is it?

3. Do partner organizations use the same terminology? If not, what are the differences?

3

Common Myths and Traps

AS PROGRAM THEORY has become more widely used, a number of myths have arisen about how to go about it and how useful it might be. In this chapter, we challenge seven myths that can get in the way of effective use of program theory. We also set out seven common traps for those using program theory and discuss how to avoid them. We provide more detail throughout the book about avoiding these traps.

SOME COMMON MYTHS

As program theory has become increasingly widely used, and even mandated, within organizations, a number of myths have been promulgated. Knowing that these are myths and being able to discuss them with colleagues and partners will reduce your risk of making common mistakes when using program theory. The myths are set out in Table 3.1 and discussed in the sections that follow.

Table 3.1   **Some Common Myths About Program Theory**

Myths	How These Are Sometimes Expressed
1. New approach	Program theory is a new approach to evaluation, and what is contained in a local guide is all there is to know about it.
2. One way to do it	There is one way to draw a logic model.
3. Not a good model	Program theory does not address all necessary aspects of evaluation and therefore is not a good model of evaluation.
4. Too much time	Developing a credible and useful program theory always requires so much time, content knowledge, and research expertise that it is usually impractical.
5. Just draw it	Developing a credible and useful program theory can be done by simply asking people to draw a logic model.
6. Can't really test it	Testing a program theory can be done only by formal experimental methods, which are beyond the scope of most evaluations.
7. Assume causality	Program evaluation should not bother with testing the program theory. If results are consistent with the theory, it can be assumed that the program has caused the results.

Myth 1: A New Approach

A surprising number of introductory guides to evaluation state that this is a new approach. These statements demonstrate an ignorance of previous developments of program theory and of the lessons that can be learned from these. Program theory has been used for over fifty years, and current practice should draw on what has been learned by this.

Myth 2: One Way to Do It

Many guides to program theory show only one way to draw a logic model—usually a version of a pipeline model. In fact, there are many ways to draw logic models, including results chains and causal matrices. Some show activities and some external factors. Some show how the program works differently for different types of participants. Some show how the program works in combination with other programs. Some should be read left to right, some from top to bottom, and some from bottom to top. In Chapter Nine, we set out a range of options and discuss their comparative advantages in particular situations.

Knowing there is a range of options and when each might be most suitable makes it more likely that the form of logic model will suit the needs of each situation. Knowing that other organizations are likely to represent logic models differently and use terms differently can make it easier to communicate with partners effectively.

Myth 3: Not a Good Evaluation Model

We can understand the term *model* in two very different ways. A model can be a heuristic—a device to help us think about things, which focuses on some key aspects. When this definition is used, combining several evaluation models in a single evaluation can be useful. Program theory, for example, can readily be combined with particular evaluation designs that address causal analysis (for example, randomized control trials or comparative case studies), particular approaches to evaluation management (for example, external evaluation, peer review, or self-evaluation), particular approaches to identifying evaluation questions (for example, utilization-focused evaluation, or discrepancy evaluation), and approaches to collecting and analyzing quantitative and qualitative data (for example questionnaires, interviews, economic statistics, physical measurements). Similarly, theories about how change occurs and models of different types of programs such as "carrots and sticks" can be used in conjunction with other processes to help us think about a program theory. We discuss these models in Chapters Eleven and Twelve.

However, the term *model* is sometimes understood as a comprehensive package—like buying a particular model of car. With this definition, an evaluation model needs to provide guidance on all aspects of evaluation. This definition has led some influential evaluation theorists (for example, Stufflebeam, 2004, Stufflebeam and Shinkfield, 2007) to dismiss program theory as a valid approach to evaluation. By itself, program theory clearly does not provide guidance on gathering the evidence for monitoring and evaluation; it needs to be combined with evaluation expertise to draw appropriately from methods for research design, data collection, and data analysis. In Chapter Fourteen, we discuss how program theory can be used to guide the development of measures and indicators and in Chapter Fifteen how program theory can be used to guide causal analysis. In all cases, those using program theory need to have, or be able to access, expertise in evaluative skills, especially in collecting and analyzing data.

Myth 4: Too Much Time

This myth appears to have arisen from observing inexperienced evaluators fall into the trap of getting stuck on developing and revising program theory, at the expense of spending time designing the evaluation, collecting and analyzing data (leading to trap 1: not actually using program theory in the evaluation). In our experience, program theory can be developed without getting stuck or taking an excessive amount of time (we discuss these in Chapter Six). There are diminishing returns on finessing logic models, so quick and simple versions can be used as a way to check understanding and facilitate thinking about evaluative criteria, indicators of progress toward long-term goals, and so on. However, assuming that program theory will take almost no time, as the next myth suggests, is a risk.

Myth 5: Just Draw It

This myth is the opposite of the previous one and may have arisen from program theory training courses where people work through the process of developing logic models without access to previous research or observations of the program. The truth lies somewhere between this myth and the previous one.

A credible program theory needs to go beyond simply documenting people's assumptions about how things work, although this might be a useful starting point. Any use of program theory, even under time and resource constraints, needs to review the program theory critically for its consistency with research and other data and its logical plausibility. Chapter Ten provides criteria for critiquing a program theory. If there is sufficient time and access to specialist content knowledge, it is possible to further develop the program theory, drawing on systematic review of previous research literature and expert review. Investing these resources is particularly important when a program theory is being developed for a large-scale or high-risk intervention. Purposeful program theory will find the appropriate investment of time in developing the program theory. Chapter Six describes the range of evidence sources that should be used to inform the development of program theory.

Myth 6: Can't Really Test It

Evaluations can use the whole array of methods for causal attribution, including experimental, quasi-experimental, and nonexperimental methods. It is not realistic to expect that any single evaluation, or any single research project, can provide a definitive proof of a program theory, but evaluations can and should provide insights into its credibility. In Chapter Fifteen, we discuss different ways to undertake causal analysis using program theory, including rigorous systematic approaches that can be used when this is a major focus of the evaluation.

Myth 7: Assume Causality

This is the opposite of the previous myth, and again the truth is somewhere in between. Converting a logic model into a series of indicators and then reporting these as evidence of the impact of the program can be tempting, but it is inadequate for evaluation. In Chapter Fifteen, we discuss different ways to undertake causal analysis using program theory, including approaches that can and should be added into small evaluations.

TRAPS TO AVOID WHEN DEVELOPING AND USING PROGRAM THEORY

Being able to draw a logic model and produce an evaluation plan does not mean that program theory has been developed and used well. This book provides advice on avoiding seven common traps, which we set out in Table 3.2 and discuss in the sections that follow.

Table 3.2 **Seven Traps to Avoid When Developing and Using Program Theory**

1. No actual theory
2. Having a poor theory of change
3. Poorly specifying intended results
4. Ignoring unintended results
5. Oversimplifying
6. Not using the program theory for evaluation
7. Taking a one-size-fits-all approach

Trap 1: No Actual Theory

Many versions of program theory, particularly those using a pipeline approach to logic modeling, fall short of having an actual theory. They simply display boxes of activities and boxes of outcomes without demonstrating logical and defensible relationships between them and the various items listed in the boxes. This setup can make it difficult to understand the causal chain and identify what measurement and evaluation would be appropriate for purposes of causal attribution.

Program activities and resources are what staff are most familiar with on a day-to-day basis. So it is not surprising that the program theories that they construct can be more preoccupied with what the program does than what it achieves in terms of outcomes. Renger and Titcomb (2002) introduced the notion of activity traps: well-intended activities that appear to address particular problems but on closer inspection do not address any of the conditions that underlie the problems. Flowing on from these activity traps, a program monitoring system may expend much effort on measuring activities of little consequence in producing outcomes and resolving the problems or issues that the program was established to address.

A frequent criticism of monitoring systems and, perhaps to a lesser extent, evaluation studies is that they place too much emphasis on measures of busyness—measures of what the program does on the assumption that what it does will produce outcomes. In other words, the measures of activity come to be used as proxy measures of outcomes. It is particularly important in these situations to be confident that what the program does is relevant to the outcomes it wishes to achieve and that the program has not become caught in activity traps.

A program theory should be able to show that each activity is relevant to achieving one or more outcomes and that each outcome is addressed by one or more program activities unless there is reason to believe that activities that achieve lower-level outcomes will propel clients to higher levels of outcomes with little further program intervention. Even if the activities are relevant, a program theory that gives excessive attention to the detail of activities and insufficient attention to outcomes can lead to monitoring and evaluation that focuses on aspects of program delivery rather than outcomes.

Strategies to Avoid This Trap

- Instead of having a single box labeled "outcomes" in a pipeline logic model, present it as an outcome chain to show the assumed relationships among the various outcomes (see Chapter Seven).

- Identify program and nonprogram factors that affect each outcome and how the program activities and resources address those factors (see Chapter Eight).

- Identify all significant program activities and resources and show their relevance to one or more outcomes and to factors that will affect outcomes, and identify activity traps and gaps (see Chapter Eight).

Trap 2: Having a Poor Theory of Change

Just because you can represent a program theory in a logic model does not make it credible or relevant. Program theories may fail to provide a credible explanation of why one would expect a higher-level outcome to flow from a lower-level one. For example, it is not unusual to see a logic model that explains that a program to change health behaviors is understood to work by providing people with information about the health consequences of their choices of diet, exercise, and smoking. Knowledge of these consequences (achieved, for example, by improving food labeling about fat and calories) might be a necessary part of changing choices, but it is rarely sufficient to achieve behavior change. A program that works only through this mechanism is unlikely to be successful. More problematic, a monitoring system that looks only at these processes will be ignoring other important processes and intermediate outcomes and either directly or indirectly discouraging them.

From some program theories, it can be difficult to envisage how the outcomes that are the main focus of a program (and are also likely to be the focus of measurement and evaluation) will make a difference to the overall problem that gave rise to the program. There may be missing links and large gaps between the intermediate outcomes that the program is designed to achieve and the ultimate outcomes to which the program is to contribute in order to reduce or resolve the problem. For example, the outcomes chain for a mass media campaign to improve awareness of HIV/AIDS would have a significant gap if the chain jumped straight from increased awareness to reduction in HIV/AIDS. Sometimes these gaps in the outcomes chain occur because the situation that gave rise to the program (the nature and extent of the problem, its causes and consequences) has been poorly conceptualized and analyzed, measured, or poorly documented and explained.

Another version of this trap is to develop a credible solution but to the wrong problem. For example, people's failure to consume fresh fruit and vegetables might not be a deliberate choice, and therefore amenable to strategies aimed at changing individuals' choices. Rather, it might be a consequence of constrained options in their local neighborhood or budget, which would require a solution at the supply end, not the demand end, or it may be a combination of both. Moreover, further analysis or social marketing may show that even if one identified impediment is removed, compelling reasons, such as personal preferences, may exist not to consume fresh fruit and vegetables. When behavior change is an objective, good evidence about what motivates different people is essential.

Failure to undertake a good situation analysis can lead to developing a solution to the wrong problem and producing a program theory that does not reflect the situation (needs, resources, problems, and opportunities) or its causes and consequences or that has large gaps in the causal chain. Inadequate formulation can lead to failure to collect baseline data for program evaluation and to routinely monitor the problem addressed by program causes and consequences.

Strategies to Avoid This Trap

- **Involve the right mix of people** to develop the program theory to ensure that adequate knowledge is brought together to develop a plausible and defensible program theory that can be directly related to the situation analysis (see Chapter Six).

- **Draw on research and previous evaluations** to identify previous theories that may be relevant (see Chapter Six).

- **Undertake a situation analysis** that identifies problems and opportunities to be addressed, causes and consequences, including baseline data and other information about the problem and its causes and consequences (see Chapter Seven).

- **Systematically critique the quality** of the program theory (see Chapter Ten).

- **Draw on wider theories of change** and theories about how different types of program work (see Chapters Eleven and Twelve).

- **Include an evaluation of the program theory** as part of an empirical program evaluation (see Chapter Fourteen).

Trap 3: Poorly Specifying Intended Results

This trap has two variations. One is that the program theory gets stuck in the direct, tangible products of the program and does not include the longer-term outcomes and impacts that form its rationale. For example, the outcomes are expressed in terms of completed cases rather than anything about the results for clients. This can come about when people believe they will be held accountable for achieving everything in the program theory, and they know that they cannot totally control the results for clients. The other variation is that the longer-term outcomes and impacts are included but expressed in narrow ways that reflect what is readily measurable rather than what is actually the intended result. For example, instead of including a hard-to-measure outcome such as well-being, the program theory shows "percentage of clients satisfied," which is an incomplete indicator of the actual outcomes sought.

Program managers are often advised to include only measurable outcomes among their objectives. In many cases, the intent of this advice is to guard against vaguely stated ambiguous outcomes, and so it is useful advice. However, this advice can also result in a tendency to include in a program theory only easy-to-measure features. A preoccupation in many measurement systems with SMART (specific, measurable, achievable, realistic, and time-bound) objectives may foster that tendency. Including only measurable outcomes in a program theory can be quite counterproductive—a case of the measurement tail wagging the program dog. It may, for example, be difficult, if not impossible, to measure the deterrent effects of some anticrime programs. However, measurement difficulties would not be a sufficient reason to stop trying to deter crime, and it would be a brave program manager or program theorist who would advise discarding a program simply because it was not measurable.

Measurability should not be the first or even a major consideration in deciding what outcomes will be admissible in a program theory. The expected role of the intended outcome in contributing to the problem to be solved should be the main consideration in deciding which outcomes are desired of the program and which features (definitions of success) those outcomes should have. And sometimes the cost of measurement will be disproportionate to the nature of the program.

Outcomes should be defined as clearly as possible and in measurable terms as far as possible, but genuine difficulty in doing so should not be cause for eliminating the outcome from the outcomes chain. Continuing effort may need to be applied to identify innovative and proxy measures of outcomes.

Outcomes in program theories are often expressed in simple summary terms for ease of communication (for example, "Knowledge of target group improves"). However, at some point, the desired features of these outcomes need to be made explicit. An example of a desirable feature of an intended outcome might be that it should be distributed equitably across a target group and that those who are least advantaged should not be further disadvantaged. Some program theories fail to identify these important features and jump straight from the broad statement of outcome to a description of performance indicators or measures for that outcome.

Some desired attributes or features of outcomes will be easier to measure than others. In particular, some quantitative features such as numbers of participants in a program may be easier to measure on a routine basis than whether the right people are participating or what they are gaining through participation. A focus on measurement rather than definition will tend to favor those features over features that are not easily definable in measurable terms. This can result in poor program design, measurement of easily measurable outcomes, or easily measurable features of outcomes only. This in turn can result in failure to measure what is important, measurement of what is unimportant, and diversion of the efforts of the program to achieving what is measured sometimes at the expense of outcomes that are important but not measured (goal displacement).

Strategies to Avoid This Trap

- Consider whether the program theory needs to include complicated aspects of the intervention, such as a sequential process that continues beyond the involvement of your program (see Chapter Five).

- Involve the right mix of people in developing the program theory to ensure that these issues are addressed during development of the program theory (see Chapter Six).

- In logic models, express outcomes and impacts in broad terms to capture the aspects that are important (see Chapter Nine).

- Address measurement issues separately and systematically rather than quickly deciding them as part of developing the program theory (see Chapter Fourteen).

Trap 4: Ignoring Unintended Results

While program theory often focuses on the intended results, proper management and ethical evaluation require attention also to unintended results. Unintended results can be negative (such as side effects) or positive (bonus effects that add to the benefits produced by the program). Unfortunately, most logic models show only intended results, and many evaluations based on program theory ignore unintended effects.

Strategies to Avoid This Trap

- Specifically include consideration of possible unintended effects when articulating the program theory, including using negative program theory—a model of the program that represents how the intervention strategy might lead to negative results (see Chapter Six).

- Specifically include unintended results in the evaluation plan, drawing on the negative program theory and also data collection strategies that will capture unintended and unanticipated results (see Chapter Fourteen).

- Include consideration of unintended results when developing an overall judgment about the effectiveness of a program (see Chapter Sixteen).

Trap 5: Oversimplifying

Some aspects of programs are simple, with clear causal sequences under the control of the program, which can be identified in advance and managed tightly. However, not all aspects of programs are like this.

Many interventions have complicated aspects. A program might be just one piece of the jigsaw needed to produce the intended results, and it will work only if other components are in place—for example, other interventions, favorable implementation environment, or particular participant characteristics. A program theory needs to identify the outcomes that others need to achieve, as well as those to be achieved by the program.

Many interventions have complex aspects where the program cannot be specified in advance. In such cases, the program is appropriately emergent and adaptive in response to needs and opportunities that arise and as understandings of what is effective develop over time. As open systems, programs need to be scanning their environment, looking for warning signs, and picking up on opportunities to influence those out-of-scope conditions that are critical to the program's success. Complex programs, in particular, need to adopt approaches that involve working with other systems and subsystems.

Treating a program as if it were a simple, closed system, when in fact it has important aspects that are complicated or complex, can lead to insensitivity to the inherent unpredictability of the contexts within which programs function and the ways in which they are implemented, the need for programs to adapt, and the likelihood, and in many cases desirability, of emergent outcomes. Although it is important to define the scope and boundaries of a program, it is also important to recognize that boundaries are often moving and that what is on both sides of the boundary, in scope and out of scope, need to be considered in the program theory. These considerations include how what is out of scope interacts with and affects what is in scope and the extent to which there is fluid movement back and forth across the boundary. It is important to monitor and evaluate the impacts on program success of out-of-scope outcomes of other programs and other factors or conditions that affect program outcomes and its capacity to make a contribution to resolving the problem. If this is not done, there is a risk of overclaiming or underclaiming outcomes. When external factors are an element in producing outcomes, then attributing the outcomes to the program may be overclaiming. However, external factors may run counter to and dilute the apparent effects of a program. In this case, ignoring them can lead to underclaiming results. The situation might very well have been worse without the program. Counterfactuals (what would have happened without the program), even if difficult to apply, should be considered a key part of developing a program theory.

Although recognizing complicated and complex aspects is important, a risk is that stakeholders may see their program as operating so much in an open system that they believe it is powerless to affect anything but that which is within their total control. As a result, they may limit their program theory to quite instrumental and sometimes banal achievements (for example, delivery of outputs). Such an approach encourages measurement of only outcomes that are relatively easy to attribute to the program and loss of perspective on the big picture context within which the program operates. The links between what the program does and solving the problem are lost.

Strategies to Avoid This Trap

- Undertake a situation analysis that identifies as many of the important causes as possible—those that the program will and will not directly address (see Chapter Seven).

- Construct an outcomes chain that culminates in impacts that address the main situation that gave rise to the program, even if the program will not be held accountable directly or fully for those ultimate outcomes (see Chapter Seven).

- Scope the program to identify both what lies within and what lies outside the program's boundaries. As part of the scoping, show which items in the full outcomes chain will be the direct and primary responsibility of the program. Scoping is not intended to narrowly define a program in ways that prevent it from seizing opportunities as they arise. The more complex the program is, the more fluid the boundaries may need to be (see Chapter Seven).

- Develop success criteria for each of the results in the outcomes chain—for example, quality, quantity, timeliness, equity, different target audiences and other features, and comparisons. Knowledge of these criteria helps with identifying factors (program and external) that will affect the achievement of various aspects of the outcomes. Different factors will be relevant to achieving different attributes. For example, achieving an outcome with many participants (a success criterion relating to quantity) will require consideration of different factors from those that would need to be considered for achieving an outcome to a particular intensity or in a sustainable manner (a success criterion relating to quality). (See Chapter Eight on how to identify success criteria.)

- Identify both program factors and external factors that are likely to affect each intended outcome and the extent to which the desired features of each outcome are successfully achieved. (In Chapter Eight, we discuss how to identify factors that will affect success.)

Trap 6: Not Using the Program Theory for Evaluation

Although one of the main reasons for developing program theory is to use it in evaluation, one of the common traps is then not actually using it to conduct the evaluation. A program theory might have identified intermediate outcomes, for example, but the evaluation gathers data only about ultimate impacts, in the same way that a black box evaluation would. More commonly, program theory is used to identify intermediate outcomes, and the evaluation gathers data about these, but the analysis of the data does not go beyond simply reporting whether the outcomes were achieved. One of the most important ingredients of a program theory and its evaluation is not just measuring the building blocks but exploring the relationships among the building blocks.

Strategies to Avoid This Trap

- Ensure that the intended purposes of program theory are clearly articulated and agreed on (see Chapter Four).
- Rather than getting bogged down in endlessly revising the program theory, use an efficient process for articulating it (see Chapter Six).
- Make sure the evaluation uses the program theory to guide data collection (see Chapters Fourteen and Fifteen).

Trap 7: Taking a One-Size-Fits-All Approach

There are two versions of this trap: one at the program level and one at the level of an organization.

For a program, the trap is to develop a program theory, draw a logic model, and then present it as if it can meet all possible needs for all time. It is unlikely that any one model will be able to simultaneously provide an overview of all important aspects and important details. It is much more likely that different versions might be useful, each highlighting a particular aspect—a stage of the program, or how it works in a particular context, or how it is viewed from a particular perspective, for example.

For an organization or an evaluation practice, the trap is to use the same method to develop the program theory, the same way of representing it, and the same way of using it to guide evaluation. This is not to deny some value in streamlining processes and making it easier to develop and use program theory. A balance needs to be found, however, to ensure that the program theories and logic models developed represent the particular features of interventions adequately and suit the particular purposes.

Strategies to Avoid This Trap

- Carefully consider alternative ways of developing and representing program theory. Use different types of representation for different purposes—for example, big picture overviews for communication and program or organizational cohesion and nested program theories for parts of the big picture (see Chapter Nine).

- Periodically review the program theories and logic models in use, and decide whether they need revising (see Chapter Ten).

SUMMARY

The myths and traps identified in this chapter frequently reduce the usefulness of program theory. Remain alert to them, especially when introducing program theory to an organization, and be prepared with strategies for myth busting and trap avoiding.

EXERCISES

1. Are any of the seven myths set out in this chapter alive and well in your organization? If so, is this likely to be a problem? How might you address it?

2. Review an evaluation report or article that has used program theory. Is there evidence of falling into any of the traps discussed in this chapter? If not, are there descriptions of the processes they used to avoid them?

## PART 2

Assessing Your Circumstances

4

Scoping Intended Uses

PURPOSEFUL PROGRAM theory begins by identifying its intended users—who will be using it, how, and when. These have implications for who should be involved in developing program theory, what sources should be used, and how it should be represented.

WHY INTENDED USE MATTERS

Different types of program theory are needed for different uses. As a communication device to explain an intervention broadly to outsiders, it should be brief and clear. As a framework for developing accountability performance indicators, it needs to be much more detailed and show external factors that influence results. In addition, program theory developed for one use might not be suitable for another use.

Considering Intended Users

Early in our careers, we worked on a project to develop performance indicators for reporting to senior management and funders. We worked closely with staff and their supervisors to develop a program theory that they thought clearly represented important aspects of the program. Unfortunately, soon after this, a new manager was appointed who had not been involved in the development of the program theory. She did not understand what had been represented and felt no commitment to revising it to address the issues she had raised. The program theory that the group had developed was set aside and ignored when performance indicators were developed, leading to disengagement and cynicism among the staff.

Those who have contributed to developing a program theory often feel much more committed to it than those who have not been involved. Identifying who will use it, and seeking to engage them in some way, is important for successful use of program theory. Since turnover will always occur, a strategy is needed for briefing and involving new staff, managers, funders, and policymakers.

Processes that can be used to develop program theory with a small group of people who share knowledge and values will not work for a large group of people with different understandings about how an intervention could, should, and does work. In the latter case, time might be needed to negotiate the different perspectives or elaborate the details of intermediate steps and standards of performance that are not commonly understood or accepted. If intended users are numerous, dispersed, or time poor, participatory processes

THE IKEA EFFECT IN PROGRAM THEORY

The IKEA effect, according to Wikipedia, is the disproportionate sense of attachment felt for a piece of furniture, such as a bookcase, that one has assembled. This effect is often evident in program theory when those who have been involved in developing a logic model both understand and value it, while other people, who also need to use it, are reluctant to engage with it and are inclined to dismiss it.

to develop program theory might need to be curtailed, adapted, or not used at all. In Chapter Six, we discuss processes that can be used to develop program theory in a way that is appropriate for the situation at hand, including some ways of involving people who are geographically separated or not available for long periods.

Considering Timing

Program theory can be developed during the planning stage of an intervention, during implementation, at the end of a cycle of implementation (for example, at the end of one year's program before the next intake of participants), or after the intervention has ended.

Program theory developed as part of the initial planning of an intervention increases the scope for its use, but reduces the sources of information that can be used to develop the theory. It is not possible to observe the program in operation or speak to participants or staff about how they think it works, as can be done when program theory is developed after implementation. Early development of program theory also increases the likelihood that it will need to be revised to reflect emerging priorities and evidence.

Considering Intended and Other Possible Uses

Some years ago, as part of the move to change the view of government as steering, not rowing, Meals on Wheels services that had been delivered by municipal councils were contracted out to the private sector. This service supplies a daily meal to elderly people and people with disabilities who need some additional help to keep living in their own homes rather than move to residential care. (We will return to this example in later chapters.) The contract documents that specified the levels of service were based on the performance indicators that had been used to manage the service: specifications about the number of meals, the nutritional content of the meals, and the timeliness of delivery, for example. These indicators had been developed from a program theory that focused on the meals component of the service, but they did not explicitly refer to two other important strands of the program: daily social contact for the clients and checking for accidents and hazards in the client's residence. Although the simple program theory did not adequately represent the complicated nature of this multistrand program, it had not been a problem when the managers and staff shared an understanding that the program was about more than delivering meals.

The situation changed when the agency was required to contract out the Meals on Wheels service, and this program theory was used to develop the performance indicators for contractual arrangements. The successful bidders implemented a service that met all the stated requirements but ignored the elements of the program that had not been made explicit. Soon there were complaints about meals being left on doorsteps during delivery, delivery schedules being too fast to allow any time for conversation, and staff who were not able to assess recipients' well-being. In the next round of contract negotiations, a revised program theory and associated performance measures were developed that addressed the three strands of the program in terms that were understandable to people from outside the organization.

It is also important to recognize that unintended uses might not always be positive for the originating organization. For example, competitors might seek to learn from it how to implement a similar intervention, and opponents might use it to plan strategies to reduce the success of the program. These possible uses should be anticipated and managed. If commitments about the confidentiality of the program theory are made to stakeholders, they must be honored. It may be appropriate to have a less detailed public version of a logic model with the more detailed version being only used internally.

USING PROGRAM THEORY

The uses of program theory can be grouped into four clusters: planning, management, monitoring and evaluation, and synthesis for evidence-based policy and planning. We discuss these different intended uses in the rest of the chapter.

Using Program Theory for Planning

Program theory can be used for preplanning an intervention by undertaking a situation analysis, planning how an intervention will address particular needs, and planning how different elements of a large initiative are intended to work together.

Situation Analysis   A needs analysis identifies the needs or problems in a community or organization. A situation analysis goes beyond this deficit focus to also identify strengths or opportunities. We strongly recommend that program theory development begin with a systematic situation analysis. If program theory development begins instead by filling in boxes of inputs, processes, outputs, and outcomes, the risk is that incorrect assumptions about needs and strengths will remain implicit and unexamined. If a program has an existing program theory, perhaps shown in a logic model, this can be the starting point for a review of the needs analysis or situation analysis to check whether it is still valid.

Jane Davidson (2006) outlined the use of a theory-based approach to situation analysis in her evaluation of a community nutrition program. The vision or ultimate intended result of the program was "well nourished, healthy Native Hawaiian families and communities." Barriers to achieving this were identified as the limited availability and expense of healthy ingredients, the perception that healthy food was not tasty, and a lack of knowledge about nutrition. At the same time, the situation analysis identified a number of strengths that could be drawn on: extensive extended family networks, respect for elders and tradition, and a tradition of meeting and exchanging ideas in the context of a meal.

In Chapter Seven, we discuss how to undertake a situation analysis to develop or revise a program theory and how to use an existing program theory to undertake a situation analysis.

Planning an Intervention   Using program theory to ask direct questions about what the intervention is trying to do, and what evidence suggests that this is likely to be successful, can help to bring together existing information or identify gaps in existing knowledge. Developing a program theory increases the likelihood that all components have been considered, that they have the potential to work together as a cohesive whole, and that all important factors that are likely to affect performance (people, systems, resources, and organizational and extraneous factors) have been addressed. This planning can be done before an intervention starts or as part of reviewing and revising an intervention. In many cases, an evaluation that uses program theory will identify gaps and inconsistencies in the planning process or changes that have occurred since then that need to be addressed.

Being able to engage program developers in revising their understanding of the program might be an important way in which program theory can have an impact through its processes, in addition to its products and empirical findings. However, while many people report that this use is a major benefit of program theory, sometimes those who have been involved in planning are reluctant to revisit these issues. Leslie Cooksy (1999), reporting on the use of program theory for an evaluation of a curriculum delivery program in the middle grades, commented that "the development of the logic model, while important for the evaluation purposes, did not appear to stimulate much re-thinking of program processes by the program developers" (pp. 135–136).

Planning Integration of Broad Strategies   Program theory can also be used for strategic planning and whole-of-government approaches. Many programs endeavor to achieve their results through a variety of projects often scattered far and wide and delivered by different groups. Without a strategic overview, fragmentation can occur, as many and diverse projects are funded in different parts of the nation. Complicated multistrand and multilevel logic models can be used to portray and design cross-agency and whole-of-government programs that require several interdependent strategies (for example, legislative, educational, and service delivery) to achieve results. For example, Adler (2002) reported on the use of a program theory that incorporated all parts of a domestic violence service system: police, residential services, counseling, and legal services.

Developing a strategic overview of a program and portraying it using a logic model has these advantages:

- Projects and their desired results can be placed within the overview.

- The overview can identify and facilitate the interrelationships among the projects to ensure that collectively, they achieve the end results over time—for example, improvements in economic and social outcomes for individual communities and for the nation as a whole.

- The overview can show what more needs to be done to build on achievements of projects so that each project or group of projects becomes a stepping-stone to further development within the community along the path to ultimate high-level outcomes.

Funnell (2000b) showed how the program theory for a national farm forestry program was used in these ways.

Using Program Theory for Managing, Engaging Stakeholders, and Communicating

Program theory can support the management of interventions both directly and indirectly, through monitoring and evaluation. It can help different stakeholders develop a common understanding of the program or recognize differences in what they value and what they believe happens. It can help disparate individuals or sections feel part of a combined endeavor. It can help communicate about the program to newcomers or outsiders. It can be used as the basis for developing performance contracts. It can engage and empower communities. And it can identify gaps in knowledge that research can help to fill. We discuss each of these in more detail below.

Developing a Shared Understanding of the Program or an Appreciation of Different Perspectives   Program theory can be used to clarify, air, and resolve conflicting views about program objectives, priorities, and modes of operation. Very different views about how a program is supposed to operate and its underlying assumptions may be brought to the surface, and sometimes these will need to be carefully and tactfully managed. In Chapter Six, we discuss some ways of dealing with divergent perspectives.

Clarifying How Individual People or Sections Contribute to the Overall Result   Program theory can be used as a team-building and morale-boosting exercise to show how each member contributes to the success

of the program as a whole. The manager of a drug and alcohol program for prisoners who was using program theory to provide a framework for developing and testing her intervention model said this:

> Participation in the process reassured staff that they are part of a program which has a commonsense, coherent and believable rationale. This reassurance provides a buffer against external criticism. . . . However our evaluation process is not simply about giving a warm inner glow to staff. By exposing the assumptions underlying the program, and making them testable, we have raised expectations that they will be tested and will continue to be tested from time to time. We certainly have a commitment to doing so [Matthews and Funnell, 1987, p. 7].

Sometimes clarification makes existing interrelationships explicit; other times it reveals a need to improve the quality or nature of these relationships.

Communicating the Intent and Rationale of Programs to Outsiders   Program theory is a useful communication tool in that it gives a schematic overview of what the program is trying to achieve and how various short-term objectives and achievements will contribute to higher-order outcomes. McLaughlin and Jordan (1999) refer to logic models as "a tool for telling your program's performance story."

Developing Program Specifications as a Basis for Contracting Out Programs or Parts of Programs   Program theory been used to develop service agreements and specifications for contracted programs. Such specifications typically identify the outcomes and outputs to be delivered. In some cases, they also identify quality assurance mechanisms required of the contractor to ensure that activities adequately address factors within control and that contingencies and risk management strategies are built in for factors outside control. The program theory matrix introduced in Chapter 8, *Developing a Theory of Action,* will also provide a rationale for the selection of performance measures by which the contractor can reasonably be held accountable. We used this in a project to develop an agreement between state-owned electricity enterprises and a social welfare department regarding community service obligations (subsidized electricity services for poor people in crisis).

Developing Programs of Research Projects and Pilot Projects   Program theory
has also been used as a framework for identifying projects to be undertaken
as part of a research program. For example, the Sydney Olympic Roads and
Transport Authority established a program theory for the successful operation of public transport for the 2000 Olympic Games in Sydney. This was
developed two years before the games and used as a basis for identifying the
components and relationships of the overall system that would need to be
put on trial and their outcomes assessed in preparation for the games. To do
this, they conducted a series of research projects using major community
events at the various sites as pilot projects. It would be drawing a long bow
to attribute the general success of the Olympics transport in Sydney to the
use of program theory, but its use certainly helped them to organize their
thinking and initiate action.

Engaging and Empowering Stakeholders   Like any other intervention, using
program theory can affect the balance of power and knowledge among
stakeholders. In its guide to using logic models, the W. K. Kellogg Foundation (2004) described the potential impact of program theory in terms
of strengthening the capacity and power of practitioners and the community:

> Learning and using the empowering tools and skills in logic mod-
> eling can serve to increase the voice of practitioners in the plan-
> ning, design, implementation, analysis and knowledge generation
> domains. . . . Developing and using logic models is an important
> step in building community capacity and in strengthening commu-
> nity voice. The ability to identify outcomes and anticipate ways to
> measure them provides all program participants with a clear map of
> the road ahead [p. iii].

Using Program Theory for Monitoring and Evaluation

Program theory can be used to guide different types of monitoring and
evaluation, including process evaluation, performance monitoring, outcome
and impact evaluation, and cost-benefit and cost-effectiveness evaluation.
Program theory can also be used to develop an evaluation framework that

covers the range of monitoring and evaluation that will be undertaken during the life of the intervention.

In Chapter Fourteen, we discuss how program theory can help to focus an evaluation, identify what should be measured or described in an evaluation, and provide a coherent framework for analysis and reporting.

Using Program Theory for Evidence-Based Policy and Practice

Program theory is an essential part of evidence-based policy and practice in order to support appropriate translation of knowledge about successful pilots and innovations to other settings.

Documenting Innovation   Evaluation can provide evidence about what works or what works in what circumstances, but without adequate documentation of innovation, what "it" is may not be clear. Program theory can provide a framework for documenting an innovation. You might use as your starting point specific instances of implementation, described in concrete detail. From these, you can iteratively develop and redevelop the program theory, gradually building up an explanation of why things are being done in a particular way and what seems to be making a difference to the end product.

In an evaluation of a pilot project that helped those who were arrested get bail (CIRCLE, 2003), important insights about program theory came from interviews where those who had been arrested, who had rated the project as very helpful, were asked what it was that made a difference. These open-ended responses from different clients built up a picture of the important elements in the project—particularly the two key features of the project officer who was approachable and clearly dedicated ("not like a public servant at all") but also made it clear she would not waste her time on clients who were not serious about actively engaging in the project. Using rich qualitative data to develop and redevelop the program theory can be important when documenting an innovation.

Supporting Appropriate Adaptation and Fidelity   Program theory is used differently if you are trying to replicate or scale up a successful program. Type II translation focuses on learning how to successfully translate interventions that have demonstrated their efficacy under controlled conditions into implementation in the community, or from one type of implementation environment to another (Rohrbach, Grana, Sussman, and Valente, 2006). How do you know which elements need to be exactly replicated and which elements can and should be varied to adapt to the new situation? Without program theory, it can be difficult to know which things can be varied.

One of the projects we worked with was a middle school prevention project that worked with at-risk children through providing mentoring, after-school programs, and improving children's relationships with their parents and teachers. The project had been seen to work effectively in a school on the East Coast of the United States and was being implemented for the first time in California. One of the features of the program was a pizza night held at the beginning of the school year. Students, parents, and teachers came together to meet socially, play games, and begin to establish relationships. One day, a call came through to the project office from the school that was implementing the project for the first time. They were ready to start the program but had a question. Did it have to be pizza?

Without program theory, we are constrained to exactly copy something that has been shown to work, like children copying all the mannerisms of their parents, even though these may not be necessary to perform the action. In an era when organizations are simultaneously encouraged to implement evidence-based programs that have been shown to work, and yet also respond to the particular needs of their clients, it is essential to be able to identify which elements must be copied exactly and which can be adapted. "Does it have to be pizza?" is a question that reveals a lack of understanding of how the project actually worked or a lack of authority to make any changes, even if these would improve the local appropriateness and accessibility of the project. There was nothing magical about having pizza rather than, say, tacos or hamburgers or other food that has connotations of a social occasion. For this purpose, program theory needs to clearly distinguish between the theory of change (develop a relationship among students, parents, and teachers in a casual, social atmosphere), which should be kept, and the theory of action (have a pizza night), which might well be adapted to suit the local situation.

Synthesizing Evidence from Multiple Evaluations   Lester Salamon, in an article on the new forms of government action that became popular in the 1980s as governments moved to steer rather than row, commented:

> Rather than focusing on individual programs, as is now done, or even collections of programs grouped according to major "purpose" as is frequently proposed, . . . we should concentrate instead on the generic tools of government action that come to be used in varying combinations, in particular public programs. . . . The development of a systematic body of knowledge about the alternative tools of public action is the real "missing link" in the theory and practice of public management [(Salamon, 1981, p. 256].

Various authors have suggested ways of identifying underlying patterns among programs to create taxonomies of programs for program theory and program evaluation. Mark Lipsey (2000) advocated the use of meta-analysis to build up research-based knowledge of particular types of programs and reported some examples of the use of meta-analysis to build and strengthen our understanding of program theory. He suggested a taxonomy built around three dimensions: change paradigm, problem area, and program area.

Anthony Petrosino (2000) also reported use of meta-analysis for different types of programs to combat recidivism. Petrosino used meta-analysis to establish that cognitive-based programs produced larger effect sizes than behavioral, individualized, or group counseling. He then focused on the intermediate outcomes or mechanisms through which the cognitive-based category of programs produced its effects.

Using program archetypes (which we discuss in Chapter Twelve) as a thinking tool for program theory development rather than as a blueprint is consistent with Pawson's (2003b) approach when he said:

> The aim should be to produce a sort of "highway code" to programme building, alerting policy makers to the problems that they might expect to confront and some of the safest measures to deal with them. What the theory-driven approach initiates is a process of thinking through the tortuous pathways along which a successful programme has to travel. What it produces and what you, dear

evaluators, should be advising is: "remember A," "beware of B," "take care of C," "D can result in both E and F," "if you try G make sure that H is in place" [p. 488].

SUMMARY

Identifying the intended uses and users of program theory is an important starting point because these will have implications for how program theory should best be developed and used. These implications are summarized in Table 4.1.

Table 4.1   **Implications of Different Aspects of Use for Developing and Using Program Theory**

Aspect of Use	Possible Implications for Development of Program Theory	Possible Implications of Use of Program Theory
Stage of intervention	Availability of data to inform development of program theory	Opportunities to collect data about the intervention
Primary intended users	Who should be involved in the process, whether to use a prescribed format, ability to explicate tacit knowledge or identify other data sources	Risk of ignoring elements not represented in the logic model, common understanding of terms used
Other possible users	Level of willingness from key informants to explicate their tacit knowledge	Need to consider gaps in their existing knowledge of the intervention
Intended uses	Whether a consistent format is needed, level of detail required, elaboration of interaction with context, need to show potential unintended effects, need to show all relevant causal strands, process use agenda	Whether the program theory is adequately understandable by all intended users

EXERCISES

1. Find an example of an intervention that has used program theory. At what stage of the intervention was the program theory developed? Who were the intended users? What were the intended uses of it? Is there evidence of other uses being made of the program theory?

2. Consider an intervention you know well. What would be a likely intended use for a program theory of it? What might be the implications of this intended use for developing and representing the program theory?

5

The Nature of the Situation and the Intervention

ONE OF THE key messages of this book is the importance of taking into account the nature of the intervention when developing, representing, and using program theory. Are many organizations involved in making a strategy work? Does a program work the same way for everyone? Will a project succeed only if other services are also available for participants? These issues all relate to whether it is reasonable to think about an intervention as simple, or whether there are multiple components or complex aspects. This chapter helps you to identify whether an intervention has important complicated or complex aspects, or whether it is appropriate to think about it as simple.

SIMPLE, COMPLICATED, AND COMPLEX

Despite our emphasis on complicated and complex aspects of interventions, we endorse the approach of Albert Einstein who, in a lecture at Oxford University in 1933, urged theorists to aim for an appropriate level of simplicity: "It can scarcely be denied that the supreme goal of all theory is to make the irreducible basic elements as simple and as few as possible without having to surrender the adequate representation of a single datum of experience" (Einstein, 1934, p. 165). Or, as it is usually paraphrased, "Everything should be made as simple as possible, but no simpler." The advice holds true for program theory: it too should be as simple as possible, and no simpler.

Conceptualizing Simple, Complicated, and Complex

Interest in considering how evaluation should address the simple, complicated, and complex aspects of interventions is increasing (Eoyang and Berkas, 1998; Sanderson, 2000; Arnkil and others, 2002; Barnes, Matka, and Sullivan, 2003; Douthwaite, Delve, Ekboir, and Twomlow, 2003; Patton, 2003, 2008, 2010; Barnes, Sullivan, and Matke, 2004; Sibthorpe, Glasgow, and Longstaff, 2004; Stame, 2004; Guijt, 2008; Mayne, 2008; Ramalingam, Jones, Young, and Reba, 2008; Rogers, 2008; Leeuw and Vaessen, 2009; Rogers, Guijt, and Williams, 2009; Rogers, Stevens, and Boymal, 2009). These terms have been defined in slightly different ways by those who have written about applying ideas from complexity science to human organizations and behavior. The approach we set out in this book draws on three ways of conceptualizing what is complicated and what is complex.

One of the most influential characterizations of simple, complicated, and complex, including referencing in the *New York Times* (Segal, 2010), was outlined by Sholom Glouberman and Brenda Zimmerman (Glouberman, 2001; Glouberman and Zimmerman, 2002) in exploring the implications of the typology for reforming Medicare in Canada. They developed a characterization with three exemplars: a recipe, a rocket ship, and a child. The distinctions among the three, shown in Table 5.1, relate to the universality of prescriptions for practice and the level of expertise required to implement them successfully.

Table 5.1 **Simple, Complicated, and Complex Problems**

Following a Recipe	Sending a Rocket to the Moon	Raising a Child
The recipe is essential	Formulae are critical and necessary	Formulae have a limited application
Recipes are tested to assure easy replication	Sending one rocket increases assurance that the next will be OK	Raising one child provides experience but no assurance of success with the next
No particular expertise is required. But cooking expertise increases success rate	High levels of expertise in a variety of fields are necessary for success	Expertise can contribute but is neither necessary nor sufficient to assure success
Recipes produce standardized products	Rockets are similar in critical ways	Every child is unique and must be understood as an individual
The best recipes give good results every time	There is a high degree of certainty of outcome	Uncertainty of outcome remains
Optimistic approach to problem possible	Optimistic approach to problem possible	Optimistic approach to problem possible

Source: Glouberman and Zimmerman (2002, p. 2)

Their main message was that interventions such as health care systems are complex, and attempts to repair them as if they were complicated are doomed to failure. Or, as Douglas Adams put it, "If you try and take a cat apart to see how it works, the first thing you have on your hands is a nonworking cat" (as reported in Dawkins, 2001).

Brenda Zimmerman (2001), building on the work of Ralph Stacey (1993), discussed a slightly different conceptualization. She represented the concepts in terms of a matrix of two dimensions: agreement about ends and certainty about means. Simple situations have both agreement and certainty, complicated situations have one of these, and complex situations have neither. Technically complicated situations have agreement about ends but uncertainty about means; socially complicated situations have disagreement about ends but certainty about means. In practice, as Patton (2010) has pointed out, "High uncertainty about how to produce a desired result fuels disagreement, and disagreements intensify and expand the parameters of uncertainty" (p. 90).

Cynthia Kurtz and Dave Snowden (2003) defined the terms slightly differently, focusing on the knowability of situations and adding two more options. They characterized simple as the domain of the known, where cause and effect are well understood, and best practices can be confidently recommended; complicated as the domain of the knowable, where expert knowledge is required; and complex as the domain of the unknowable, where patterns are evident only in retrospect. Their framing also included chaotic (no patterns were evident, even in retrospect) and disordered (the classification was not clear) (Snowden and Boone, 2007).

Implications for Program Theory

We have drawn on this previous work of the concepts about simple, complicated, and complex and adapted it to focus on the most salient features and implications for using program theory in evaluation (Table 5.2). For thinking about program theory, we have found it useful to consider complication in terms of multiple components—multiple and competing objectives or causal strands, interventions that operate on multiple levels or involve multiple implementing agencies, and multiple contributors to outcomes—and complexity in terms of adaptation and emergence—evolving focus, activities, and collaborations. It is important to realize that classifying an intervention as simple does not mean it is trivial or necessarily easy to implement.

We find it useful to think about these in terms of what interventions look like—their focus, governance, and consistency—and how they work—their sufficiency, necessariness (that is, their state of being necessary), and change trajectory. An intervention can appropriately be thought of as simple if it looks simple and works in a simple way. It looks like a discrete, standardized intervention with an agreed set of objectives that is implemented by one agency and carried out the same way everywhere. It is both necessary and

Table 5.2 **Distinguishing Simple, Complicated, and Complex Aspects of Interventions**

	Simple Aspects	*Complicated Aspects*	*Complex Aspects*
What interventions look like	Standardized intervention activities, implemented by a single organization	Multiple components or implemented by multiple identifiable organizations in predictable ways	Nonstandardized and changing, adaptive, and emergent Implemented by multiple organizations with emergent and unpredictable roles
How interventions work	Pretty much the same everywhere	Differently in different situations and for different people or in different implementation environments	Generalizations rapidly decay Results are sensitive to initial conditions as well as to context

sufficient to produce the impacts of interest, has a constant dose-response relationship, and works the same for everyone in every situation. Simple interventions can be adequately evaluated in terms of what works.

In practice, few, if any, interventions meet these criteria exactly. In some cases, it might be reasonable to treat an intervention as if it were simple, and to develop and use program theory suitable for simple interventions. But if there are important complicated or complex aspects, recognizing these when developing, representing, and using program theory is a better choice.

An intervention will likely have elements of each of these categories: some parts that are conceptually simple but logistically difficult, others that comprise multiple components, and some degree of emergence. For example, in some respects, it might be useful to think about vaccination as a simple intervention. For many diseases, preventing illness from that disease is both necessary and sufficient; there is no other way to achieve immunity, and vaccination by itself is enough to cause immunity. However, there might well be ways in which it is best to think of it as complicated—for example, in terms of different organizations needed for the logistics of implementing it across a country—and ways in which it might be helpful to think of it as complex— for example, in terms of ways to encourage people to participate in voluntary vaccination. Therefore, this classification is best used as a heuristic to help in thinking about interventions and their evaluations and to consider the implications of seeing them in different ways rather than as a classification system where interventions can be absolutely classified into a single type.

In the remainder of the chapter, we explore these distinctions in more detail with respect to focus, governance, consistency of implementation, necessariness, sufficiency, and change trajectory and discuss their implications for developing, representing, and using program theory throughout the book.

FOCUS

Program theory explains how an intervention contributes to its intended or observed outcomes. It can therefore be a challenge when intended outcomes are multiple or emergent.

Simple

A simple intervention has an agreed set of objectives in terms of intended outcomes and processes. This makes the process of developing program theory easier because the end point of the intervention is agreed. It also makes monitoring and evaluation easier because of agreement about the evaluative criteria for outcomes, even if there is disagreement about the standards to be used. For example, in an adolescent pregnancy prevention project, if there is agreement that the intended outcome is a reduction in the number of adolescent pregnancies and that this should be achieved through preventing conception rather than implementing terminations, there may still be disagreement about how much of a reduction would constitute success: reduction to zero, reduction compared to the previous year, or reduction compared to similar communities where the project was not implemented.

Complicated

An intervention has complicated aspects when different individuals or organizations value different impacts, the intended impacts are multiple and competing, or impacts are needed at different levels of a system.

Different Valued Outcomes
Partner organizations, or even individuals within the same organization, do not always have the same agenda or share the same view on intended impacts. For example, the WagePause program, which provided training and work experience for long-term unemployed people in Australia, involved a federal employment department (that provided funding), a state technical education department (that delivered training), and local government (that provided work placements). These different partners had different legitimate accountabilities and different needs in terms of what were measured as desirable impacts: employment after leaving the program, completion of qualifications, or completion of local projects such as flood mitigation. These different impacts could be in conflict with each other. For example, a participant who left the program halfway through a placement to take up a permanent position would be a success in terms of the first of these (employment) and a failure in terms of the other two.

It can be important to recognize when there are different legitimate views on the objectives of an intervention and to address this issue effectively. For different intended outcomes of a program, it may be difficult to develop an agreed program theory because agreement about what should be identified as the intended impacts may be lacking.

The best option is probably to develop a program theory that includes all of the identified objectives of the different parties. It can sometimes be helpful to have an external person assist in this process. Where different objectives are addressed in a single program theory, these may be in tension and need to be handled in terms of the issues we discuss in the next section. In Chapter Six we also discuss the use of multiple program theories.

Another option is to develop different program theories for the different perspectives. It cannot be assumed that a single evaluation can meet partners' differing needs. For example, a review of evaluations of economic development programs found that joint evaluations, where the one evaluation is undertaken of a program with multiple partners, may end up being of little use to either partner because the consensual focus of the evaluation ends up meeting no one's actual information needs (Toulemonde, Fontaine, Laudren, and Vincke, 1998). However, this approach can come with risks; the partners in an intervention need to bear in mind the needs of the others when planning, managing, and evaluating the intervention to ensure successful continuation of the cooperation.

Multiple Competing Imperatives   Sometimes an intervention has multiple strands and are all needed to produce the intended outcomes. Monitoring or evaluation that focuses on only one of these is likely to be dysfunctional. When procurement, accountability, or arguing for resources will be based on the program theory as articulated, all the causal strands need to be included. Recall the story in Chapter Two about a Meals on Wheels program where the program theory failed to identify all the important outcomes; the result was dysfunctional management and performance monitoring.

Activities Needed at Different Levels of a System   Sometimes interventions are complicated in that they need to operate at different levels of a system in order to bring about change (Figure 5.1). For example, a new program might seek to change the way schools operate so that teachers can change the way they operate, so that students can engage in learning in different and better ways. The intervention might have activities at each of these levels. Or different organizations might be involved in the activities that relate to different levels of the system. A programming implication of socioecological theory, which we discuss in Chapter Eleven, might be the need to work at different levels of the systems that surround program participants.

Complex

Interventions with important complex aspects are dynamic and responsive to changing needs, opportunities, and challenges rather than following a path

Figure 5.1   **A Complicated Program Theory for a Multilevel Intervention**

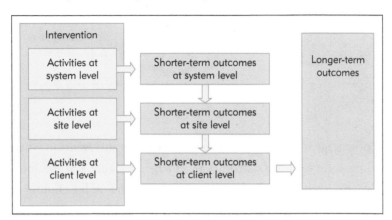

that has been tightly defined in advance to achieve tightly specified objectives. Planning is bounded to some degree, because it is done with an overall vision in mind. Nevertheless, it acknowledges that the intended impacts are likely to change as different people become engaged, opportunities and threats emerge, and the program unfolds.

There are two different version of emergent outcomes: one where intermediate intended outcomes are changeable and emergent and one where even the long-term intended outcomes are emergent.

In the first of these, an intervention has a clear statement of its intended ultimate outcomes, but the intermediate outcomes are emergent (Figure 5.2). Community capacity development programs often look like this. We worked on the evaluation of a community-strengthening program that encompassed hundreds of community projects. All of the projects were intended to develop community capacity in terms of human capital, social capital, organizational capital, and economic capital, but the specific intermediate outcomes, and the activities to achieve these, were not specified in advance; rather, they were decided in consultation with local communities. They had an enduring long-term goal, but within this, they had varying and sometimes changing short-term outcomes, such as creating a community garden, establishing reciprocal child care, and training volunteers.

Those who have always worked within the framework of clearly stated program objectives can find it challenging to consider the second type

Figure 5.2   **Emergent Focus of the Intervention: Intermediate Intended Outcomes**

of emergent outcomes and conceptualize an intervention where even the long-term intended outcomes are emergent (Figure 5.3). Yet for some interventions, the nature of the challenges and the understanding of these are constantly changing, so this is an appropriate way to conceptualize them. A project might start with a quite limited aim; as issues and opportunities emerge and a deeper understanding of what is needed and what is possible is developed, the ultimate goal is reframed. When a program theory is revisited and revised, any substantial reframing of goals should be transparent, explicit, and acknowledged. We have experience with several programs for which staff and others have developed a shared understanding that the program has moved in a different direction, but this movement has never been formally recognized and documented. As a result, there have been problems when an audit is conducted that takes the original goals of the program as its point of reference. The program may be judged in terms of something it is no longer trying to achieve, and what it is trying to achieve may be overlooked.

In their book *Getting to Maybe* (2006), Frances Westley, Brenda Zimmerman, and Michael Patton describe several programs that have evolved over the years, not only in terms of what they do but in terms of how they understand their overall mission. For example, the Hope Community in Minneapolis, Minnesota, began as St. Joseph's House, a service for homeless women and children. As the crack epidemic worsened, landlords abandoned their properties, and families hid their children inside rather than risking the violence on the streets. The organization changed its mission to one of neighborhood revitalization and changed its name to the Hope Community, starting a process of gradually changing what it did and how it operated in reaction to the changing environment.

Figure 5.3   **Emergent Focus of the Intervention: Long-Term Intended Outcomes**

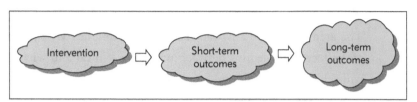

Dynamic and emergent interventions present a challenge to conventional linear processes of developing an evaluation, implementing it, and reporting the findings. Dynamic interventions change substantially over time, and their specific impacts cannot always be identified in advance. For example, community capacity-building projects often work by bringing together community members to identify local issues, develop a plan for addressing them, gather resources and implement the plan, and then review progress and begin planning again. Within a community-based health promotion program, some projects might focus on under-age smoking and some on diabetes management. Even different projects that have a similar desired outcome are unlikely to implement the same activities.

The solution is not to make logic models increasingly complicated. In fact, it is not possible to develop a blueprint for the intervention. What is needed in situations of complexity is an agile heuristic that can be revised and combined repeatedly. Many central funding programs have these characteristics; examples are comprehensive community initiatives in the United States (Weiss, 1995; Connell and Kubisch, 1998), community-based prevention programs, Health Action Zones in the United Kingdom, and many grant-making programs such as the European Union's structural funds.

An evaluation of the program that provides funding to these projects cannot specify SMART objectives (specific, measurable, achievable, relevant, time-bound) in advance, such as building a certain number of playgrounds or providing a certain number of hours of training, since these will vary for each project, will emerge over time, and need to be developed by the community rather than be imposed.

Ricardo Wilson-Grau (2010), an evaluator based in Brazil and the Netherlands who uses complexity concepts in his evaluation practice, has described his experiences when organizations have insisted on SMART objectives for complex interventions:

> The arena of public advocacy is rife with complexity—the cause and effect relationships between what you will do and what you will achieve are simply unknown. In fact, defining what you want to achieve so precisely is usually counterproductive, because of the wasted time and resources in attempting to predict the unpredictable, or because of the blinders you set for seeing the unexpected, or the discouraging disappointment of not achieving what is so meticulously planned.

Sometimes it is possible to use high-level outcome measures that are likely to be common across projects, such as various social indicators of individual and family well-being. But this is feasible only when projects have a clearly defined geographical focus that matches the regions for which social indicator data are available, and when there is sufficient time for these outcomes to become evident during the life of the evaluation. This is relevant only in terms of learning about the final impact. It does not address the need for evaluation to learn about improving performance during implementation.

Dynamic and emergent interventions are therefore likely to need an element of dynamic and emergent evaluation design and measures: "Developmental programming calls for developmental evaluation in which the evaluator becomes part of a design team, helping to monitor what's happening, both processes and outcomes, in an evolving, rapidly changing environment of constant feedback and change" (Patton, 1994, p. 313). This might involve a process of iteratively identifying and answering important evaluation questions. Alternatively, it might involve progressively more specific questions during the life of the project.

GOVERNANCE

By *governance,* we mean the organizations that are making decisions about an intervention. John Donne famously wrote that no man is an island, and, similarly, few interventions are now completely separate. Rather, they encompass a number of organizations as partners, funders, referrers, regulators, or advocates. It is therefore important to include these different organizations in the process of developing program theory and to ensure that the program theory addresses their contributions.

Simple

A simple intervention is undertaken by a single organization, with agreed objectives and procedures. This provides a common reference point for identifying the intended outcomes of the intervention, makes it easier to ensure that all relevant parties participate in program development, and makes it easier to ensure access to data for monitoring and evaluation.

An intervention that is implemented by a single organization but has different departments that do not share objectives, procedures, or data is not simple. Conversely, if a number of organizations have been working together for so long, and so effectively, that they share common objectives, procedures, and data, it is more appropriate to consider this as simple. The litmus test for using the classification is whether it is likely to make an important difference to developing, representing, or using program theory.

Complicated

In many interventions, multiple stakeholders work together. These arrangements can be considered complicated when there is agreement about what will be done and how—often in the form of contracts, letters of agreement, or lists of deliverables. The process of developing program theory needs to ensure that it includes these different partners sufficiently and uses the agreements to define intended activities and outcomes.

Complex

The difference between complicated and complex governance is the issue of emergence. Complex governance involves an emerging list of partners and emerging relationships and responsibilities. It is perhaps best illustrated by a quote from *The Dish,* a 2000 film about the cooperation between the Australian radio telescope and the *Apollo 11* moon mission: "You remember that night at my place? Trying to sort out the contract with that fella from

NASA? . . . Two hours, and you finally speak. 'Gentlemen, this should be the contract. We agree to support the Apollo 11 mission.' That was it—one sentence."

CONSISTENCY

Consistency refers to the types of activities that are undertaken and how they are undertaken.

Simple

Simple interventions look the same even when different people at different sites implement them. Fast food companies have derived much of their success through standardizing the preparation and serving of food and beverages. This standardization ensures consistency of quality and also makes it possible to employ staff without specialist skills. A similar strategy is used in some approaches to evidence-based teaching, social work, and counseling, where considerable effort is put into ensuring fidelity to already-developed curriculum or protocols.

Whether this is an appropriate strategy depends on whether the intervention actually works in a simple way—that is, the same for everyone. If the intervention works better if it is adapted to meet the specific needs of participants, then conceptualizing it as complicated or complex is more effective.

Complicated

Complicated interventions have been adapted to meet the different needs of groups of participants. Even fast food companies are increasingly adapting their menu to meet the specific needs and preferences of customers in different locations, with McDonald's producing a vegetarian burger in India and a coffee bar in Australia.

In human services, service delivery often is adapted to meet the needs of a particular demographic group or a specific site. For example, a maternal and child health program that we evaluated worked largely through home visits but used different strategies for teenage mothers and Indigenous mothers. A special center was designed for teenage mothers that would encourage the development of social networks and access to other services. A visiting service to a community-based Indigenous health service was provided to encourage trust and engagement among Indigenous mothers, many of whom have understandable concerns about having government officials visit them at home, given the history of forcible removals of Indigenous children.

The appropriateness of this approach depends on the accuracy of the needs of the group and their homogeneity, as well as the appropriateness and legitimacy of providing different services to different groups. It can be inappropriate to make assumptions about people's needs, literacy, and preferences on the basis of their demographic characteristics.

Complex

Complex interventions take the notion of adaptation further. Rather than having a standardized intervention for a particular group of participants, the intervention is adapted for each individual participant, at least in theory. The "individual" could be a person, a family, a group, an organization, a community, or even a nation. What makes the case individual is the uniqueness of their situation and the responses needed to it. For example, an effective maternal and child health nurse who is providing home visits and clinic sessions for new mothers will customize the advice she provides and how she provides it to match the assessed needs and preferences of a specific client, and she will monitor and follow up the client's response to this, including changing strategies as needed. In such cases, the focus of the intervention is fixed, but the specific activities undertaken and how they are undertaken is highly adaptive. The case management and community capacity building archetypes, which we discuss in Chapter Twelve, are useful for these types of programs. Figure 5.4 represents how a varied and changing intervention might still be intended to contribute to fixed outcomes.

Figure 5.4   **Emergent Activities to Achieve Stable Intended Outcomes**

Complex interventions can also change their activities considerably over time. For example, community capacity-building programs have many emergent aspects, as they work by identifying needs and strengths and then iteratively planning and implementing specific actions as priorities and opportunities change. A health promotion program that seeks to bring about behavior change through providing information might use a range of activities (brochures, Web sites, stalls at community fairs) and respond opportunistically to new possibilities. It is not possible to report in terms of what works because that is constantly changing.

NECESSARINESS

Necessariness is the first component of the classic causal conceptualization of a cause as being necessary and sufficient.

Simple

A simple intervention is essential to achieving the intended impacts—for example, if participation in a training program is the only way to develop particular knowledge or skills.

Complicated

Many interventions are not the only way to achieve the intended impacts. For example, students in a statistics course might learn the course content by reading books, attending other courses, or getting private tutoring instead of attending classes. This aspect of complication has particular importance when undertaking causal analysis. If there were no differences in achievement between students who attended or did not attend class, a naive causal analysis might conclude that the classes have been ineffective. A proper evaluation would require a comparison of the costs and benefits associated with the alternatives. Perhaps the classes were very efficient for participants, and those who could not attend had to spend more time and money to undertake alternative activities. We discuss this issue in more detail in Chapter Fifteen.

SUFFICIENCY

Sufficiency relates to the second component of "necessary and sufficient": can the intervention produce the intended impacts without assistance from other interventions or favorable contexts? Logic models often imply that the intervention by itself is sufficient to produce these impacts. If this representation is not accurate, however, it can lead to misleading causal analysis.

Simple

A simple intervention is sufficient to produce the intended impacts. This means that an intervention works equally well for everyone.

Complicated

An intervention is often only one part of a causal package, and other interventions (previously, concurrently, or subsequently) or other factors (such as a favorable implementation environment or participant characteristics) are needed for success. Recognizing and addressing these types of complications when they are present is essential for effective management and accurate causal analysis.

Contributions of Previous, Concurrent, or Subsequent Interventions   A common challenge in using program theory is that the organization implementing the intervention does not have control of the whole causal chain to the final intended outcomes and is dependent on previous, concurrent, or subsequent interventions. For example, training programs aim to develop participants' skills and knowledge, but the real benefit of the training depends on whether they are able to apply these skills when they return to the workplace. Often trainers have no control over this. In agriculture, basic scientists produce knowledge that applied scientists use to develop new varieties or breeds, which extension programs communicate to farmers and other end users, who then implement the new technology to produce the intended outcomes. At any point along this multistage intervention, the causal chain could break.

An evaluation can itself usefully be thought of as a multistage intervention (Figure 5.5). One of the standards for judging the quality of an evaluation

Figure 5.5   **An Intervention Dependent on the Contribution of a Subsequent
Intervention**

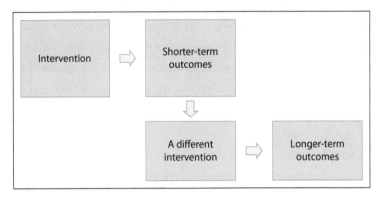

(Joint Committee on Standards for Educational Evaluation, 1994) is utility (the others are accuracy, propriety, and feasibility). Evaluations with high utility make a difference to decisions and actions that contribute to improved programs, better selection of programs, and better resource allocation. But the actual use of evaluations is usually outside the control of an evaluator or an evaluation team. They can (and usually should) seek to encourage the utilization of an evaluation report, but in the end, it may not be their fault if an evaluation is not used.

Simple logic models that have all the inputs and processes at the front produce a misleading representation of complicated interventions, because additional processes (and resources to produce these) will be needed at different stages. Sometimes the outputs from one intervention will be the inputs to another intervention, which then produces the ultimate outcomes and impacts (Figure 5.6).

Interventions that depend on other interventions to succeed present challenges to notions of accountability. If someone in a relay race drops the baton, whose fault is it? Those involved in the first stage are often adamant that it is not reasonable to hold the implementers of the first stage accountable for achievement of the final outcomes, since they do not control them. But if there is no success in moving from the first stage to the final outcomes, continuing to implement this first stage has little point. In Chapter Nine, we discuss outcome mapping, a type of program theory that focuses on the

Figure 5.6   **Outputs from One Intervention Forming the Inputs to a Subsequent Intervention**

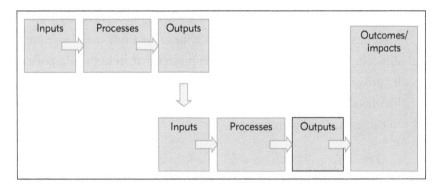

Figure 5.7   **A Complicated Program Theory for an Intervention That Works Only in Conjunction with Other Interventions**

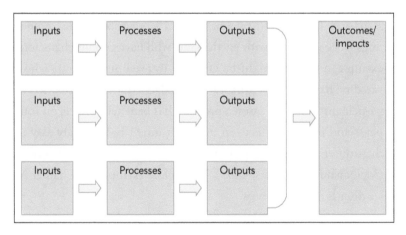

handover between two or more organizations that need to work together to produce the intended outcomes.

The success of complicated interventions can also be dependent on concurrent interventions (Figure 5.7). For example, Meals on Wheels is intended to help support elderly people and people with disabilities to remain living in their own homes rather than move to institutional accommodation. But Meals on Wheels can achieve this only if clients' other needs are being met by other services to provide house cleaning, home maintenance, and possibly visiting nurse services. If these other services are not in place or not working well, the outcomes and impacts might not be achieved, even if Meals on Wheels is working well.

If these concurrent interventions are being implemented by other organizations, getting information about them for a full causal analysis can be difficult. For example, an agriculture project that seeks to help poor farmers by providing better crop varieties might be successful in improving rural livelihoods only if other programs provide support to farmers to enable them to harvest and sell their crops.

Differential Effects Depending on Implementation Context   The implementation context refers to the physical, social, and other characteristics of the implementation site. These can sometimes have an effect on the results achieved. For example, a substance abuse prevention program will work differently at a site where geography helps control access to substances.

Differential Effects Depending on Participant Characteristics   Some interventions work well only with participants who have certain characteristics—for example, a teacher might be effective in teaching students who are motivated or have the prerequisite knowledge—and be ineffective with other participants. In some cases, a program has beneficial effects on some participants and negative effects on others. It might be relatively easy to identify these differential effects, or it might require systematic market segmentation (McDonald and Rogers, 1999). We explore examples in Chapter Nine when we discuss realist matrices.

CHANGE TRAJECTORY

The change trajectory describes the pattern of impacts over time or the relationship between the intervention and the impacts.

Simple

In a simple intervention there is a constant relationship between cause and effect and a constant change over time that can be readily understood by anyone. An example of a simple relationship would be where a small increase in police numbers leads to a small decrease in crime, and a larger increase in police numbers leads to a larger decrease in crime.

Complicated

A complicated change trajectory requires expertise to understand. It might be curvilinear; for example, performance is best in situations of medium stress, with both low stress and high stress producing poor performance. It might be so complicated that it needs expertise to model it—for example, through systems dynamics modeling of the interactions among many components, including reinforcing loops (a vicious or virtuous circle) or balancing loops. An example of a complicated relationship would be where rehabilitation services appeared to have little benefit until the threshold of achieving a return to work was achieved (Batterham, Dunt, and Disler, 1996).

Complex

In complex interventions, even experts cannot predict results because of the changing nature of the relationship between cause and effect or the many factors affecting it. Only in retrospect can these can be understood.

Complex interventions might have tipping points where a small difference in a variable can have an unpredictable disproportionate effect on results. An example of a complex relationship would be where an advocacy program achieved a critical mass of active supporters and was able to produce information resources that went viral, growing exponentially. Figure 5.8 represents such an intervention, where in a slightly different situation, quite different outcomes result.

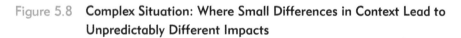

Figure 5.8   **Complex Situation: Where Small Differences in Context Lead to Unpredictably Different Impacts**

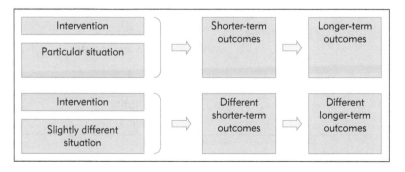

SUMMARY

The ways of thinking about simple, complicated, and complex aspects of interventions are summarized in Table 5.3. The framework should be understood as a heuristic rather than a set of recipes to follow for different situations.

Table 5.3 **A Framework for Addressing Simple, Complicated, and Complex Interventions**

	Simple	*Complicated*	*Complex*
What it looks like			
1. Focus	Single set of objectives	Different objectives valued by different stakeholders Multiple, competing imperatives Objectives at multiple levels of a system	Emergent objectives
2. Governance	Single organization	Specific organizations with formalized requirements	Emergent organizations in flexible ways
3. Consistency	Standardized	Adapted	Adaptive
How it works			
4. Necessariness	Only way to achieve the intended impacts	One of several ways to achieve the intended impacts	
5. Sufficiency	Sufficient to produce the intended impacts Works the same for everyone	Works only in conjunction with other interventions (previously, concurrently, or subsequently) Works only for specific people Works only in favorable implementation environments	

	Simple	Complicated	Complex
6. Change trajectory	Simple relationship that is readily understood	Complicated relationship that needs expertise to understand and predict	Complex relationship (including tipping points) that cannot be predicted; understood only in retrospect
7. Unintended outcomes	Readily anticipated and addressed	Likely only in particular situations; need expertise to predict and address	Cannot be anticipated

EXERCISE

1. Choose an intervention you know well. What are some aspects of it that could usefully be considered simple? Are there some aspects that would be better to consider as complicated? If so, which ones are important? Are there some important aspects that would be better to consider as complex? What are these?

## PART 3

Developing and Representing Program Theory

6

Processes to Identify or Develop a Program Theory

MANY METHODS and approaches can be used to identify the theory that underpins an existing program or develop a program theory that can be used to design a new program or redesign an existing one. There is no one best way to develop or identify a program theory. This chapter discusses factors to consider when making decisions about processes to develop a program theory. Rarely should the development of a program theory be a one-off exercise. The program theory should evolve along with the program.

PROCESS OPTIONS FOR DEVELOPING PROGRAM THEORY

Purposeful program theory makes deliberate choices of developmental methods to match the desired use(s) of the program theory. Some common purposes were discussed in Chapter Four, for example, use of program theory to develop an evaluation framework, reach consensus, and communicate with stakeholders about what a program is trying to achieve and how it will do so. Purposeful program theory also identifies other purposes or desired by-products of the process of developing the program theory and chooses processes that will be appropriate for those purposes. For example, a program theory development process may be used for team building. Some processes are more suited to team building than others. Some methods foster theories about what the ideal program should look like. They can be useful for judging the quality of an existing program theory or generating a program theory from which to design a program. Other methods generate theories about how key stakeholders think about the program. These methods are useful for consciousness raising and communication with stakeholders about what the program is all about and how it will achieve results. Yet other methods give rise to program theories that reflect what happens in practice. These methods are useful for testing what actually happens in practice when compared with formally stated principles or ideals about how a program is expected to function and achieve results. They may lead to judgments about the fidelity of program implementation, the validity of its underpinning principles, or closer scrutiny of what works best in different situations.

Purposeful program theory also considers the nature of the program for which program theory is to be used. Processes for developing program theory need to be matched to type of program, including whether it has simple, complicated, or complex aspects or some combination of these. For example, simple service delivery programs may benefit mainly from drawing on staff and client perspectives. Program theory development for a program with complicated aspects benefits from the participation of all key partners who are responsible for delivering various parts or interrelated streams of the total program. Some may be outside the organization that has the main responsibility for a program. The complexity of a program can be a reason for developing multiple theories. Smith (1994) concluded that "complexity and uncertainty suggests a need for multiple working theories of program impact, attention to conditions as well as causes" (p. 86).

This chapter describes three approaches to describe program theory: articulating mental models and deductive and inductive approaches that can be used on their own or in conjunction with each other. In general, it is better to use a combination of approaches. The balance among the approaches will vary from one set of circumstances to another, including time and resources available and intended use of program theory. The emphasis may also differ according to differences of cultural context, such as the relative emphasis of different cultures on group participation and achieving group consensus. We discuss six decisions to make about how to develop a program theory:

Decision 1: Who should be involved in developing a program theory, and what should be their role? Who takes the lead role in developing a program theory, and who else should be involved?

Decision 2: What is the appropriate mix of approaches for developing or eliciting the program theory (stakeholder mental models, deductive, or inductive), and how should various approaches be used, for example, iteratively?

Decision 3: How might workshops and interviews be used in developing program theory?

Decision 4: As challenges arise, how should they be addressed? Under what circumstances should a program theory development process be terminated or the effort be wound down?

Decision 5: How much effort and resources should be invested in a program theory exercise?

Decision 6: When is it time to revisit a program theory?

Although we discuss a range of approaches that can be used, we have a general preference, based on experience, for including, whenever feasible, stakeholder articulation of mental models and collaborative approaches in the mix of approaches used.

DECISION 1: WHO SHOULD BE INVOLVED IN DEVELOPING A PROGRAM THEORY, AND HOW?

An initial decision needs to be made about who will take the lead role in developing a program theory and who else will participate—for example, staff, program clients, and funders. It should be noted, however, that as further questions are addressed about the most appropriate approach, the lead role may change, as may the numbers and types of stakeholders who should be involved and their roles. The following topics are addressed in this decision:

- A program theory can be developed by funders, management and staff, clients, or an evaluator or program theorist within or outside an organization, working alone or collaboratively.

- Decisions need to be made about what other stakeholders should be involved and in what capacity: as a source of information or participant in construction.

- Consider the benefits that can be gained by involving staff, program clients or intended beneficiaries, policymakers and funding agencies, partner or related organizations, and subject matter experts.

- Roles of various stakeholders are likely to differ according to whether deductive, inductive, or stakeholder mental model approaches are being used to develop the program theory.

- The nature of the relationship between those in the lead role and other participants needs to be determined. Will it be directive and authoritative? Collaborative? Facilitating? Who has authorship of the theory?

Development of Program Theories by Staff, Funders, and Others

Increasingly staff in organizations running programs are becoming familiar with program theory processes and are able to develop their own program theories, especially when they are relying primarily on program documentation and staff experience. This approach is useful when the intention of the program theory development is simply to document current organizational beliefs about the program. However, these program theories, and hence their programs, are less likely to include and benefit from external perspectives and experience that go beyond that of the organization. The absence of these perspectives can be a particular disadvantage when program theory development is being used to design a new program or refine the design of an existing program.

Another model that is popular in some organizations is program theory construction by program staff, with a theorist or evaluator with program theory expertise exercising a peer review function as the theory emerges. Professional subject matter experts can also contribute. This approach serves the dual purpose of providing feedback and contributing to program theory capacity building through mentoring.

Facilitation or Development by Program Theorists and Evaluators

Program theorists and evaluators can also take a more active role as facilitators. During the facilitation process, the program theorist or evaluator can add value for purposes of program development and improvement by exercising two other roles: devil's advocate and key informant. In some situations, it may be possible to pull together the bare bones of a program theory as part of the facilitation process, such as the immediate product of a workshop. However, when evaluators facilitate a program theory development process, they may find that it falls to them to pull the program theory together in a document. In performing these various roles, they need to make sure that they do not unilaterally or excessively impose their own view of what the program should look like. Checking with key stakeholders such as the party in the designated lead role in the program theory development process, managers, staff and funders, and generating stakeholder buy-in are critical. As program theorists, we have found to our detriment that it can be all too easy to slip into enjoying the challenge of applying our expertise as program theorists without adequately bringing along with us key stakeholders unaccustomed to thinking in terms of program theory.

Program Theorist or Evaluator as Devil's Advocate   The specialist may play this role by raising questions about the legitimacy of stated and unstated assumptions and identifying gaps and inconsistencies in the emerging program theory. In exercising this role, program theory specialists can make a constructive contribution to refining the design of the program.

Program Theorist or Evaluator as Key Informant   Unless the program theory specialist is intent on eliciting only the group's theory uncontaminated by other considerations, the facilitation processes will usually be underpinned by prior reading of relevant documentation, selected stakeholder interviews, and knowledge of other programs and other program theories. In Chapters Eleven and Twelve, we explore the ways in which evaluators can add value to the process by drawing on theories of change and program archetypes.

Program Theorist or Evaluator as Developer   Constraints on time and resources of stakeholders such as staff may lead to a request that a program theorist develop a draft program theory document for discussion with stakeholders. This can be a useful process if it is accompanied by active processes such as a hands-on workshop to refine the draft so that it adequately includes stakeholder perspectives, information, and insights; develops stakeholder understanding of the documented program theory; and generates commitment to it.

Decisions About Other Stakeholders to Involve

Choice of stakeholders and decisions about the nature of their involvement in developing or identifying a program theory depend on the intended use of the program theory.

When the purpose of the program theory is to develop an understanding of what is needed to make the program work, involving program clients or intended beneficiaries and partner agencies is important. When program improvement and development are the main objectives, involving staff and management is critical. If the program theory is to be used as a framework for determining what routine monitoring information will be collected by staff or to design an evaluation, then it is generally helpful to get staff understanding and acceptance of the program theory. Involving a person with expertise in monitoring and evaluation is also useful.

When a program theory is being developed for an external evaluation whose main purpose is accountability, buy-in by the client or commissioning agent for the evaluation is important to sign off on the program theory that is to be used for the evaluation. Even so, it would be unusual and generally undesirable to exclude staff from such a development process.

Their participation can result in a more robu⸤
on staff experience about what is realisticall⸀
cooperation in the evaluation if they con
achievement are to take into consideratiᴄ.
Participation can also build staff capacity to deveᴛ.
ory. They may be able to apply that capacity to other progᴛ.

When program theory is being developed to guide a high-proᴛᴜ
nal evaluation, it may be important to involve critics or potential critics
of the program who otherwise might argue that important outcomes or
desirable program processes (especially unintended ones from a program
perspective) had been overlooked.

INVOLVING PROGRAM CRITICS IN THE DEVELOPMENT OF A PROGRAM THEORY

A program theory development process was used to design an evaluation of the piloting
of new legislation concerning eviction of public housing tenants in the event of repeated
antisocial behavior. The development of the program theory included group discussions
with various advocacy groups (tenants in general, people with mental health issues,
and people with disabilities), some of whom opposed the legislation and the piloting
processes. The groups were to articulate their views about what the program should
look like if it were functioning well. This process helped to ensure that their concerns
were addressed by the evaluation and also set those concerns within an overall theory
of change.

Experts in the field can also make a valuable contribution to identifying important features that should be included in the program theory of an
effective program. They can be included among interviewees.

DECISION 2: WHAT IS AN APPROPRIATE MIX OF APPROACHES FOR DEVELOPING OR ELICITING THE PROGRAM THEORY?

Program theory can be constructed in many ways. We start by describing
three approaches in their pure form—deductive, inductive, and articulating
mental models:

- *Articulating program stakeholder mental models* by working with various stakeholders in groups or as individuals to draw out their mental models of how they understand the program to work or how they would like to see it work: what the program would look like if it was successful and why that type of success is important in terms of contributing to important and often high-level outcomes.

- *Deductive development* from formal and informal documentation about the problems the program is addressing, the causes and consequences of the problem, and wider research literature and professional experience that is relevant to the program and effective practices. Deductive development also includes logical analysis.

- *Inductive development,* which involves inferring the program theory from how the program operates in practice in the field based on observation; interviews with staff, clients, and other relevant stakeholders; and use of other relevant information. It includes how opportunities are being used as they arise.

In practice, it is usually helpful to have at least some elements of each of these. They favor different processes and sources of information, each of which can contribute to program theory in different ways and for different purposes.

Program theories that are based on stakeholder mental models and deductive approaches often are used to design a program evaluation, known in such circumstances as a theory-driven evaluation. Program theories that arise from inductive approaches may well be the product of a completed evaluation. These emergent program theories may shed a different light on how the program works than appears to be the case from program theories arising from deductive or stakeholder mental model approaches.

This classification of methods for identifying program theories is somewhat similar to Patton's (2008) comparison of user-focused, deductive, and inductive approaches. However, our category of articulating stakeholder mental models, while it includes user-focused approaches, also acknowledges that many stakeholders in addition to those who will use the program theory have mental models about how the program should work. Understanding and responding to the different mental models can be critical to the success of a program. There are many legitimate perspectives on how a program does, could, or should work and different perspectives about how it relates to other efforts and the wider context in which it sits.

Our use of *deductive* differs slightly from Patton's in that we suggest as sources of information from which theories can be deduced not just the academic literature but also program documentation. We also include the experience of relevant professionals concerning other models outside the program that might be obtained through interviews, the experience of evaluators and other program theorists, and processes of logical analysis.

Articulating Stakeholder Mental Models

Mental models for programs are about how various stakeholders believe a planned or existing program will achieve what it is designed to do. In other words, what are the mechanisms (psychological and other) by which a program is believed to achieve its results? Usually this involves articulating an entire causal model or key links in it. Anderson and others (2006) used a different approach by asking stakeholders to generate ideas about the purpose and function of a program and then to cluster these into groups, summarizing these results in the form of concept maps, and only then labeling them as inputs, activities, or outcomes.

Among the key stakeholders whose mental models might be elicited are program staff and management, policymakers and funders, target groups and intended program beneficiaries, and key partners who have a role to play in the program.

Engaging a range of stakeholders in the program theory development process can assist with identifying the potential for unintended outcomes, whether positive or negative. However, it is often not feasible to work directly with stakeholder groups other than program staff. Getting informed and experienced members of staff to role-play different stakeholder perspectives can be an alternative approach. For example, in developing a program theory for a Meals on Wheels service, different participants in a group can be asked to take on the role and perspectives of the clients of the service, their families, the staff, the management, the clients' neighbors, the local health service, and others. Their different interests can then be incorporated in a more comprehensive program theory for program improvement than would be available by considering the perspective of just one group or another.

STAKEHOLDERS WHOSE PERSPECTIVES SHOULD BE INCLUDED IN MENTAL MODELS FOR A PROGRAM THAT PROVIDES CASH GRANTS TO VICTIMS OF DOMESTIC VIOLENCE

A simple (perhaps overly simplistic) mental model for a program that provides cash grants to victims of domestic violence might be that the cash will encourage victims to report domestic violence to the police by removing one of the current impediments to doing so: their reliance on financial support and accommodation from the perpetrator. Encouraging victims to report the crime makes it more likely that perpetrators will be caught, prosecuted, and punished, and victims' future safety and well-being will be more secure.

It would be helpful to elicit not only the mental models of staff delivering the program (the most likely intended users of the program theory) but also of victims and their families, perpetrators, and police, who are important to making the program work. We would also need to obtain the perspectives of victims for whom financial considerations are not a deterrent to reporting. The mental model of each of these types of stakeholders is a piece to add to the jigsaw.

Mental models of staff are often tacit rather than explicit and may operate as a series of propositions or guiding principles without having been woven together into a cohesive theory. Articulating these mental models is a useful means of bringing them to front of mind so they can be questioned. The process of articulating a model may encourage more in-depth thinking about it. Involving stakeholders is a powerful means of improving the understanding and plausibility of the program theory.

Among other things, stakeholders may question the plausibility of the program theory, the likely achievability of intended outcomes and objectives, whether the objectives are focused on outcomes rather than delivering activities, and the feasibility of program implementation. In the case of program staff, policymakers, and funders, this questioning may also lead to program redesign, program improvement, revision of intended outcomes, new choices of alternative measures of success, and so on. If the mental models prove to be quite robust, then having them at front of mind can help them to play a stronger role in guiding decisions and behavior.

Eliciting a program theory quite often reveals that objectives were set at an unrealistic level. For example, they may refer primarily to ultimate impacts, such as changes in the population's social and economic indicators that may arise from a combination of many programs and other factors. Such aspirational objectives have a place in providing a compass for the overall direction of the program. However, they are problematic if they are not accompanied by objectives about intermediate outcomes to which the intervention can make a more direct and discernible contribution. Using program theory to identify what is achievable can make it more likely that a theory-driven evaluation will measure those achievable outcomes and, in the process, more readily give credit for what has been achieved. It can also explain why those outcomes are important by showing their expected or research-based links to further, and ultimately more important, outcomes and impacts.

Mental Models of Program Participants   These models can help with identifying the outcomes that will be important to the participants and the conditions in their lives that will affect whether those outcomes are achieved—for example, what will affect whether they will use the services in the manner intended. Participant mental models may also identify how they would like the program processes to operate.

MENTAL MODELS OF PROGRAM PARTICIPANTS

In the domestic violence cash grants example, intended beneficiaries may identify the conditions under which those grants are and are not likely to work, some of which will relate to the grants themselves (for example, their size, how people qualify for them, and the extent to which victims need them or are financially independent) and some of which will relate to external factors. The external factors can include how police respond to victims of domestic violence, the adequacy of the justice system in relation to domestic violence, the probability that the victim will be protected from future violence from the person he or she has reported, and the adequacy of other support systems for victims to assist them to live away from the perpetrator. Of course, there are many other issues that factor into domestic violence and that we have not considered in this simplistic analysis, such as victim–perpetrator relationships and the fact that financial considerations will not be relevant to many victims.

Victims of domestic violence have expressed preferences about how the program treats them: they want to be treated in a way that empowers them rather than controls them, respects them rather than judges them. Whether these preferences are addressed may well influence whether potential clients use the service and benefit from it. Their preferences would need to be incorporated in the theory of change.

Mental Models of Partners Other Than Program Staff   Mental models of partners such as other organizations that may facilitate or inhibit effective delivery of the program and achievement of its outcomes can also be critical, especially for complicated programs that incorporate a range of codependent strategies. Partners may have a range of agendas that are outside the program and may or may not be consistent with its requirements.

The process of developing an overall mental model that shows not only what the program is doing but also what other stakeholders are expected to do by, such as putting the names of particular stakeholders against particular outcomes, can provoke discussion and negotiation. Assumptions that various partners have made about what others are doing can be tested and may well prove to be false. In Chapter Seven on developing a theory of change, we argue for developing an overall road map theory of change that includes not only elements that relate to the program in question but also key outcomes for others to achieve. This draws attention to its interdependencies with other aspects of its operating context.

Including the behaviors of partners and other key stakeholders in the identification of mental models can lead to ideas about what the program can do to influence those behaviors or manage the risks associated with them. Techniques such as outcome mapping and social network analysis that show roles, relationships, and communication patterns among various stakeholders can be used to assist with developing mental models that show partners' roles. In particular, these partners can be instrumental in achieving key immediate or intermediate outcomes needed to progress to longer-term outcomes. For example, they may play a gatekeeper role that affects whether the target groups for the program are ever referred to it. This can help with overcoming miracle thinking and missing middles in program theory: large gaps in the middle of a program theory that require a leap of faith from immediate intended outcomes to ultimate outcomes. We discuss these issues further in Chapter Seven.

WHY MENTAL MODELS OF PROGRAM PARTNERS CAN BE IMPORTANT INCLUSIONS IN A PROGRAM THEORY

For the domestic violence cash grants program, the ways in which the police exercise their role will be critical—for example:

- What are their standard procedures for responding to and treating victims who report domestic violence to them?

- What is the impact of their processes on victims and on the probability that they will report the crime?

- How do the police document cases to increase probability of successful prosecution?

The program may need to influence the police to reconsider their mental models and in practical ways change how they work with these victims.

Engaging other partners can help to develop a deeper understanding of the context within which the program operates. A program that seemed to be simple might emerge as having complex aspects that need to be addressed if it is to be effective. Conversely, engagement of partners may help to overcome any apparent complexity, unpredictability, or sense of powerlessness by identifying simple steps for removing obstacles to success and increasing certainty of outcomes.

For example, partner organizations in the mental health arena could be better engaged in care coordination and referral processes that make the transition between clinical and community mental health services more seamless so that clients are less likely to fall through the cracks. This in turn might decrease the likelihood that they will suffer setbacks that can result in any one or more of a host of uncertain, idiosyncratic, and undesirable outcomes. Some aspects of early intervention and prevention, such as those relating to better referral processes, may be relatively simple compared with treatment and management of the interrelated flow on effects of family breakdown, homelessness, unemployment, and so on.

This example is not intended in any way to suggest that addressing mental health issues is a simple matter, only that programs and the situations they are addressing can seem more complex than they need be. Introducing the relatively simple process of more effective referrals could make the challenges of addressing mental health issues and their flow on effects more manageable.

Deductive Development

Deductive development of a program theory uses formal and informal documentation about the program and the needs it is intended to address. It may draw on wider research literature and experience that are relevant to the program. It can reveal whether there is a documented theory, the absence of which could raise questions about whether anyone had an actual theory (documented or undocumented) about how the program would work.

LOGICAL ANALYSIS APPLIED TO THE DOMESTIC VIOLENCE CASH PAYMENT PROGRAM

If victims of domestic violence are assured that they will receive financial assistance after reporting incidents to the police, then they will be more likely to report incidents. This is based on the assumption that one of the barriers to reporting is the fear of some victims that financial dependence will force them to return to the same domestic situation and confront the perpetrator. Among the assumptions that need to be met and tested are that the financial assistance will be sufficient to make return to the domestic situation financially unnecessary (a feature of program design) and that the desire for the person to leave the domestic violence situation outweighs competing desires to stay (external factors). If these assumptions cannot be shown to be met, then the program theory may be inadequate.

Amongst the formal and informal documents that might be considered are program guidelines, past reports, performance information collected for the program, and previous evaluations of a program. Desired outcomes can be identified by extracting from documentation all statements about program outcomes. They can then be put together as a logical argument or as several strands of a logic argument: a series of *if–then* statements accompanied by statements about the assumptions concerning mechanisms for change (for example, psychological or sociological) that lie behind them and about the conditions that need to be in place in order for the *if–then* relationship to work. Simply working through the *if–then* statements can help to identify gaps or anomalies in the documents.

Deductive development could also turn to academic research and theoretical literature, avoiding the trap of having a theory that is not supported by or is contradicted by research. In Chapter Eleven, we discuss how research-based theories of change can be used to develop and improve a program theory by strengthening its research base. Research is about what works for particular types of programs in the relevant policy area—for example, health, education, or corrections—or about relevant policy tools—for example, carrots and sticks—that can also be useful. In Chapter Twelve, we discuss program archetypes that relate to various policy tools.

These theories can also be used as templates against which to compare the documented theory for the program, and once again this can be used to identify potential difficulties with the theory. However, because each program has unique features and contexts, and we know of no theories of change that have been developed to cover all possible variations, these theories of change should be used as a heuristic device for raising issues for discussion rather than as a gold standard.

Research-based theories of change and program archetypes can be a source of ideas about what to look for in program documentation and how to organize findings in a logical manner. These theories can provide shortcuts to improve the efficiency of the process, reduce the amount of time needed, and avoid the trap of theories that are not supported by or are contradicted by research.

Program documentation and wider research literature could be used separately or in combination. Use of research in combination could improve an existing program theory. Program documentation also contextualizes the more general research literature to the problem at hand. When relevant program-specific research evidence is available to underpin a program theory, then it is important to seek out and apply that evidence. Types of research evidence that will be relevant include evidence about the nature and extent of the problem being addressed by the program; the problem theory, that is, causes and consequences of the problem; and what is known about effective practices or treatment theories. These may or may not be directly related to treating the causes, the problem theory (Lipsey, 1993). They may be about treating consequences instead. A strengths-based approach can also provide information about how other programs effectively use opportunities.

Previous evaluations of a program, in whole or in part, or of its predecessors or similar programs can also be valuable sources of information for developing a program theory. The lessons learned from previous evaluations often include factors that are likely to affect how well the outcomes will be achieved in different circumstances. As we show later, a program theory can be improved by including key factors that are likely or are known to affect its success. Many of these factors can be identified from research. Others, including those relating to the actors who are involved in some way with a particular program or affect the program outcomes, can be identified through consultation with program staff, participants, and others.

Deductive development can be particularly useful for designing a new program or comparing the design of an existing program with an ideal program theory that might emerge from research evidence. Deductive approaches can also be used simply to extract a program theory that is embedded in program literature but has not been articulated or explained in a cohesive, logical manner.

Intended outcomes that are not clearly stated in program documentation can sometimes be inferred from documented statements about program activities. This approach should be supplemented by further questioning of stakeholders since it can sometimes give rise to incorrect assumptions about the program theory. Theories that have been deduced in this way may appear to be implausible because they are built on assumptions and incorporate objectives that were never in the minds of the program designers. Therefore, it is important to ask about intended outcomes rather than to assume them from the statements about activities. For example, inclusion of training as an activity in a program description is often used to infer an intended outcome or objective, such as "improved knowledge and skills of participants." However, the purpose of training might instead be as a tool for bringing people together to build social capital, and any formal learning could be almost incidental. Or it may be that training is being provided because of a need to meet employment conditions requiring a certain frequency of training rather than to achieve particular learning outcomes.

Deductive approaches, especially those drawing on program documentation, can be particularly useful for identifying the espoused program theory (see Argyris and Schön, 1974). This is the stated position about how the program is supposed to work. Espoused theory can be the most important theory to elicit if it is important to portray the official policy position about how the program ought to work or even the public statements of staff and others about how they believe the program works. An evaluation may be being undertaken for administrative accountability purposes to check compliance with the theory. Is the program operating in the expected way? Are the intended outcomes occurring? Does the theory as it is enacted correspond to the espoused theory, and if not, why not? In our discussion of workshop processes later in this chapter, we revisit the question of whether the processes are used to elicit the espoused theory, a preferred theory, or the theory of how a program operates in practice.

Inductive Development

Inductive development involves observing the program in action and deriving the theories that are implicit in people's actions when implementing the program. The theory in action may differ from the espoused theory: what people do is different from what they say they do, believe they are doing, or believe they should be doing according to policy or some other principle.

The program in action could be observed at the point of service delivery in the field that is closest to the clients of the program. It could include observation of the program in action, including through participant observation. Interviews can be conducted with staff about how they implement the program and about why they undertake some activities that may appear to be at variance with the program design or omit parts of the program design. Program participants can be interviewed about how they experience the program (or have experienced it, if data are gathered through exit interviews; Unrau, 2001), and how they would like to experience it. The analytical approach of grounded theory can be used to make sense of this ethnographic data (Goertzen, Fahlman, Hamption, and Jeffery, 2003).

Theory in action could also be identified from looking at how program managers have interpreted the program, as indicated by the types of practices they adopt. However, it is important to confirm that inferences drawn are correct. For example, what they consider to be important about the program and how they interpret the program's intent might be inferred from their choice of particular performance indicators and how they use them. This inference would need to be confirmed with program managers, since the selection of indicators may have been imposed on management and staff as, for example, part of national nonprogram specific monitoring requirements. Or the indicators may have been selected simply because they were available and easy to measure and report, but not necessarily considered by staff to be meaningful.

Many of the stakeholders involved in inductive development processes are the same as those involved in the process of articulating mental models of stakeholders. The stakeholders would be sources of information, but the program theory developer would have the main role of putting together a program theory, which might then be discussed and contested.

Inductive development is also about recognizing that programs will be implemented very differently, in different sites and by different people. Therefore, there may be many significantly different interventions, all operating under the same program banner. Many of these differences may reflect quite appropriate adaptations to local contexts such as is often required by programs with complex aspects operating in complex situations. Understanding the theories that are operating in these different contexts can help to enrich the program theory by identifying what works for whom, under what circumstances, and why various adaptations are useful.

Sometimes these variations are so great as to beg the question of whether there really is a program as such (see Patton, 2008). Inductive development of program theory has strong crossovers with implementation evaluation, which shares similar processes of assessing what is actually happening. Questions might be raised as to whether the variations and the ensuing program complexity can be accounted for by quite functional adaptations of a program to local conditions and perhaps changing situations. Or perhaps the variations arise from faulty, sloppy, or unregulated delivery of what should be a simple program responding to relatively simple circumstances—for example, failure of a building program to adhere to building codes.

Choosing an Approach: Articulating Mental Models, Deductive Approaches, or Inductive Approaches?

For conceptual purposes, it is useful to differentiate among the approaches and their consequences for program theory. However, program theory development processes often benefit in practice from drawing on elements of all three approaches. Thus, the choice of approach entails deciding on the degree of emphasis on each of the various approaches, taking the following into consideration:

- How the program theory process and product are to be used—for example, to design an evaluation or use an evaluation to develop a new theory

- The type of theory you are trying to portray—official theory, espoused theory, theory in action, stakeholder preferences, or academic theory

- Practical considerations such as availability of time, resources, and stakeholders

Combining Approaches   The process of eliciting the mental models has an impact on the mental models themselves, just as conducting an opinion survey has the capacity to affect the very opinions it seeks to measure. So if the process of eliciting models also shapes them, it is helpful to do so in a constructive way.

The different approaches and the sources of information that are typically associated with them can be used sequentially, iteratively, or interactively. For example, the experience of program theory specialists, although identified as part of the deductive approach, could also be applied during facilitation approaches used to elicit and articulate stakeholder models or to shape emerging theories.

AN ITERATIVE PROCESS COMBINING ALL THREE APPROACHES TO PROGRAM THEORY BUILDING

A program theory for an extension program for farmers was developed when that program had been under way for some months. The purpose was to refine a draft monitoring and evaluation framework. The following processes were used in the order shown:

- *A review of the literature* relating to effective extension practices and review of supporting documentation for the program. [Deductive approach]

- *Workshops with funding agencies, partner agencies, course coordinators, and deliverers* to review and refine a draft program theory identified by the evaluator in the light of literature, extensive evaluator experience with these types of programs, and program documentation. In this case, the evaluator provided some direction in the form of working hypotheses because of lack of time to start from scratch. Workshops were used to identify stakeholder understandings about what the program would look like if it were successful and how the program would work. [Articulating mental models]

- *Participant observation and interviews* with participants and staff to test the program theory so that the evaluator could fine-tune the theory as the program progressed and evolved over several years. [Inductive approach]

A richer and more nuanced program theory can emerge from drawing on many sources of information. A case example for a high-profile national partnership program working with ethnic communities affected by particular drug problems is illustrative.

SOURCES OF INFORMATION USED TO IDENTIFY THE PROGRAM THEORY FOR A PARTNERSHIP PROGRAM

Deductive Approach

Review of Program Documents Including

- Senate report calling for the continuation and refinement of a pilot program

- Description of the program, its background, and its rationale
- Budget statements for the establishment of the program

Review of Academic, Research, and Other General Literature on the Topic

- Relevant literature around this particular drug problem, including a Senate report and the research evidence cited in that report and coroners' reports
- Transcripts from Senate hearings with stakeholders about approaches to drug abuse
- Previous evaluations of predecessor and related programs
- Information about monitoring of prevalence of drug abuse in the relevant communities over several decades
- Research literature on drug abuse and treatment
- Literature on whole-of-government and partnership approaches to developing and implementing difficult and complex public policy issues

Inductive Approach

- Interviews with key stakeholders from across four federal government agencies and three state government agencies involved in the strategy about how the program was working and intended to work, future directions, data that were being collected, and how the data were being used
- Workshop presentations by various stakeholders, including staff, about what was actually being done (as distinct from what was on paper) to address each of the important outcomes that they had incorporated in the model identified through the stakeholder mental model process
- Interviews with staff who had extensive experience with the day-to-day operation of the program and the needs and experiences of participants (site visits were not feasible for this program theory project)
- Program monitoring information, strategic plans, and information about the implementation of work plans
- Performance contracts with service providers that identified specific activities to be undertaken, performance measures to be used, reports to be prepared, and so on

Articulating Stakeholder Mental Models

- Workshops with key stakeholders from across the government agencies involved in the strategy—for example, an ideas writing workshop concerning what success would look like, how life would be different for communities if the program was working well, the use of flip chart paper, and sticky notes

- A discussion and adaptation of a draft program theory prepared by the evaluator drawing on documents and stakeholder interviews

- A redrafted model to include conditions that were outside the scope of the program but were critical to its success so that they would not be overlooked during program redesign, implementation, and evaluation

Being Selective About Approaches and Sources of Information   Ideally the best aspects of all three approaches would be used, along with as many sources of information as possible, especially if the program theory is to be used to design a new program or redesign an existing one. However, it is not always feasible or desirable to use all of these approaches and sources of information. Selections need to be made based on important considerations—for example:

- *How the program theory process and product is to be used.* We have commented on some of the different benefits of stakeholder mental models, deductive approaches, and inductive approaches, especially for stakeholder ownership and understanding of the program and for program development and improvement. Any of the approaches can be compatible with monitoring and evaluation purposes, depending on what type of evaluation is being undertaken and whether the evaluation is largely internal or external.

- *Whether the program theory needs to be a direct reflection of policy or whether it can accommodate other perspectives.* Different approaches are more suited to representing the theory on paper: the espoused or official policy position, the theory in action, how various stakeholders would like to see the program operating, or theories that might be preferred based on the academic research literature.

- *Whether the program theory identified should be the one that applied when the program was first documented or the current program theory, documented or not.* Typically there is a time lag between changes in stakeholder and practitioner theory and documentation of those changes. Thus, the deductive approach, in drawing on documentation, is more likely to reflect older or original theories, whereas stakeholder mental models and inductive approaches are more likely to reflect current theory.

• *Whether external validation of the program theory is important at the time of identifying the theory.* This relates to a program theory's evidence and research base, external stakeholder perspectives, values, contextual factors that affect the operation, and outcomes of the program. Sometimes the purpose of the program theory development process is to postulate a theory for testing empirically through research or evaluation. In this case, the process is really about stating hypotheses in testable ways. On other occasions, where testing is less important, a program theory that simply provides a conceptual framework will suffice (see Margoluis, Stem, Salafsky, and Brown, 2009).

• *Budget, human resource time and regular availability, and whether the process is on a fast track.* Availability and willingness of relevant stakeholders to commit time and effort to the process is a relevant consideration. However, regardless of resources available, there may also be a need to work quickly. In other situations, it may be possible to accommodate successive iterations of an emerging program theory over several occasions.

• *Staff ownership.* How important is it to obtain staff ownership and understanding of the program and build staff capacity in relation to the program or to develop expertise in program theory?

• *Engagement and momentum.* Consideration should be given to whether it will be possible to maintain staff and other stakeholder engagement and momentum if consultative and workshop processes are used.

• *Whether conflict is to be avoided or creative tension encouraged.* This could influence decisions about whether or how to use group processes and whether to have homogeneous or heterogeneous group membership.

Table 6.1 summarizes some of the considerations for choice of approach (deductive, inductive, or articulating stakeholder mental models) and for choices of specific methods and sources of information for preparing a program theory.

Table 6.1 Considerations for Choice of Approach, Methods, and Sources of Information

Note: Checkmarks in cells show the types of sources and methods that are appropriate when particular considerations are applied. Additional text in a cell qualifies the conditions under which a source or method would be useful or provides further explanation.

If These Considerations are Important:	Main sources for deductive theory building			Main sources for inductive theory building			Main methods for articulating mental models		
	Review of program documents	Review of research literature	Use PT specialist experience	Interviews with program staff, policy makers	Interviews, surveys, etc., with program users	Site visits; observation	Group work: single occasion	Group work: multiple occasions	Interviews and other work with individuals
External validation		✓	✓	✓	✓	✓ If sites are carefully selected			All key types of stakeholders are included
Fidelity of PT to written policy	✓			✓					
Reflect program in practice				✓	✓	✓	✓	✓	✓
Small budget, little staff time for Program Theory	Good for small budget	Relatively small budget	If PT is done as a desk job	If with small number of key informants; by phone	If just with key informants	If only a small number of sites	Lower budget	Higher budget	if with small number of key informants; by phone

			Interviews with key informants only	
Potential to fast track	✓	✓ If PT is done as a desk job	✓	✓
Staff ownership, understanding, and capacity building	✓	✓	✓ More grounded theory	More likely than with single meeting
Maintain momentum and engagement with process				✓ Multiple meetings may lead to fatigue; disengagement
Encourage creative tension	✓			✓
Avoid conflict	✓	✓ but may just delay conflict		✓

Implications of Choice of Approach for Who Should Be Involved

Earlier in this chapter, we provided some suggestions about deciding who should be involved in identifying or developing a program theory. Decision 1 concerning choice of stakeholders and decision 2 concerning choice of approach are clearly somewhat related.

From our discussion of the three different approaches, it is evident that the development processes range from being largely driven by the evaluator or program theorist, to collaborative, with a strong emphasis on active stakeholder participation.

Deductive and inductive approaches are more likely to be led, and may even be completely undertaken, by an evaluator or program theory specialist. However, such approaches could also be conducted by program staff, funders, and others as part of the process of program development and management. Articulating stakeholder mental models is always a collaborative process. It needs to be facilitated and is often done so by an evaluator (internal or external) or other program theory specialist.

As program theory experience in organizations develops, all three approaches could be done by program staff and others. Indeed, some organizations now include program logic among their core business processes. Nevertheless, we believe that evaluators, program theorists, program subject matter experts, and experts in particular policy tools (such as how to effectively use incentives) can each add value to the process by bringing different perspectives to program theory development.

DECISION 3: HOW MIGHT WORKSHOPS AND INTERVIEWS BE USED IN DEVELOPING PROGRAM THEORY?

Using workshop processes and interview questions to elicit mental models and derive program theories is both widespread and challenging. Workshops have the potential to serve a wider range of purposes than reliance on documents, observations, or interviews. With help and advice from a skillful and well-prepared facilitator, a workshop can draw on the best elements of those other approaches. Structured interviews can also be used, and there are standard questions that can be useful for eliciting program theories.

Whether it is important to develop a convergent theory or it is useful and acceptable to encourage diverse perspectives on a program theory will affect not only whether workshops are an appropriate process in a given context but also what facilitation techniques should be used. We discuss the practice of deliberately fostering different mental models, developing mental models for consensus building and conflict resolution, and how to deal with different perspectives.

Workshop Processes

The development of program theory with groups of staff and other relevant stakeholders works best by having them work together in a face-to-face situation.

Sometimes it can be helpful to provide participating stakeholders with program documentation from which they extract key ideas for inclusion in the program theory. This can be especially helpful when the group's familiarity with the program is low or variable. These types of workshop processes are particularly useful for generating ownership among staff or other stakeholders and better understanding of their program, what is needed to make it work, and stakeholders' roles in making it work. The hands-on processes help them avoid getting bogged down in theoretical arguments. They are also useful for capacity building in program theorizing and can help participants bring a different mind-set to other programs.

Simply asking various stakeholders for their mental models is unlikely to be productive. Among useful prompting techniques are ideas writing workshops for eliciting participant views about the important features of a program. Logic modeling using practical aids such as electronic whiteboards, large sheets of paper, sticky notes, and index cards work particularly well in face-to-face group situations; computer software is also available for these purposes (see Chapter Thirteen). There is also potential to use online discussions and to review various iterations of program theory either online or through circulating hard copy to reach consensus (if consensus is needed).

In a group situation, the twin processes of drawing out the program theory and representing it in a logic model converge and often operate in an iterative manner. Which approach to use as the starting point (articulation of program theory in narrative form or development of a logic model diagram) depends on many factors, including the cognitive sophistication of the group. Either way, the main purpose is to elicit their beliefs about the cause–and-effect relationships in the program.

For more sophisticated groups with extensive program knowledge, it can be useful to start with their thoughts as to what they think the program is about, why it is needed, and how they think it will work (its mechanisms for change). These thoughts can then be portrayed in an easily readable and recognizable form that participants agree would reflect their narrative theory: a logic model. Techniques such as getting participants to start by showing an *if–then* causal chain either in text form or using a sticky note for each step in the chain with arrows linking them can be used to tease out their program theory and cause them to think about aspects they may have left out or not thought through properly.

Referring to the simplistic example of cash grants to victims of domestic violence that we discussed earlier, the steps in the *if–then* causal chain, showing mechanisms for change that relate to developing financial independence, increasing immediate safety of victims, and reducing fear would be:

1. If cash grants are available to cash-strapped victims of domestic violence, then they will be better able to be financially independent of the perpetrators.

2. If victims are financially independent of the perpetrators and able to live away from them, then they will be less fearful of the consequences of reporting them because they will be better able to escape possible retribution and further attacks.

3. If victims are less fearful of the consequences of reporting the perpetrator and believe that they can be safe from further harm, then they will be more likely to report the crime to police.

4. If victims report the crime to the police, then the police will be able to remove the victim from immediate danger and apprehend and prosecute the perpetrators.

5. If police apprehend the perpetrators, then steps can be taken to prevent them from further harming the victims, and the victims will be less subject to domestic violence and its consequences.

Laying out the program theory in this way enables us to identify, question, and test the assumptions that need to be made about what else needs to occur for the *if-then* chain to progress as planned. For example, one assumption is that the judicial system can effectively protect victims of domestic violence once perpetrators have been identified; another is that victims believe that the system will be effective. These assumptions would need to be seriously questioned; if they do not hold, then it is possible that the theory is a poor one that will not work in practice.

Program theory is largely the theory of how the program contributes to outcomes, so program theory must clearly articulate the intended outcomes. However, staff may be more used to thinking in terms of processes than outcomes. One approach for identifying the outcomes is to start with the familiar by asking what the program actually does (its activities) and what will be achieved by each activity. For example, if selection of participants is identified as an activity, then the immediate outcome of a well-executed selection process would be that the right people participate in the program. One way to put the task to stakeholders is to ask them to tell their story of how they think the program works.

Questions for Use in Workshops and Interviews

A program theory facilitator can ask a series of questions to elicit stakeholders' understanding of the problem addressed by the program, its causes and consequences, the program's contribution to addressing the problem, given its causes and consequences, and what else is needed. Some authors (Gugiu and Rodríguez-Campos, 2007) have suggested standard sets of questions that can be used for interviewing in order to construct a logic model. However, we have found that in general these work better as research questions than as interview questions and that they need to be adapted to particular interviewees to elicit the information. Some useful lead questions for interviewing are shown in Exhibit 6.1.

From the answers to the questions in Exhibit 6.1, a facilitator works with a group or groups or with individuals to construct their *if-then* story of what the program will achieve in terms of outcomes for program clients and others. Participants are then asked why they would expect one thing to lead to

EXHIBIT 6.1 QUESTIONS FOR DRAWING OUT PROGRAM THEORIES

- Can you give me an example of where this program is working really well? Why did you choose that example? What do you think is making it work well? (You can also ask about examples that are not working so well.) If the answers are about program processes only and not outcomes, then extend the questions by asking why those processes are important for program clients and outcomes for clients.

- How would life be better for participants or intended beneficiaries if this program worked well?

- What are the current barriers to a good life for program participants? (You could explore this in relation to particular domains such as health, employment, or social participation.)

- How would you see this program overcoming those barriers?

- What is it about the way the program operates that would or could make life better for participants or intended beneficiaries?

- What does the program currently do that helps to make it work and what is not working so well?

- What else needs to happen?

- Who else needs to be involved, and how?

- Does the program try to influence those other parties, and if so, what would you expect them to do differently?

the other. These questions about why they would expect an *if–then* relationship to hold are a means of drawing out important information:

- The mechanisms for change—for example, if some victims do not report domestic violence because they are financially dependent on the perpetrator, a program might endeavor to remove or reduce this financial dependence.

- The assumptions that need to be met about how the program will operate—for example, that the financial assistance to domestic violence victims is sufficient to significantly reduce dependence.

- Awareness of the external conditions and other factors that may affect whether the financial assistance will work—for example, whether competing motives to stay with the perpetrator will override any impacts of reducing financial dependence, whether alternative, satisfactory, long-term accommodation, and family caring arrangements are available

The *if–then* stories that emerge from these processes, along with the stated mechanisms for change, the assumptions, and the external conditions and other factors, are all part of participants' mental models program theories.

Fostering Divergent or Convergent Theories Through Workshops and Other Processes

Workshops can be used to foster different ways of thinking about and addressing social issues and to facilitate the development of different program theories.

Choosing workshop participants and workshop processes that foster or at least allow divergence can discourage groupthink and avoid premature closure on program theories. However, processes that foster divergence do take more time. This is especially so if participants are required to engage in a series of workshops and it can be difficult to convince participants of the benefits of this extra time unless they are truly committed to using program theory processes to critically review the development process (for example, for program improvement purposes). We address this issue further in our discussion of challenges in developing program theory.

Group processes and selection of group members for workshops in particular can be used to encourage either convergence or divergence depending in part on the nature of the group and the facilitation processes. Parallel and heterogeneous groups can be an effective means of encouraging divergence and avoiding artificial convergence. Parallel groups representing particular perspectives can also be used, but it will then be important to make sense of the different program theories that emerge. For example, it will be important to identify whether they are conflicting perspectives or just different but compatible perspectives, perhaps focusing on different aspects of a program. We address this issue further in our discussion of challenges in developing program theory.

For a group that has a tradition of working together and is relatively homogeneous or cohesive in outlook, then having that group workshop about a program theory may reinforce the agreed understandings about the way things are done. This may be a useful reminder for group members and a means of communicating with outsiders, but it may also discourage reality testing, critical analysis, and innovative thinking about the program. A different but equally cohesive or homogeneous workshop group might come up with a different program theory. A heterogeneous group may also put more program theory variations on the table for discussion.

Different groups might represent different disciplines that will operate from different mental models. For example, health professionals could be expected to produce medical models or program theories; social workers might be more likely to produce community development or individual case management program theories; and economists might view the program and construct its theory through a public and private costs-and-benefits prism. Sometimes workshop group processes that bring together these different professional and other interests can be useful where a purpose of undertaking the program theory project is to bring simmering conflicts to a head and resolve them or simply give due recognition to the multifaceted nature of the issue or problem that the program is addressing.

Another way to elicit different types of theories during workshops is to invite participants to describe the theory from one or more of a number of different perspectives, which can be roughly sorted into these categories:

- Espoused theories—theories about how the program is supposed to work and the way the program theory would be publicly presented (whether by policymakers, program managers and staff, or others)

- Theories in action—what participants believe actually happens in practice

- Preferred theories—how various stakeholders would like the program to operate or think it should operate

Some espoused theories, especially those inherited through political decisions, may lack credibility with staff for a variety of reasons, including inadequate resourcing or conflicting directives. Espoused theories may or may not be underpinned by research evidence but may be ideologically driven.

Theories in action are more useful for portraying staff experience of what happens in the field; why staff and others implement a program in a particular way; their beliefs as practitioners about the nature, causes, and consequences of the problems being addressed; and the strategies that they believe to be effective. Theories in action are often more realistic than espoused theories.

Preferred theories relate to how various stakeholders would like to benefit from the program or see others benefit, how they would like the program to operate, and some of the barriers to effective operation. Preferred theories could also be those that are preferred in the light of research evidence about effective programs. Preferred theories can be particularly useful for program design and improvement.

Even when stakeholders are asked to focus on one or another type of theory (espoused, action, or preferred), they can sometimes confuse the various types either in their own minds or when representing them to others. Their program theories may be a combination of what they really believe and how they operate, on the one hand, and what they consider to be politically or socially desirable, on the other hand. It can also be quite challenging for a program theory facilitator to disentangle the various types.

DECISION 4: AS CHALLENGES ARISE, HOW SHOULD THEY BE ADDRESSED?

There are many challenges in identifying and developing program theories but also some useful strategies for dealing with each of these. We discuss a selection of these challenges and some strategies that we have sometimes found useful (the challenges are summarized in Exhibit 6.2).

Dealing with Different and Sometimes Conflicting Perspectives on a Program's Theory

A natural consequence of articulating the different mental models of different stakeholders is that different theories may emerge that may or may not be compatible or complementary. Differences can also emerge through lack of clarity about whether the mental models are to be about espoused theories, theories in action, their own preferred theories, theories preferred by the research literature, original theories, current theories, or some hybrid of one or more of these. Similarly, if the three approaches to deriving program theory were conducted independently in their pure form for the same program, different program theories might emerge. These theories could be

---

EXHIBIT 6.2 SOME CHALLENGES WHEN DEVELOPING A PROGRAM THEORY

- Different and sometimes conflicting perspectives on a program's theory

- Addressing unintended outcomes

- Lack of research evidence to underpin the development of a program theory or support an existing program theory

- Reluctance to disclose the program theory

- Lack of interest and lack of engagement or faltering engagement in the process, including inability or unwillingness to commit time and resources to do it

compatible, complementary, or conflicting. On occasion, there will be merit in reaching agreement about a theory that would take the best of all three types of program theory as a basis for refining the program for the future or repositioning it.

Sometimes the differences that arise, whether they are from different mental models or different processes for developing program theory, can create conflict. For example, an espoused program theory arising from a deductive approach brought by an evaluator to the table for discussion during stakeholder meetings may not ring true with the experience of practitioners and beneficiaries, who may be cynical about it. Staff may consider that the theory based on written policies reflects a superficial understanding of the nature, causes, and consequences of the problem to be addressed or has little practical application. They may wish to move on to something more useful rather than be saddled with a program theory that they consider to be unworkable, unrealistic, indefensible, or no longer current.

In another situation, policymakers may feel uncomfortable with a theory in action that is at variance with the stated policy and seems to take the program in a direction other than intended by the policy. They may believe it reduces their capacity to meet public commitments and accountabilities. Perhaps they undertook to deliver so many widgets. An analysis arising from a theory in action, a discussion of stakeholder mental models, or a review of the research literature might conclude that widgets were not useful. Despite this conclusion, policymakers and evaluators may still have a formal or contractual obligation to report on the number of widgets delivered.

Sometimes the differences among stakeholders may be mainly around the presentation rather than the substance of the program theory. Issues concerning presentation such as those discussed in Chapter Nine are usually relatively easy to resolve, and on occasion there can be merit in producing different types of program theory models for different audiences. For example, for one program, we found it useful to portray exactly the same program theory concepts in two different formats for different audiences: a vertical linear portrayal and a series of concentric circles. However, in some cases, there may be quite different theories about how change occurs, different worldviews and values, or different stakeholder interests, including some that may relate to unintended positive or negative outcomes. Where substantial differences emerge, a variety of different approaches can be used:

- Compare the different theories using logical argument or external supporting evidence (if available) with a view to showing that one is better than the other.
- Negotiate or facilitate the development of a common path.
- Develop an amalgam of theories.
- Allow different theories to stand as alternatives.

Decisions about which course of action to take in the face of differing theories of change will depend on such considerations as how extreme the differences are, how committed the stakeholders are to their position, and how the program theory is to be used. Negotiating an agreed position can be useful for team building, as a touchstone for further program development and implementation, and for external communication.

Sometimes the theories of different stakeholders contribute a particular dimension not found in the other types of program theories—for example, different stakeholder interests or disciplinary backgrounds such as psychology, sociology, or economics. In such cases, it may be useful, for purposes of future program development and for designing an evaluation, to take the best of all the theories. This is particularly feasible and useful when the theories are compatible with or complement each other.

Tension that arises either within an organization, or between that organization and its stakeholders, or between a program theory specialist and the organization about different program theories can be creative as long as it is well managed. Real differences need to be surfaced and discussed rather than papered over. We have learned that negotiating a common path when very fundamental differences remain can lead to a program theory to which no one relates in any practical way and is not a useful touchstone for either program implementation or program evaluation.

Sometimes resolution of differences strikes at the heart of an organization's philosophy or differences with its stakeholders. The process of doing so may be protracted and fall outside the scope of work of the facilitator.

CONFLICTING PROGRAM THEORIES

In a state government agency, a statewide program manager and a regional program manager participating in a training program on program evaluation independently developed logic models for the same program: a land management program using community development processes.

The divergence in their logic models that emerged during training sessions reflected deep-seated discrepancies in their perspectives on the program. Differences appeared with respect to ideology, intended outcomes, and views about acceptable, appropriate, and effective activities and processes. The facilitator worked with them initially to develop sufficient common ground and then encouraged them to work together over an extended period to address their conflicts and toward a more cohesive program rationale rather than denying their points of conflict. Subsequently, a joint logic model provided a better basis for moving forward with the process of designing the evaluation.

A decision may be made to allow conflicting theories to stand because it is deemed useful and worthwhile to do so until sufficient evidence is available to support one or another. Sometimes conflicting theories provide stimulus material for discussing and resolving deeper points of philosophical, methodological, or values conflict.

When it has been impossible to reach agreement among stakeholders, and the disagreements are around which approaches are most effective, there may be value in retaining the various program theories for testing, as in classical research methodology that compares different treatments. However, this could add significantly to the costs of the evaluation.

Addressing Unintended Outcomes

Sometimes the theories that emerge from different stakeholders incorporate outcomes that were not explicitly or specifically intended by the program. In this regard, differences in perspective and even conflict can make a significant contribution to program theory and should not be seen as necessarily negative. A sound program theory should also identify any anticipated unintended outcomes and, as far as possible, either capitalize on them if they are positive or manage risks associated with them if they are negative. Overlooking unintended outcomes creates problems for program design, implementation, and achievements. It also becomes more likely that unintended outcomes will be overlooked in evaluation design.

Amalgamated theories and logic models showing parallel or interacting theories of different stakeholders can certainly be accommodated in both the narrative description of a program and its corresponding logic diagram. Sometimes these different paths simply represent the different interests that the various stakeholders have in the program. They may be intended and central to the program's effectiveness or simply spin-offs from the program.

Negative program theories can also be identified (what might happen if each intended outcome was not achieved or the opposite was achieved) and can stand alongside positive program theories. Allowing negative and positive program theories to stand side by side is another way to ensure that the program theory addresses some possible unintended outcomes.

Examples of amalgamated program theories and negative program theories follow. They are useful for addressing conflicting or simply different stakeholder theories and the issues of unintended outcomes.

Amalgamating Theories That Represent Different Stakeholder Interests   List (2004) gives an example of a program to introduce radio talk shows in Vietnam where the primary purpose was to expand democracy by giving the general population a voice. A potential spin-off from that program was that radio would become stronger as a business. The radio business was a definite stakeholder whose interests were somewhat different from those of either the government or the people of Vietnam. In one sense, this outcome could be considered unintended in that it was not central to the purpose of the program.

Although the sustainability of radio was not a specific objective of the program, its survival was certainly required to make this particular program theory and program work. In Figure 6.1 we have added a feedback loop to show this connection between the two strands of the program theory and the importance of the unintended outcome to the sustainability of the program and its impacts.

Competing program theories may spring from simplistic ideas about how change occurs, with each stakeholder advocating one solution that will fit all situations. Narrow perspectives may overlook the need for different solutions

Figure 6.1   **Amalgamating Different Stakeholder Theories**

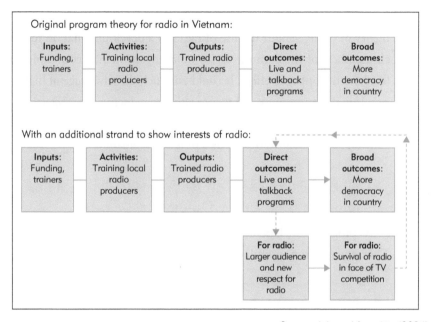

Source: Adapted from List (2004)

to address different mechanisms for change (for example, that change is more likely to result from hope or fear, or that it results from being better educated, or that it results from using better price signals) and different policy tools to activate these mechanisms (carrots, sticks, education, and so on). (We discuss these program types in Chapter Twelve.) However, many programs need to combine various policy tools in order to be effective. Carrots work with some members of the target audience under some circumstances, sticks and education under others, and often all will need to work together. The program theory needs to address this complexity rather than ignore it. Combining apparently alternative program theories and showing how they relate to one another is one way to approach this complexity.

Showing Competing Theories That Include Negative Program Theories and Unintended Outcomes in a Logic Model   Negative program theories posit that the program will have the opposite or counterproductive effects to the ones desired by those who constructed the program or to the way the program is portrayed. They are another way to capture unintended outcomes

and, in particular, negative unintended outcomes and to ensure that those
that can be foreseen are not overlooked.

Figure 6.2 extends the Vietnam radio example to show how an initial
theory of change about the use of radio in a developing nation to empower
all members of the community to hold government accountable could in
fact have the opposite effect. It could entrench the advantage of higher-status
people in the community and reduce rather than increase democracy. Figure 6.2
shows alternative pathways: a positive pathway and a negative pathway. Both
pathways could be investigated by a program evaluation.

What to Do When Little Research Evidence Is Available as a Basis for Developing a Program Theory

Some programs are based on little prior research or theoretical investigation.
They emerge from popular wisdom about what works or a decision to try

Figure 6.2   **Positive and Negative Program Theories**

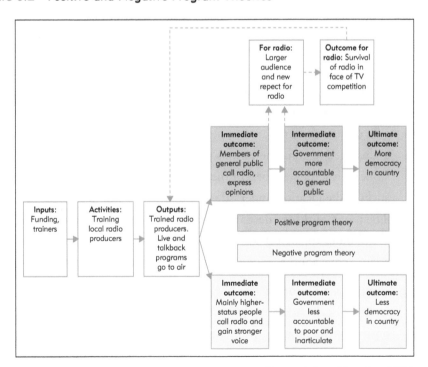

Source: Adapted from List (2004)

something new. The absence of an evidence base may weaken the credibility and external validity of a theory, but it certainly does not prevent one from identifying a program theory. If only program theories with a strong evidence base were permitted, then theories might never be developed or tested for innovative programs. Moreover, if the program is going ahead, and someone has a rationale for adopting it, then undoubtedly some kind of theory, even if home grown, is underpinning the program.

For many reasons, it can be helpful to make this theory transparent. One important function of program theory is to bring theories to the surface so that they can be analyzed and tested, not just to come up with the perfect theory. There are also occasions on which programs that appear to be poorly constructed or lacking a sound research base will nevertheless achieve worthwhile results. Also, as we discuss in later chapters, a theory that was applicable at one time may no longer be applicable. Transparency makes it easier to recognize when changes may be needed.

Reluctance to Disclose Theories Underpinning Programs

Key stakeholders who need to be involved in the process of describing program theory may be reluctant to discuss their theories of change openly. Sometimes this is because they recognize that the program is not a rational one, is in some way indefensible or deficient (see Fraser, 2001), or is pursuing objectives that might be questioned if they were publicly disclosed. Sometimes nondisclosure is a means of avoiding conflict, and sometimes it is a means of holding onto power. We discuss some of these reasons in more detail:

- *There is no theory to speak of: A theory does not exist at all or, if it does exist, stakeholders may be embarrassed about its lack of coherence.* This can happen for reasons that have nothing to do with what the stakeholders have done themselves in designing a program. Program staff may have inherited what they recognize as a faulty program. They may consider it a waste of time to look for the theory yet are loath to make public that there is no theory. Even so, sometimes those who recognize that they have been given the stewardship of a poorly conceptualized program may want to reshape it into something useful that will achieve valuable outcomes. It may be possible to entice participants to do so through the program theory development process. Or managers, staff, and others recruited to identify a program theory for a well-established program may have had a great deal to do with knowingly setting up a program that began with a poor theoretical basis. They may have established a program simply to meet some political imperative, but without much expectation of or commitment to its effectiveness. Program theory development may be a waste of everyone's time and energy in that situation.

•   *"Don't mention the war": Concerns exist that efforts to identify a program theory will bring to the surface and disclose points of division among various stakeholders in a confrontational manner.* There can be an unwillingness to air the dirty linen concerning points of conflict to an outside facilitator. There can also be a desire to avoid conflict and maintain internal comfort levels by not discussing some points of contention. Ambiguous terminology is one approach that stakeholders might use to avoid conflict. They may resist the efforts of program theory facilitators to define their terms more precisely. As George Bernard Shaw (1944) said:

> As long as we describe the virtues we are to practice and the vices we are to eschew in abstract terms every sane thinker from Confucius and Moses to Jesus and Muhammad . . . agrees with us wholeheartedly. But the moment we come down to tin tacks, the agreement vanishes . . . smokers and non-smokers cannot be equally free in the same railway carriage [p. 5].

For example, stakeholders for educational programs might agree that equality of opportunity in education is a desirable outcome. For some, that might mean equal access to particular educational facilities or every school having the same pupil–teacher ratio or the same number of science laboratories per one hundred students. For others, equality of opportunity might mean equality of outcomes across students regardless of background. Intervention strategies, intended outcomes, and evaluation strategies may well differ depending on perspective. Resistance to clarification, while it may avoid some conflicts and other difficulties, can contribute to uncoordinated and dysfunctional program behavior.

•   *The professional knows best: Nondisclosure is a means of holding onto power by arguing that the professional knows best and should not be questioned.* Logic modeling in particular has the potential to demystify programs and make inherently complex programs appear simple—perhaps simpler than they really are. And sometimes programs that are touted as being complex really are quite simple once their program theory has been unpacked.

•   *The elephant in the room: The program has hidden agendas (or restricted access) that, if exposed, would be revealed as encapsulating some values, ideologies, or priorities that stakeholders would prefer not be made public.* For example, there might be political reasons that the program favors the interests of some target groups over others, competing agendas that result in poor program trade-offs, or the purpose of the program may be politically incorrect or unacceptable to the population or to partners. Intended outcomes that are about influencing the behaviors of partners may be considered by those partners to be presumptuous or mechanistic social engineering.

A variation is that so-called minor agendas such as generating public support for government, getting a seat at the table or currying favor with various organizations, or not alienating particular groups are really the main agendas. And they eventually take priority over resolving the problems that the program is ostensibly about. Achievement of the publicly stated but incomplete agenda without also achieving the other unstated agenda would lead to the program's being considered unsuccessful by those who have set its agenda and are responsible for the program.

•   *The emperor's new clothes: Management and staff recognize that their program has limited potential to have an impact and so resist making the program theory explicit.* Given the size and nature of the problem at hand, they fear that their efforts, no matter how well intentioned and exercised, will fall well short of being sufficient to make a substantial impression on the problem. Eliciting the program theory may make staff feel helpless. Undertaking the program theory process is not a happy one for them unless it can be used as a means of identifying, measuring, and giving credit for more realistically achievable outcomes and also for identifying opportunities for greater success.

• *Talking up the potential of a program to obtain or retain funding.* In this case, eliciting the program theory might disclose to others that descriptions of the program have inflated its potential to have significant outcomes and sustainable impacts. The program theory might lead to queries being raised in relation to potentially specious claims about the contribution of a program to various changes that have been observed (such as observed changes in behaviors, social problems, or educational outcomes that may or may not have anything to do with the program). Funds may be reduced to a level that would be commensurate with the extent of program impact.

Cummings, Stephenson, and Hale (2001) give an example of a logic modeling project for a Local Management in Schools Project (LMS) in which

> nearly all stakeholders . . . were initially uncomfortable with the idea of explicitly stating the assumptions that underpin the LMS. This appears to be due to reluctance to be asked to explain how the LMS was expected to work (perhaps in case it didn't actually work as expected), and to an unwillingness to expose the LMS process to formal critical review by colleagues and others. At this stage in the project it was evident that stakeholders did not share a mutual understanding of LMS and lack a common language for talking about it [p. 33].

Being aware of these types of motivations and resistances can help with identifying how much commitment there is to develop a program theory and with decisions about how far to push the theory identification and description process. There may also be implications for who should be involved in the process and how to deal with situations where stakeholders bring too many unhelpful motives to the process.

There is always the potential for evaluators to raise questions about efficacy and legitimacy of the program theory when it is difficult to elicit a coherent or otherwise defensible theory. However, as Fraser (2001) concluded,

Across all the conflicting schools of methodology evaluation has always operated on the fundamental principle that letting in a little sunshine can only do good. Where perversity is at play, this expectation becomes a lot less secure. The evaluator finds himself in the realm of the un-acknowledgeable and the un-discussable, where too much clarity can be seriously unwelcome [p. 25].

Given all of these potential points at which the program theorizing process may be derailed, an evaluator or program theorist needs to make a professional judgment about whether the development of a program logic and the evaluation project as a whole is useful or is a pseudo-evaluation, and then act accordingly.

Lack of Interest, Nonengagement, and Faltering Engagement

Many factors can affect willingness to engage in the process of identifying program theory, additional to those already discussed about reluctance to disclose program theories. A key factor is an inability to see how program theory can help them in their work. This may lead to failure to use program theory once it has been developed. We have found that it can be helpful at the start of a program theory development process to show participants a simple example of a program theory and how it was used to design an evaluation.

Some program theory development can be quite complex and require considerable mental agility and a willingness to think strategically and explore alternative scenarios. It is not a task that suits the abilities or inclinations of all people. Some people do not see the point of program theory. Others do not enjoy engaging in processes that use systematic logical approaches or analyses of causes and effects. These approaches may be at variance with their own methodological orientations, and they may decide that program theory stifles rather than liberates their thinking. They may not be prepared to commit to the hard thinking and documentation that program theory can require. Cognitive style and personality differences such as preference for operating intuitively versus thinking analytically may well affect a propensity to engage in program theory.

Occasionally it becomes evident to a program theory facilitator that the stakeholders who have been nominated to participate in the program theory identification process for a complex program lack the strategic thinking skills or commitment to be effective. This situation may require some subtle maneuvering to involve more appropriate people.

Support and leadership for the process from senior management is important. If senior management is not persuaded from the outset that the exercise will be useful for them and the organization, then the value of proceeding with the theory of change processes should be seriously questioned. Giving examples of where program theory has been useful to other managers can be persuasive. When managers see similarities to their programs or to the issues that they confront, they may see the significance. This is particularly true for those who have come to the exercise looking for how program theory can help their organization.

Even so, it can be difficult to persuade senior managers that the work of fleshing out the theory of change is sufficiently strategic to engage their attention and get them to commit resources. Competing priorities for senior managers' and staff time can lead to a preference for employing someone else to undertake or lead the program theory development process. When there is no possibility of more active engagement of senior managers, then engaging them for a relatively short time, typically at the beginning of the process to set the agenda, can be more effective than trying to engage them throughout. However, it is important to check with them in an engaging manner from time to time to ensure that progress is on track and the program theory is being understood.

Engagement can falter when program theory development processes are protracted. Workshop processes that require participants to engage in a series of workshops can make it particularly difficult to maintain engagement. Laycock (2005) provides some useful practical ideas for maintaining the engagement of participants:

- Making efficient use of group time to apply incremental processes, such as having meetings that are long enough to make substantial progress or at least including some longer meetings at strategic points in the process to refine the logic model and reach agreement

- Maintaining enthusiasm by making the connections between the theoretical and the practical by incorporating the discussion of program theory in the discussion in implementation planning

Laycock found that having the evaluator incorporate comments from the meeting in the logic model meant that the committee became even more isolated from the process of developing the model. Undertaking more of the revision processes during the meeting rather than after may help if time is available.

Another practical issue arising from the development of program theories and logic models over several sessions is that it is rarely possible to get the same people in the room on each occasion, so each replacement person brings a different perspective.

Longer, fewer meetings with less turnover of participants may make life easier for the program theorist and other stakeholders in terms of reaching agreement about a program theory. However, it must be recognized that this is a consensus limited to those participants, and for more controversial programs this may be a fool's paradise. The belief that "the program theory has been done and we can all go home!" may be a misplaced one.

DECISION 5: HOW MUCH TIME AND RESOURCES SHOULD BE INVESTED IN DEVELOPING OR IDENTIFYING A PROGRAM THEORY?

Identifying program theory can undoubtedly be a time-consuming and resource-intensive process. However, there are ways to make the process more efficient. Also, sometimes perceptions about inordinate amounts of resources and time to undertake program theory development processes arise from a lack of appreciation of the benefits of doing so.

Convincing Stakeholders to Invest Enough Time

It can be difficult to persuade funders of evaluations and other stakeholders involved in the program theory process that this will be time well spent. Some key advantages can be brought to their attention concerning investment in program theory as part of a program evaluation process:

- Their involvement in program theory development can give more control over what is investigated in an evaluation.

- Developing a program theory can produce some early evaluation findings relating to the design and feasibility of the program. These findings can contribute to refining program design and implementation.

- Developing a program theory can be the basis for deciding whether it would be appropriate to undertake an empirical impact evaluation without first refining the program design or its implementation. A decision of this type was made in relation to the antidrug program for ethnic communities described in this chapter. The use of program theory for this purpose can save valuable resources from being unnecessarily or prematurely directed to conducting an evaluation of outcomes.

- Developing a program theory can reorient an evaluation toward focusing on achievable outcomes rather than on unrealistic objectives.

Some Considerations for Decisions About Resourcing Program Theory

We have argued that program theory can be useful for all programs: simple or complex, large or small. However, decisions need to be made about how much time, effort, and rigor should be devoted to the process. A number of factors can be considered when making these judgments. Our discussion has noted that many methods can be used either singly or in combination to develop a program theory. In general the greater the number of methods used, the more protracted and resource intensive the process will become. Note the following considerations relevant to resourcing program theory:

- *Which one or more of the range of purposes the program theory process is to be used (such as program design, evaluation frameworks, team building, or communication)?* Some single purposes, such as use of a technical process to develop an evaluation framework, may require less time than others (such as team building) or combinations of purposes. Also, in many cases, all that is needed is a program theory that is good enough. Much time can be spent on refining the details of a program theory and its presentation, but with diminishing returns in terms of usefulness. It is always important to consider whether the program theory in its current form is sufficiently fit for purpose rather than whether it is perfect, bearing in mind that theories rarely are perfect, have a limited lifespan, and that in many respects, the processes of thinking through them are more important than the finished product.

• *Size, complexity, and budget of the program* (Truman and Triska 2001). In general, more time, resources, and more rigorous methods would be applied to larger, more complex programs. This is not to say that program theory should not be done for small or simple programs. However, the level of resourcing of the program theory should be commensurate with the size and simplicity of those programs. In addition, some programs, regardless of size, will already have a more cogent rationale than others, and less time and effort will be required to lay out and document the program theory: it will fall into place relatively simply.

• *Size of the assumed impacts of the program* (Leeuw, 2003). The larger the assumed impacts are, the more rigorous and resource intensive the program theory methods should be. In some situations, a program such as a pilot program has small potential impacts and perhaps costs little on its own, but it would have large potential impacts and cost a lot were it to be adopted widely. In those situations, more resources for program theory development might be justifiable.

• *Level of risk with regard to impact and side effects of the program.* The larger the risks in the program, the more rigorous the processes for identifying the program theory should be (Leeuw, 2003).

Getting Smarter About Resourcing Program Theory

Renger and Titcomb (2002) have described the processes they use to teach students how to develop program theories as part of planning an evaluation. They estimate that the identification of just one key aspect of a program theory (what they refer to as the identification of the antecedent conditions of the problem to be addressed) can consume about 75 percent of the time typically allocated to planning an evaluation. It is important to continue to look for more efficient ways to identify program theory.

Sometimes stakeholders have genuine interest in and are committed to having a useful program theory but lack the time to engage actively in all aspects of the development process, especially for complex and multifaceted programs. Employing a program theorist to develop a draft program theory from documentation and interviews can help to start the process and save time. One approach is to have a skilled program theorist develop a draft program theory based on the best information available at the time: program documentation, research reports, knowledge of other similar programs and of research-based theories of change, and a small selection of interviews. This draft can be brought to the table for discussion with stakeholders.

The program theorist needs to manage the process carefully to avoid overcommitment to her or his own theory during discussions and to foster ownership by relevant stakeholders. Introducing a third party to facilitate the discussion and amendments has been a useful process on some occasions.

When group processes are used over an extended period of time, with the identification of a program theory being an incremental process, the time required for rework can easily expand to a level that threatens the practicality of doing so and engenders frustration or disengagement of stakeholders.

Again we wish to emphasize that we see that the main value of program theory development is as being a thinking tool rather than the capacity to produce elegantly documented theories.

DECISION 6: WHEN IS IT TIME TO REVISIT A PROGRAM THEORY?

Program theories should evolve as part of ongoing organizational learning. However, from time to time, it will be important to undertake a more systematic and comprehensive back-to-basics review of a program theory. A clear indication that it is time to do this is when the program's context or the problem or situation that initially gave rise to the program has changed. Changes in the research base and knowledge about what makes particular types of programs more or less effective can also suggest a need to revisit a program theory. Changes of personnel and other stakeholders can also mean that it is useful to revisit the program theory to develop understanding and commitment and get everyone back on track or to benefit from new perspectives. Sometimes we have been commissioned specifically for the purpose of undertaking relevant research and developing a new program theory for an existing program whose context has changed, as, for example, when it has become part of a wider strategy that has different priorities from those originally embedded in the program.

The fact that different stakeholders can come up with alternative program theories or changes over time to an agreed program theory reinforces the fact that most program theories reflect the views of particular stakeholders at a particular point in time and the particular sample of stakeholders engaged in the program theory process. As such, the generalizability of any program theory may not reach far beyond those members. Similarly, theories inferred from site visits (inductive theories) represent the particular sites observed at a particular time through the lens of the particular observer. Deductive theories based on program documentation certainly need to be checked from time to time to ensure that the documentation is still considered relevant or, if it is based on research, that the research is consistent with current knowledge.

Implications of the Fact That Program Theories Are Often Relative

If program theories are relative (to people, time, and so on), they need to be tested not in terms of absolute truth but in terms of usefulness for particular purposes. For example, do they help us to think more systematically about what underpins a program, leading us to question and improve the program? Do they help us to communicate the essentials of what the program is about? Do they help us to ask sensible and useful evaluation questions and choose more appropriate performance measures?

Many programs and their contexts are dynamic rather than static. Key players change. New values come to the fore, or the balance among values changes or is renegotiated. Contexts change, and a program that was once central becomes peripheral. Factors that were once critical to the success of a program are no longer so, and those that once appeared to be irrelevant can suddenly assume great importance. Accordingly, a program theory that seemed eminently plausible and appropriate at one time may rapidly lose validity.

Programs with complex aspects are especially susceptible to these types of changes. Recognizing that a program has complex aspects or is in response to a complex situation can be a prompt to explore the legitimacy of program theories that may have been developed and documented in the past. The program theories for programs with complex aspects also need to incorporate some features that are responsive to the changing nature of the program rather than straitjacketing the program concepts. For example, the program theory needs to be described in a way that allows emergent outcomes and multiple causal paths, whether complementary or alternative.

Programs and the ways in which they are viewed by staff and others may also be rapidly changing—not because the programs are necessarily complex but simply because they may be relatively new and program staff are still finding their way around them, reconceptualizing them as they go. This can be a source of enormous frustration for a program theory facilitator, especially when the purpose of the program theory is to provide the framework for designing an evaluation. At some point, the group (or the program theory facilitator) needs to drive a stake into the ground. It is possible to agree that a theory is workable and useful for the purpose of designing the evaluation and still keep an open mind about that theory as the evaluation proceeds. This would include iterative evaluation processes that allow emerging paths to be explored and receptiveness to unintended and unanticipated outcomes.

Documenting the Program Theory in an Ever-Changing World

Because programs and their contexts are forever changing, it is helpful to clearly document both the program theory developed at a particular point in time and the methods used to develop the theory. Documentation assists with putting on record the origin of the program theory for program and institutional memory purposes. It will make assessing the continuing relevance of the theory easier in years to come. Documentation of methods for developing the program theory, sources used, and whose perspectives are being represented assists program theory developers as reflective practitioners. Documentation of changes to program theory as they occur is also important

Figure 6.3   **Linear Theories with Branching Structures**

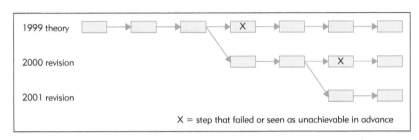

Source: From Davies (2004)

but often overlooked or not done at all because stakeholders are only vaguely aware that their theories have changed; they have not been reflective practitioners in relation to program theory.

Documentation of changes can be in simple narrative form or in pictorial form. Davies (2004) gives an example of how changes in a program theory arising from organizational learning could be portrayed (Figure 6.3).

Documentation of a program theory and the processes used to develop it also helps uninvolved stakeholders who may wish to understand, reflect on, critique, or apply the program theory. Knowing about the evidence base and how well the program theory has been constructed from it is likely to be particularly important when key social policy advice is being given, such as that about choice of policy tools and implementation strategies.

SUMMARY

In this chapter, we have discussed various methods for identifying a program theory and why it is important to approach the process with some deliberation. We have emphasized that the choice of methods for developing a program theory should be governed to a large extent by its purposes. Multiple purposes can require hybrid approaches. We have discussed methodological issues, identified common challenges when describing program theory and provided some ideas on how to address them, and have identified some signals for determining whether and when a program theory should be systematically and comprehensively reviewed.

EXERCISE

Think about a program theory exercise in which you have participated or with which you are familiar and identify:

a. The sources of information that were used and what other sources of information could have been used to develop the program theory.

b. What general approach was used to develop the program theory: articulating stakeholder mental models, deductive approaches, or inductive approaches, and what other approaches might have been feasible.

c. How the program theory might have looked different with the addition of other sources of information or other methods of developing the program theory.

d. Challenges that arose in developing the program theory and how those challenges were addressed. What other approaches could have been taken to address those challenges?

7

Developing a Theory of Change

I N THIS CHAPTER, we identify three key features of a program's theory of change that can improve the contribution of program theory to monitoring and evaluation, as well as to program design. Although these features are not included in all approaches to program theory, we have found them useful for improving the quality of a theory of change.

We present the three features of the theory of change, followed by the three features of the theory of action discussed in Chapter Eight, as a sequential process for identifying key elements of a program theory. However, the process may in fact be an iterative one, especially if program theory is being used to design a program. For example, the process of identifying program activities and resources, as part of describing the theory of action, may lead one to revisit parts of the theory of change and to reconsider what would be reasonable outcomes for the program and which would be out of scope or borderline. In addition, although it is helpful conceptually to differentiate a theory of change from a theory of action and to ensure that important aspects of both are incorporated within an overall program theory, in practice, work on the two aspects often takes place at the same time and is seen as a single program theory development task.

FEATURES OF A THEORY OF CHANGE

For each of the three features of a program's theory of change shown in Table 7.1, we explain what that feature is and why it is important, and we give some suggestions about how to identify it.

In broad terms, identifying these features of a program's theory of change can assist in these ways:

- Showing how the program's intended outcomes are expected to contribute to addressing the situation that gave rise to the program and to the outcomes that are ultimately sought by this program (often in concert with other programs or other factors)

- Identifying the baseline data that should be collected in relation to the ultimate outcomes even if these are beyond the direct influence of the program

- Clarifying program boundaries while simultaneously recognizing where the program sits in, interacts with, influences, and is influenced by its wider context

Table 7.1   **Components of a Program Theory**

Theory of change—the topic of this chapter	Situation analysis: identification of problem, causes, opportunities, consequences	Focusing and scoping, setting the boundaries of the program, linking to partners	Outcomes chain: the centerpiece of the program theory, linking the theory of change and theory of action
Theory of action—the topic of Chapter Eight	Desired attributes of intended outcomes, attention to unintended outcomes	Program features and external factors that will affect outcomes	What the program does to address key program and external factors

- Ensuring that the outcomes chain gives adequate attention to outcomes that are beyond the direct influence of the program but that are critical to the program's success and its potential to make a difference to the main problem

We show how effective attention to the features of a theory of change that we discuss in this chapter can avoid common traps such as those we identified in Chapter Three. Table 7.2 provides a summary of those traps to which the various features of a theory of change are particularly relevant. We also note links to the different approaches to developing a program theory that were discussed in Chapter Six: articulating stakeholder mental models, deductive approaches, and inductive approaches.

SITUATION ANALYSIS: UNDERSTANDING THE PROBLEM, ITS CAUSES, AND ITS CONSEQUENCES

Developing an appropriate theory of change begins with an accurate analysis of the existing situation.

What a Situation Analysis Is

A situation analysis identifies the nature and extent of the problems or opportunities to be addressed by the program. It describes the various features of the problem. Who is affected by it directly and indirectly? What evidence is available about the size of the problem, its history, and whether it is changing over time with respect to its nature and extent? How good is the evidence?

The situation analysis also identifies the known causes of or causal pathways to the problem and the known consequences of the problem (why the situation is problematic and worth addressing). Preventive programs will primarily address the causes of a problem—the iceberg. Treatment programs will be about addressing the consequences of the problem—often the tip of the iceberg.

A program may be trying to address all three aspects of the situation: the problem, its causes, and its consequences. It will try to reduce the problem through preventive measures that may also reduce recurrence of the

Table 7.2 Features of a Theory of Change for Addressing Common Traps

	Features of a Theory of Change That Can Help Avoid This Trap		
	Situation Analysis	Focusing, Scoping, and Boundary Setting	Complete Outcomes Chain as a Centerpiece of the Program Theory
Program Theory Inadequacy			
Trap 1: No actual theory about how the program will contribute to outcomes			
Failure to show how particular activities will contribute to particular outcomes, especially outcomes that are further along the chain of outcomes			✓
Excessive focus on activities and insufficient emphasis on outcomes: activity traps			
Trap 2: Poor theory; implausible theory			
Inadequate formulation of the problem to be addressed—its causes and consequences	✓		
Large gaps in the theory (miracle thinking)			
Indefensible connections among outcomes in the chain of outcomes			✓

Trap 3: Poorly specifying intended results

Excessive focus on what is measurable and exclusion from the program theory of any outcomes that are too difficult to measure, which can lead to goal displacement ✓

Trap 4: Ignoring unintended outcomes

Outcomes chains that are insensitive to unintended outcomes that lie beyond the immediate boundaries of the program ✓

Outcomes chains that include only the main intended outcomes and ignore important outcomes for other stakeholders ✓

Trap 5: Treating programs as simple when there are important aspects that are complicated or complex or vice versa

Failure to recognize that programs rarely function as closed systems

Insensitivity to the inherent unpredictability of the contexts within which programs operate ✓

Failure to attend to what is outside the scope of the program as well as what is within scope ✓

Overclaiming or underclaiming observed outcomes ✓

problem following treatment. At the same time, given that prevention is rarely 100 percent effective, the program may also include or link in with some treatment strategies to mitigate the impacts of any residue of the problem that the program has not successfully addressed.

By *situation,* we do not mean the particular event that triggered the program, which might have been a critical incident or a political imperative arising from public complaints. Rather, we mean the problem, set of problems, or opportunities that the program is to address and the context within which the program operates. However, sometimes analyzing a critical incident or gaining an understanding of a trigger event for a program can help with a situation analysis that will inform the program theory.

USING A TRIGGER EVENT AS A STARTING POINT FOR DOING A SITUATION ANALYSIS

A program to provide crisis financial assistance to those having difficulty paying their electricity bills began as an immediate response to a death caused by disconnection of a debtor's electricity. Because of the disconnection, the debtor had resorted to using candles. A candle caused a fire that led to his death.

Concern to ensure public safety by providing continuity of access to electricity was an initial driver of this program. As the program took shape, it became apparent that there could be many other adverse consequences of having electricity disconnected (for example, health related), and these also needed to be included in the program theory.

Neither the trigger event nor its consequences or implications for intended program outcomes such as safety were explicitly included in program documentation. Discussions with staff who were employed at the time the program was introduced unearthed this information, which subsequently was confirmed by other less easily accessible documentation.

A good situation analysis goes beyond a focus on problems and deficits to identify strengths or opportunities and may reframe perceived problems as opportunities. An antigraffiti program for which we prepared a program theory converted the problem of graffiti into an opportunity by developing a strategy to support creative street artwork that would be an alternative outlet for youth. For some of these youth, the artwork became a source of self-esteem, satisfaction, and income while delivering aesthetic benefits to the community, helping to mainstream disaffected youth into the community, and changing their leadership base from one that revolved around crime to one that revolved around community contribution.

Other examples of opportunities are those that arise from new legislation, noticeable improvements in political and community support for particular types of objectives or ways of meeting them, emergence of new enabling technology, and new evidence about practices that do or do not work.

The following example is a situation analysis for an employment program for mature-age workers that we will be following throughout this chapter and others to demonstrate the development of various components of a program theory.

SITUATION ANALYSIS FOR AN EMPLOYMENT PROGRAM FOR MATURE-AGE LONG-TERM-UNEMPLOYED PEOPLE

THE MAIN PROBLEMS

The main problem to be addressed is long-term unemployment of people over the age of forty-five. Older age groups (those over age forty-five) have a higher share of long-term unemployment, reflecting the fact that these older workers, once unemployed, have a high risk of becoming unemployed for the long term. Employment statistics show that this discrepancy between older and younger workers has been increasing over time. (A complete situation analysis would cite or include the baseline and trend data.)

Research evidence at the time this program began showed that the longer people are out of work, the harder it is for them to get a job. A person unemployed for less than three months had more than twice the chance of finding work than a person unemployed for between one and two years and four times the chance of a person unemployed for two years or more. Moreover, when older unemployed people do secure a job, they face a higher risk of not sustaining that employment beyond twelve months. (A more complete situation analysis could include data about such other dimensions of the problem as whether, within this group, there are differences relating to gender, ethnicity, age, location, type of employment, educational background, and other characteristics.)

CAUSES, CONTRIBUTING FACTORS, AND OPPORTUNITIES

Long-term unemployment increases the probability of even longer-term unemployment because the longer a person is out of a job, the more likely she or he will have:

- Fewer contacts with the labor market and so less opportunity to find out about job vacancies

- Reduced skill levels

- Loss of confidence and sapped morale

Considerable evidence indicates that the earlier that employment programs are able to engage with older people after they become unemployed, the less their motivation (including work habits) and skills have deteriorated and the higher the probability is of their being able to secure a job.

Also many employers believe, rightly or wrongly, that the longer a person has been out of work, the less employable he or she is. In relation to mature-age workers, this problem is compounded by the fact that some employers have negative views about the skills and attitudes of older workers, their capacity to adapt and to work as team members in a younger workplace, and their endurance on the job.

Counter to these negative views is a considerable amount of research that shows particular benefits that older workers can bring to the workplace. Some employers have experienced the benefits of employing older workers and are a useful resource that the program can draw on to promote the cause of older workers.

Other factors as well contribute to their higher levels of long-term unemployment:

- Mature-age workers are particularly vulnerable to long-term unemployment during times of high unemployment levels in the community as a whole.

- The nature, size, and scope of work within a community—for example, small rural communities compared with larger urban ones—can have differential effects for older workers.

- In some cases, the structure of welfare benefits for people nearing retirement age makes it less attractive for those who have become unemployed to obtain employment. This can affect their motivation to obtain a job.

- Physical or mental disabilities, or both, that may be associated with aging for some people can limit the range of types of employment for which they are physically suited and for which they have the necessary education, skills, and experience.

- Overqualification relative to available jobs can also create barriers to employment.

CONSEQUENCES

The research on this topic has identified several consequences of long-term unemployment for individuals, their families, and the community. These include poorer health, social and economic disadvantage for the individual and his or her family, economic disadvantage to the community because of inefficient use of its skills base, reduced productivity, reduced income tax from mature-age workers, and increased costs through welfare payments.

Why a Situation Analysis Should Be Done

Correctly understanding the situation that the program is to address is fundamental to understanding and portraying the program theory such that it presents plausible solutions to a correctly identified problem or maximizes opportunities. Situation analysis helps to show how the program's intended outcomes will contribute to the ultimate outcomes sought by the program and the problem that it is to solve.

A situation analysis gives proper attention to all known significant causal factors and their interrelationships that are contributing to a problem and to the consequences of the problem that show why it is important to address it. The full array of significant causal factors, relationships, and consequences is rarely completely known, and changes. Some factors are within the scope of the program to influence, and some external conditions also need to be in place. A good situation analysis recognizes elements of the situation that are especially difficult to predict because they are fluid and unstable, such as the emergence of epidemics or political volatility. In some cases, a situation analysis will be about reframing a problem as a potential opportunity, as in the case of the antigraffiti program.

As far as possible, a situation analysis includes data (not just statements) about the problem and its causes and consequences. These can be used as baseline data for evaluating the continuing need for a program or its effectiveness. Where possible, projections about how the problem will evolve, also an important part of baseline data, should also be included.

Among the information included in a situation analysis for a program to combat volatile substance abuse by youths, for example, were baseline data concerning prevalence and incidence, patterns and trends of use, as well as known health and law-and-order consequences. The situation analysis identified weaknesses in the data and improvements that needed to be made to improve their usefulness for future program design, monitoring, and evaluation. Other information for the situation analysis came from quantitative and qualitative research data and from experienced practitioner wisdom. The information related to such factors as ease of access to and supply of drugs, circumstances of individuals that predisposed them to use drugs, community contexts, and community capacity, including variations across communities, coordination or lack thereof among relevant agencies, and state and federal boundaries of responsibility.

Assessing the continuing need for and relevance or appropriateness of the program and various aspects of it is an important type of evaluation that is sometimes overlooked. Information about whether the nature and size of the problem are changing can have implications for whether the need for some type of program is increasing, decreasing, staying the same, or changing in some other way. For example, the problem may be moving to a different location or a different target group. The nature of the problem may be changing as, for example, when there is displacement from one type of crime to another, from one form of drug abuse to another. Information about whether the various causes of the problem are changing can also be useful for deciding whether the program remains the right type and whether its emphasis is right for addressing the problem.

Programs sometimes get stuck at working on a particular causal factor long after the need to do so has passed. Getting stuck can happen for many reasons. For example, a program may get stuck at concentrating on the stage of raising community awareness for many reasons: resources required for public communication and awareness raising may be less costly as a behavior change strategy than approaches needed to effectively enforce regulation; staff may be more comfortable and capable with respect to community awareness campaigns, and this may discourage them from moving to alternative approaches; and lack of monitoring data about the problem, its causes and consequences, may result in failure to recognize that the awareness problem has been largely overcome.

MONITORING CAUSES AND RESPONDING APPROPRIATELY

Particular types of causes may have been nearly or fully addressed or have been addressed to the point where they need only occasional booster shots. For example, by the end of the twentieth century, HIV/AIDS awareness campaigns might already have largely hit their mark in some countries. The awareness-raising effort may have been able to be generally scaled down, but from time to time since then, a top-up has been needed as awareness fades, the issues become less salient, new issues emerge, or complacency sets in. Monitoring information is vital for tracking the changing levels of awareness both up and down.

Data can tell us whether a particular cause and its impacts seem to have been reduced regardless of whether the progress is attributable to the program. These data can provide insights that are relevant to decisions about whether it may be time to move program effort to another possible cause and to achieve different outcomes related to that cause. However, as the above example shows, it is important to revisit the situation from time to time rather than assume that the cause has been addressed for all time.

When and by Whom a Situation Analysis Should Be Done

Like our answers to many other questions, our answer to this one is: it depends—on intended use, context, and so on. Ideally situation analysis should be done by program designers as part of the program design process. An evaluator or program theorist might also be called in to assist, given his or her research skills and program theory concepts.

The situation analysis might then be revisited sometime later (even years later) by designers or formative or developmental evaluators to check its continuing validity and provide advice about whether redesign is needed. Alternatively, it might be revisited by an evaluator at the time of doing a summative evaluation and as part of the process of making judgments about whether the program was well founded and appropriate. At that point, an evaluator may choose to collect additional data that would validate, question, or refine the original situation analysis. Sometimes new information emerges that was not available to the program designers at the time of designing the program, and this needs to be noted. This might also be the point at which an evaluator might recommend a different type of program design for the future.

Questions for Preparing a Situation Analysis

We suggest the following questions as a basis for preparing a situation analysis. It may not be possible to answer fully all questions for all programs if information is lacking. But even some incomplete information and carefully assessed perspectives from experienced practitioners can be better than none:

The Nature and Extent of the Main Problem That the Program Addresses

1. What is the problem, issue, or opportunity? Describe the nature and extent of the problem or opportunity.

2. For whom does this problem exist: individual, household, group, community, society in general, particular locations?

3. What is the history of this problem, and what projections are there about its future?

Causes and Contributing Factors

4. Why does this problem occur? What are its causes? Are some causes more important or influential than others? Are there known causal pathways—that is, successions of causes?

5. What, if anything, is known about what has and has not been effective in addressing this problem?

Consequences

6. Why should this be considered a problem? What are the consequences of this problem for those who are directly affected by it (for example, the effects of drug abuse on the users)?

7. What are the consequences of this problem for those for whom the problem exists indirectly (for example, families of drug abusers or communities affected by the drug induced violence)?

8. Alternatively, if the focus is on opportunities and strengths rather than on problems, why is this opportunity worth pursuing? What benefits will it deliver? What can contribute to successfully taking up this opportunity?

How the Use of These Questions Might Vary with Choice of Program Theory Development Process

If the situation analysis were being used with the articulating stakeholder mental models approach, then this set of questions could be used to draw out important features of the theory of change of stakeholders during group sessions and interviews. However, these questions are posed as research questions and may need to be rephrased to elicit the desired information from stakeholders.

If the inductive approach is being used, this set of questions can structure observations and follow-up questions to assess the accuracy of the observer's interpretations. With a deductive approach, this set of questions could be used to guide a research literature review and document analysis and to identify the relevance of wider theories of change and program archetypes (discussed in Chapters Eleven and Twelve).

Ideally the process will be an iterative one in which each process builds on the other: stakeholder perceptions might be checked for consistency with the research literature; the situation analysis identified through documents can be cross-checked with stakeholder mental models and their reasons for holding those models; it can also be cross-checked with what is observed in practice through inductive theory development processes.

As we have already noted, data play an important role in developing, confirming, or negating the accuracy of the situation analysis. This could come through any or all of the three processes of developing a program theory.

During the design of program, many processes can be drawn on to identify causes and select appropriate policy tools, including, as the following example shows, custom-tailoring a research project around a series of working hypotheses about the causes of a problem.

USING RESEARCH TO IDENTIFY CAUSES AND SELECT STRATEGIES

A taxation department commissioned an independent and carefully designed research program to investigate the validity of the assumptions about taxpayer behavior on which its tax audit program was based. Among the matters examined were the reasons that taxpayers submitted inaccurate tax returns. Was it a result of ignorance? Willful evasion? Clerical error? How did taxpayers perceive tax evasion? Did they see it as a serious crime? Did their perceptions of its seriousness affect their propensity for tax evasion?

The information from the research program helped the department to review its public communication programs, including those that related to the formation of community attitudes. It also shed light on its procedures, processes for detecting fraud, and the way in which it would distribute its effort over the various strategies (communication, education, detection, prosecution, and fines) in order to manage the risks most effectively.

Assessing Complication and Complexity

A situation analysis helps to identify whether the situation is a relatively simple one or one that has complicated and complex elements. When the situation clearly has features that characterize so-called wicked issues—those that are highly resistant to resolution and characterized by chronic policy failure—this may suggest that the theory of change, and possibly the program, will have many complex and uncertain aspects and will need broadly stated objectives that allow emergent outcomes. The Australian Public Service Commission (2007b) has identified some features of wicked issues or wicked problems:

- They are difficult to clearly define with respect to nature, extent, causes, and solutions, and different people have different views. There usually is no one correct solution and no point at which the problem can be said to have been solved. They can be managed on an ongoing basis.

- They have many interdependencies, are often multicausal, and may have internally conflicting objectives in part because addressing one cause may have perverse consequences for addressing other causes.

- They are often addressed by strategies that have unforeseen consequences.

- They are often not stable and are evolving at the same time as research evidence is being collected and policies and programs are being developed to address them.

- They are typically socially complex, and it is their social complexity rather than their technical complexity that makes them difficult; they generally require changing behavior.

- They are rarely the responsibility of a single agency and typically require multiagency cooperation and coordination.

Features of wicked issues are often more apparent when complex strategy development and wide-ranging change management and reform programs are being envisaged than when simple initiatives or projects are being designed, monitored, and evaluated.

FOCUSING AND SCOPING

While consideration of complicated and complex aspects of the situation can lead to an ever-expanding boundary of what a program might usefully address, setting boundaries around the program by systematically scoping and focusing the program theory is important.

Defining Focusing and Scoping

Focusing and scoping the program is about identifying which aspects of the problem, its causes and consequences, a program is to focus on directly and primarily; which are important, perhaps within its reach and capacity to influence, largely beyond the direct focus of the program but within its scope; and which are far beyond its scope. Focus is a subset of total scope. Some aspects that are not the main focus of the program may be within its scope. For example, a program to deliver better mental health services to clients may recognize that in order for it to be more effective, the practices of partner organizations need to change. The program may include among its strategies various activities to promote this change. Focusing and scoping is not intended to straitjacket a program in ways that prevent it from taking opportunities or addressing issues as they arise. The more complex the program is, the more fluid the boundaries should be to accommodate emergent needs, opportunities, and outcomes.

Although focusing and scoping may draw on much information, ultimately it is about setting priorities and making judgments about what should and should not be included in the main activities of the program and what should be able to be in some way addressed by the program. Scoping and focusing statements may be developed as part of designing a program or as part of describing an existing program for purposes of evaluating it.

SCOPING OF THE MATURE WORKERS EMPLOYMENT PROGRAM

Which causes and consequences of long-term unemployment does the program address? (Refer to the situation analysis earlier in this chapter for full list of causes.)

The program aims to assist midlife and older persons to reenter and remain in the workforce by focusing on:

- *Overcoming the effects of separation from the labor market* by helping them make positive contacts with the labor market so that the labor market knows of them and they have opportunities to find out about job vacancies

- *Overcoming skill deficiencies that stand in the way of their getting a job* by equipping them with skills that will boost their prospects of finding permanent employment

- *Overcoming lack of confidence and sapped morale* that reduce their job motivation and job-seeking effectiveness by implementing activities to restore self-confidence and morale and to restore work habits for those who have been unemployed for a long time

- *Generating the support and cooperation of targeted employers* to employ mature-age people by drawing on research evidence and recruiting as allies employers who have had positive experiences with mature-age workers

- *Helping employees placed by the program to overcome some of the factors that can reduce retention in the workplace* by incorporating follow-up support processes

The program uses the following strategies to address these causes:

- Case management work with long-term unemployed people using a range of custom-tailored activities such as computer training and job interview techniques

- Advocacy on a limited scale with prospective employers

Some causes and consequences of long-term unemployment are outside the main focus of this program, but some may be within its scope:

- Anti-aged-worker attitudes across the wider population of employers. Limitations on resources and responsibility have meant that changing employer attitudes on a large population scale is beyond the scope of the program. Instead, the program focuses on some targeted receptive employers. However, the program will take opportunities to influence other employers when they arise (for example, through service provider engagement in community activities).

- Health issues of older workers that may be an impediment to getting and retaining certain types of jobs. Effecting widespread improvements in the health status of older unemployed people is beyond the scope of this program. But the program does seek to help mature-age unemployed people to manage any health issues and seek alternative employment opportunities, upgrade their skills, and so on. It also provides some limited mental health support through counseling and some referrals for other types of assistance. This limited assistance is within scope for program participants.

- Changes in the general levels of employment and reduced employment opportunities in smaller rural communities are beyond the scope of the program's intended outcomes. However, levels of employment will need to be monitored to assist with making judgments about program impact and to make decisions about the nature and size of the program, including its approach to targeting unemployed older people.

- A welfare system that creates employment disincentives for some unemployed people who are nearing retirement. Addressing this is outside the scope of the program. The program has to work within this context but is unlikely to be able to influence it.

Why Focusing and Scoping Are Important

Focusing and scoping are to encourage recognition of the importance of what happens on both sides of the program boundary and, where the boundary is fluid, allow movement back and forth and encourage interactions across the boundary; the important role of external factors and other actors in influencing program outcomes; and the importance of neither overclaiming nor underclaiming outcomes.

In this section, we explain why it is important to identify outcomes that are in-focus, out-of-focus but within scope, and out-of-scope for a program. Differentiating between them at particular points in time is useful, but under some circumstances, what was out-of-scope may become in-scope and even in-focus and vice versa, as the program adapts to its environment and context and as responsibilities change. For example, a social welfare organization delivering a mental health program might fill a vacuum left by a departing partner organization or respond to emerging crisis needs such as homelessness (strongly associated with mental illness) in the face of a real estate market collapse. What will be important in such circumstances is that choices relating to scope and focus and changes of direction are made strategically and in line with overall purpose rather than haphazardly and reactively.

Focusing and scoping a program is an important part of program description. Failure to define what is within and outside the scope of the program and achievable by the program can lead to confusion about what should and should not be measured as outcomes of this program.

How to Focus and Scope

For an existing program, focusing and scoping will be mainly about documenting decisions that have already been made or actual practices. To focus and scope an existing program:

1. Determine or describe focus by identifying the main strategies or policy tools that the program uses or will use.

2. Determine or describe scope by identifying desired outcomes or desired conditions that are within scope or program reach even if not its direct focus.

3. Determine or describe what needs to happen at the boundaries, iden-
    tifying which other actors are expected to contribute to outcomes
    that are beyond the direct focus of the program.

For designing a new program, the tasks are quite similar to those for
an established program, but quite difficult decisions will need to be made,
taking into consideration a range of factors such as comparative advantage
of the organization running the program, likely efficacy given its resources,
and information about what other partners or prospective partners are doing
or likely to do. At the program design stage, scoping decisions are likely to
precede focusing decisions.

**To Determine or Describe Focus, Identify the Main Strategies or Policy Tools
That the Program Uses Or Will Use**   Examples of policy tools are education, regulation, incentives, provision of services, and development of
infrastructure. Identifying the main policy tools will help to clarify which
causes of the problem (ignorance, noncompliance, poor access to services)
the program is trying to address directly. Identifying the main policy tools
helps to determine the boundaries of the program and the focus of the program: what lies within the boundaries.

From time to time, the focus of a program may change. In particular,
programs that have complex aspects will need to be responsive to emerging conditions that may require different policy tools. They may need to
use a range of tools simultaneously or focus at a particular time on some
policy tools with a view to moving to other policy tools as the need arises.
Or the different tools might be planned for different stages of the life
of the program—perhaps education and capacity building followed by
regulation.

Scoping decisions can also be about what is in scope for a particular time
period. Sequencing can help to determine what outcomes might be achievable (within scope) within what time frames. For example, outcomes relating
to institutionalizing regulations or policies may be the focus of the program
at a particular time, with an expectation that the program would later focus
on outcomes relating directly to better service delivery. Preparatory work
may be under way for later outcomes while in the process of achieving earlier

ones. Also the program may want to seize opportunities to achieve in-scope outcomes as they arise rather than waiting until direct delivery of those outcomes becomes the main focus. Interventions ahead of time are sometimes used for piloting purposes before scale-up.

To Determine or Describe Scope, Identify Desired Outcomes or Desired Conditions That Are Within Scope or Reach of the Program Even If Not Its Direct Focus Some out-of-boundary outcomes may need to be included in the theory of change even though they may be identified as not being the direct responsibility of the program. As part of the scoping, show which intended outcomes will be the direct and primary responsibility of the program and which are at the boundaries within its scope but beyond its direct focus.

The broadening of the scope of a program and its theory beyond its immediate focus is a contentious issue. Some practitioners when scoping prefer to concentrate only on what is in focus and within control or able to be strongly influenced by the program. We have experienced resistance from some program theory participants to the inclusion of out-of-focus outcomes on the basis that their inclusion will require them to be measured and that this may lead to a program's being held accountable for outcomes over which it has little influence. Certainly limiting a program theory to in-focus outcomes simplifies it considerably, and there is only so far that a program theory can go in terms of incorporating out-of-focus outcomes without becoming excessively reductionist, detailed, and potentially demotivating. However, on balance, we believe there is merit in including at least some very important out-of-focus outcomes, especially those that are close to program boundaries and crucial to its success within the scope of a program theory. We argue the case on several grounds, including referring to various program theory traps that we identified in Chapter Three—in particular the traps of lacking a theory about how the program will contribute to ultimate outcomes, not taking account of complexity, and ignoring unintended outcomes. (They are summarized in Exhibit 7.1.)

EXHIBIT 7.1 ARGUMENTS FOR INCLUDING IN THE SCOPE OF THE PROGRAM SOME OUTCOMES THAT ARE NOT ITS FOCUS

- The program theory may not extend beyond instrumental and sometimes banal achievements (for example, delivery of outputs).

- The program theory may no longer show the hypothesized links between what the program can achieve directly and the alleviation of the original problem and its consequences that gave rise to the program.

- The program theory may fail to recognize that the program has complex or complicated aspects and is operating as part of an open rather than a closed system.

- The program may fail to recognize what it needs to do to interact productively with its context and, in particular, with other actors.

- The program may lose valuable opportunities to keep its finger on the pulse of the issues that it is addressing, including causes of the problem and actors who are addressing various aspect of it that lie outside the direct influence of the program.

- The program may underclaim its results if it includes in its program theory only outcomes over which it has complete control and fail to recognize that it is in fact making a contribution.

- Unintended outcomes may be overlooked through adopting a perspective that is too inward looking with respect to the program.

- *The program theory may not extend beyond instrumental and sometimes banal achievements.* For example, a theory of change for a public communications program concerning HIV/AIDS that did not show some hypothetical or theoretical links (even if only partial) between changing public awareness (the main focus of the program) and reducing HIV/AIDS (the wider scope of the program and the reason for its existence) would be of limited use and would beg the question as to why it was important to change community awareness.

- *The program theory may no longer show the hypothesized links between what the program can achieve directly and the alleviation of the original problem and its consequences that gave rise to the program.* That problem will in most cases have had multiple causes, only some of which the program may be addressing. If the program theory stops well short of including reference to the ultimate consequences of the problem it is trying to directly address, then the program can become self-serving, losing sight of its original purpose. It may also become difficult to justify the program to funding agencies and others as institutional memory is lost about why the program was established.

- *The program theory may fail to recognize that the program has complex or complicated aspects and is operating as part of an open rather than a closed system.* Most programs are open rather than closed systems. Their results are affected not only by what the program does and how well it does it but by many other external factors as well, including outcomes achieved by other programs and partners. A program theory must identify outcomes that need to be achieved by others, as well as those to be achieved by the program. As open systems, programs need to be scanning their environment, looking for warning signs, and picking up on opportunities to influence out-of-scope conditions that are critical to the program's success. Programs with complex aspects in particular need to adopt systems approaches that involve working with other systems and subsystems.

- *The program may fail to recognize what it needs to do to interact productively with its context and, in particular, with other actors.* If outcomes that are to be achieved by other programs or in conjunction with others are left out, the program may not recognize what strategies it might need to use in relation to outcomes that are largely beyond their direct or primary control but whose achievement is ultimately needed for the potential of the program to be realized. Examples of strategies that the program might use to influence these external but important outcomes are partnerships or other cooperative interfaces with other programs and other stakeholders, advocacy and enlisting community support to stimulate action and funding by others, and risk management strategies to minimize harmful or deficient outside actions. Unless some of these outside elements are in place, the program's efforts may well be wasted, and resources could have been better applied elsewhere.

- *The program may lose valuable opportunities to keep its finger on the pulse of the issues that it is addressing, including causes of the problem that lie outside the direct influence of the program.* Monitoring information about the original problem would continue to be useful for making judgments about whether the program's strategy continues to be relevant and likely to be effective and for making program design and implementation choices. Examples of relevant types of information about the problem to be solved are information about who is most affected by the problem, whether this is changing, and the implications for whom the program should target with its activities. This type of monitoring is about monitoring the situation but not necessarily the impact of the program.

- *The program may underclaim its results if it includes in its program theory only outcomes over which it has complete control and fail to recognize that it is in fact making a contribution to other outcomes.* More often than not, concerns are expressed about programs overclaiming what they have achieved. However, underclaiming can also be a problem. A program that may not want to claim responsibility for a particular outcome may have made an important contribution to that outcome, perhaps by opening the door to opportunities that would not otherwise be available to target audiences or other stakeholders or other programs. The role of the program in facilitating these types of outcomes needs to be included in the program theory and incorporated in monitoring and evaluation. Contribution analysis and other techniques can assist with identifying the valuable contributions that programs make to outcomes in situations where the program may be one player among many (see Mayne, 2001, 2008; Korvojs and Shrimpton, 2007). We discuss these issues further in Chapter Fifteen.

- *Unintended outcomes may be overlooked through adopting a perspective that is too inward looking with respect to the program.* Failure to recognize either positive or negative unintended outcomes can rebound on the program and have implications for its future effectiveness. In Chapter Six, we showed how unintended outcomes, both positive and negative, can be shown in the logic model for a program theory.

This discussion shows advantages to including, and in some cases monitoring, even those outcomes in the program theory that cannot be directly attributed to a program and therefore may not be outcomes of the program as such. However, in the course of doing so, it is nevertheless important to differentiate between those that will be considerably and directly achieved by the program, those that might be influenced even if only to a small degree by the program, and those that are largely beyond the direct influence of the program.

Out-of-focus but within-scope outcomes may become apparent when questions are asked about what else will be needed to move from one outcome to the next: for example, from raising community awareness to changing community behavior. Some of the out-of-focus causes that need to be addressed relate to higher levels in an outcomes chain than those that can be achieved by the program at any time or by the program at this time. But unless something is being done or will be done by the program or others to achieve those outcomes, including preparatory work in readiness for working on those outcomes, then the lower-level outcomes realized by the program itself at this time could be futile. The program may reach a dead end. There are, of course, occasions in which achieving the lower-level outcomes can be a stimulus to action by others to achieve higher-level outcomes by, for example, creating community demand. The following example illustrates the usefulness of identifying those higher-level beyond-focus outcomes.

THE IMPORTANCE OF INCLUDING HIGH-LEVEL BEYOND-FOCUS OUTCOMES IN A PROGRAM THEORY

One of the objectives of a public communications campaign was to improve public understanding that HIV/AIDS can affect anyone, not just particular groups in the community. Ultimately the program sought to be a step on the way to reducing the prevalence of HIV/AIDS and its impacts on individuals and the community. The full outcomes chain for the program had these ultimate impacts as the highest level of the chain.

This public communication campaign was unlikely to substantially and directly affect higher levels in an outcomes chain such as safer sex and other practices since those messages were not a significant part of the campaign. However, a full outcomes chain helped to place the awareness-raising campaign in the context of the longer-term and ultimate changes that needed to occur not as a direct consequence of the campaign but by laying the groundwork.

The messages of the campaign were crafted in a way that would generate widespread public pressure on other programs and on government to fund actions that might be needed to facilitate safer sex and other practices, such as better education, supply of condoms, and safe injecting facilities. Having achieved public awareness, campaigners then set about working with government and nongovernment groups to persuade them to take the next steps to move up the outcomes chain.

Changing practices on a wide scale may have been beyond the immediate reach of the public communications strategy. However, formulating these higher-level outcomes was an important driver of the targeting and messages of the communications campaign. Had these out-of-focus outcomes been excluded from the program theory for the communications campaign, the campaign might have taken on a very different form and, we would argue, quite possibly a less effective form in terms of reducing HIV/AIDS.

Out-of-focus determinants could also include lower-level outcomes in the chain that are actually addressed by others but whose achievement is a precondition for achieving immediate or intermediate outcomes of the program. They may create significant roadblocks for the program that, if not overcome, need to be worked around. They can occur at any point in the chain of outcomes. The next example identifies out-of-focus outcomes at various points in a results chain that proved to be critical to sustainable achievement of in-scope outcomes.

INCLUDING INTERMEDIATE- OR LOWER-LEVEL OUT-OF-FOCUS, OUT-OF-SCOPE OUTCOMES IN A PROGRAM THEORY

The program theory for a program to combat volatile substance abuse among young people in particular ethnic communities included an outcome that related to generating community support and capacity for addressing the drugs issue. Generating this support and capacity was a key in-scope activity of the program that several pilot studies had identified as critical to the success of the program for sustaining any gains and reducing the likelihood of transfer to other drugs once access to this drug had been limited.

There were several barriers to developing community support and capacity. The community had a profound distrust of government interventions arising from a history of short-term initiatives followed by withdrawal. Also the community lived in a state of crisis for a range of other reasons: inadequate housing and employment opportunities, violence, and crime, among others. Unless the members of the community received some relief from these day-to-day crises, it was difficult for them to focus on the capacity building and other developments that were the focus of the program's activities to address the drugs issue.

Although the remit of the program did not deal directly with issues such as housing and employment, getting around these issues was important for creating a space within which the program could work effectively. Accordingly, program staff saw themselves as playing a minor but significant role in facilitating housing and employment developments and working with relevant agencies to address some of these issues in a coordinated and timely manner.

For this reason, improvements in housing and employment options were included in the theory of change as necessary if the more direct aspects of the program were to work. Their inclusion legitimized expenditure of some program effort on making the necessary links with other agencies and providing some immediate practical assistance where this was feasible. Doing so helped to remove roadblocks standing in the way of the antidrug program that was working effectively with its clientele.

The program could not claim to be either the sole or even an important contributor to the housing and employment outcomes. It would judge its success not in terms of the extent to which it improved housing, employment, and so on in those communities as a whole but in terms of the extent to which what it did achieve in those areas could be shown to be useful in creating some space for the community to gain an interest in addressing the youth drug issues, gaining community trust and support, getting a foot in the door with the community, and developing community capacity to address the main program issues.

Color coding and explanatory notes were used on the logic diagram for these housing and employment outcomes to show that they were not the main game of the program but were nevertheless important and that the program staff was justified in giving some attention to them to make their program work. Although they were beyond the main program activities, they were an integral part of its theory of change.

To Determine What Needs to Happen at the Boundaries, Identify Which Other Actors Are Expected to Contribute to Outcomes That Are Beyond the Direct Focus of the Program   Obviously this can be done only in discussion with those other actors. Developing a joint theory of change across all key actors can be a basis for developing a shared vision and an integrated delivery plan. In many cases, a key part of the program will be to influence these other actors so that their efforts are aligned with that of the program. Conversely, ensuring that the program is aligned with the efforts of others is important, provided that those efforts are compatible with program objectives.

Tools such as outcome mapping can help with identifying program boundaries and what needs to occur at the boundaries. Outcome mapping (Earl, Carden, and Smutylo, 2001) recognizes that most impacts and many outcomes are the result of the efforts of many partners and that in some cases, the main purpose of a program may be to influence those other partners. Outcome mapping provides systematic processes for addressing such questions as these:

- Who are your boundary partners?
- What do they need to do for the program to work?
- How would you know if they were doing that?
- What are you doing to influence or encourage them to do this?

How Focusing and Scoping Might Vary Depending on Approach to Developing a Program Theory

When the articulating mental models approach is applied, some of the focusing and scoping will be about gaining stakeholder engagement in setting priorities around the causes to be addressed by the program and explaining why certain priorities have been selected for the program. Examples are causal importance in relation to the problem, sequencing (what has to be done first, second, and so on), resourcing (what we can do with what we have got), theories about how policy tools will address causes, and the principles of operation that they identify as important.

When the deductive approach is applied, scoping and focusing will be based on what is identified in the program documents. However, the research literature might also help to identify important conditions and other strategies that appear to be out-of-scope. This can open the way for discussion with stakeholders about what their program does in relation to these out-of-focus and even out-of-scope conditions—for example, whether it liaises with other service providers or programs and whether the program boundaries or priorities should be changed. This process can contribute to program development and improvement.

When the inductive approach is applied to developing the program theory, it can assist with identifying what in practical terms actually occurs—scope in practice as distinct from scope in theory. Priorities as they are observed and played out in practice can be identified and compared with what stakeholders and documents say are the priorities for the program. Reasons for any differences between policy and practice can be explored to ascertain what kinds of considerations are affecting these discrepancies: beliefs about effectiveness, resources, the urgent pushing out the important personal preferences, management issues, and so on. It is relatively easy to identify from program documentation and interviews what policy tools are the focus of the program. Checking in the field will sometimes identify other strategies that staff are using even though they may technically be beyond the focus of the program. In the program for youth affected by drugs, identifying that staff were working on housing issues even though housing outcomes were not specifically initially included in program objectives and activities was an example of how the inductive approach could contribute to program scoping and re-scoping. Inductive approaches may also find that some in-focus outcomes are not being pursued and some in-focus activities are not being used at all.

OUTCOMES CHAIN

At the heart of a program theory is an outcomes chain, sometimes referred to as an outcomes hierarchy (Funnell and Lenne, 1990). It links the theory of change discussed in this chapter and the theory of action discussed in

Chapter Eight. Here we use "outcomes chain" because "outcomes hierarchy" confuses some people.

What an Outcomes Chain Is

An outcomes chain shows the assumed or hypothesized cause-and-effect or contingency relationships between immediate and intermediate outcomes and ultimate outcomes or impacts (both short and long term). We include the concept of a contingency relationship as one in which the achievement of one outcome is a precondition for achieving a higher-level outcome—a necessary but rarely a sufficient or causal condition. For example, some programs need to achieve a certain level of client contact before they are in a position to be able to influence those clients. One would not say, however, that contact with clients was actually causing the program to influence clients. Moreover, as we noted in our discussion of programs with complicated and complex aspects, some causal paths will be alternative paths—either completely parallel or splintering and perhaps rejoining at various points. In some cases, a precondition may be one of several alternative preconditions.

We can also think of the outcomes chain as a ladder, though reaching a particular rung in the ladder does not *cause* the climber to move to the next one. Additional effort or mechanisms (program and nonprogram factors) propel the rise to the next rung. Equally, application of counterforces might push the climber down the ladder or cause the climber to fall off the ladder. Given these considerations, it is not always appropriate to portray the different levels in the outcomes chain as a series of *if-then* statements that hold under all circumstances.

For convenience, the term *outcomes chain* is used to refer not just to outcomes but also to impacts in relation to reducing the problem or taking maximum advantage of the opportunity that the program is addressing. The term *outcome* is a relative one. An outcome is an outcome of something. Typically when we refer to the outcomes chain for a particular program, we mean the outcomes of that program. However, there can be merit in incorporating outcomes of other programs that are key to the operation of the causal chain as long as we can identify them as being primarily the outcomes of other programs or actions.

Figure 7.1   Outcomes Chain for the Mature Workers Program

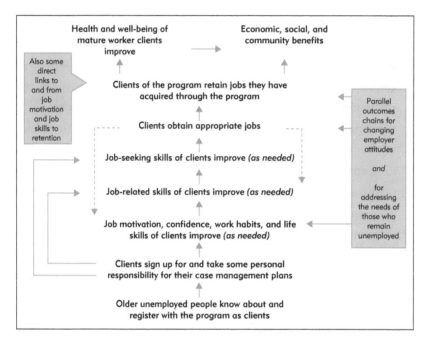

Figure 7.1 shows an outcomes chain for the mature workers employment program that we are using as our case example in this chapter. Note that because the employment program is a case management program, different clients will require different types of outcomes and services to different degrees. Some will require job motivation, skills, and job-seeking skills; others will require only one or maybe two of these three different types of outcomes. (Program theories for case management programs are discussed in more detail in Chapter Twelve.)

The arrows to the left of the diagram show that some clients may bypass some outcomes. Other arrows could also be inserted, for example, from acquiring job motivation directly to job-seeking skills or obtaining jobs. We would not expect acquisition of job motivation to be important for all participants, and that is implied in the outcomes chain by the use of bypass arrows, for example, the arrow from "clients sign up" to "job-seeking skills of clients improve," to show that some participants can skip some outcomes. These bypass arrows have measurement and evaluation implications. Program statistics that simply reported on the percentage of clients who reported improved motivation through contact with the program would be less important than information about the numbers and proportions of those who needed to acquire motivation and of these the proportion that actually did acquire motivation.

Also we have used dotted arrows for a few of the feedback loops that show either virtuous or vicious circles that might apply to a program such as this. For example, as people become employed, their self-confidence and skills are likely to improve, and these can help them to retain their jobs. If they are unsuccessful in the workplace, the reverse may occur.

As people become better at their jobs, confidence + job satisfaction improves. — Less retainment issues.

Why an Outcomes Chain Is Important

Improved confidence + better at job. ↓ able to mentor new DIS

The outcomes chain, rather than program activities, is the main device for thinking about how the program will function to achieve results and address the situation. This is a position that emphasizes that program activities are largely instrumental to achieving outcomes rather than ends in themselves and that they will be conducted differently depending on what sorts of outcomes need to be achieved. The outcomes chain is (or should be) the centerpiece for developing all other aspects of the program theory. When a program is being designed, it should provide the primary rationale for choice of program activities and resources and a structure for the development of success criteria and selection of measures of performance that will reflect those criteria. Building the rest of the program theory around an outcomes chain is an alternative to simply presenting outcomes as another box in a production line. Doing so reinforces that activities are needed not just at the beginning of the chain of outcomes (immediate outcomes) but that different activities will generally be needed for all or most outcomes in the chain.

A frequent criticism of monitoring systems and, perhaps to a lesser extent, evaluation studies is that they place too much emphasis on measures of busyness, that is, measures of what the program does—its activities—on the assumption that what it does will produce outcomes. The measures of activity come to be used as proxy measures of outcomes. It is particularly important in these instances to be confident that what the program does is relevant to the outcomes it seeks.

Using the outcomes chain as a touchstone for program design, planning, management, and evaluation can help to avoid excessive focus on activities and getting stuck in activity traps that give insufficient attention to outcomes.

Program theory is a tool that can be used to unpack the relationships between activities and intended outcomes. However, many versions of program theory, and in particular pipeline approaches to logic modeling (see Chapter Nine), fall short of realizing this potential. Often these models simply display boxes of activities and boxes of outcomes without demonstrating logical and defensible relationships between them and the various items listed in the boxes. The links between activities and the immediate outcomes may be clear, but the connections are not made explicit.

How to Develop an Outcomes Chain

Getting the outcomes chain right is vital because of its central role. A complete outcomes chain—one that goes high enough up, low enough down, and has no serious gaps relative to what the program does and the contribution of others in achieving outcomes—can help in a number of ways:

- Clarify how the program's intended outcomes (immediate, intermediate, and ultimate) will contribute to overcoming the problem that gave rise to the program. Including the high-level ultimate outcomes (or impacts) in the outcomes chain does not mean that the program will necessarily be held accountable for showing that it alone is achieving them. This needs to be carefully communicated when developing and presenting a program theory. Including the ultimate outcome does, however, remind program designers to monitor the problem and retain focus on the overall purpose of the program.

- Ensure that while the focus of the outcomes chain is on what the program seeks to achieve, proper consideration is also given to key enabling outcomes that need to be achieved by others as preconditions, in parallel, or as postconditions of the program's outcomes in order for progress to be made toward addressing the problem's causes and consequences.

- Portray the outcomes in a way that makes it easier to identify which program activities will be relevant to which outcomes, where gaps exist, where there are activities whose relationship to any particular outcomes seems to be missing, and which outcomes do not require any specific additional activities but depend instead on the achievement of prior outcomes.

As for developing a program theory as a whole, an outcomes chain can be developed in a number of ways. We have found the following approach useful for developing a first cut at an outcomes chain. It can be done by an evaluator or program theorist as a desk job using deductive approaches, especially if stakeholders have limited time to participate or specifically desire an external perspective. It can also be developed by groups of program staff or other stakeholders with or without the assistance of a program evaluator or program theorist but the basic principles remain the same. The steps are summarized in Exhibit 7.2 and discussed in the following sections in detail.

EXHIBIT 7.2 STEPS FOR DEVELOPING AN OUTCOMES CHAIN

1. Prepare a list of possible outcomes.

2. Cluster the outcomes to reduce the number of outcomes to a manageable number and give a working label to each cluster of outcomes.

3. Arrange the working label outcomes in a chain of if-then statements or multiple parallel if-then chains.

4. Identify any feedback loops among the outcomes, but as far as possible avoid fully circular representations in which everything is related to everything else.

5. Validate the outcomes chain with a range of different stakeholders and by seeing whether the outcomes chain tells a coherent story.

Step One: Prepare a List of Possible Outcomes   Possible outcomes can be drawn
from the situation analysis, derived from a problem etiology, or drawn from
sources such as those identified in Chapter Six and by using workshops and
interviews. Typically a combination of approaches is useful.

Assuming that a sound situation analysis has been prepared, the outcomes chain should reflect key aspects of what is known about the nature
and extent of the problem—its causes and consequences. This helps to
ensure that the outcomes sought will be part of a plausible strategy for
addressing the problem.

For programs that address problems (as distinct from those that focus on
maximizing opportunities), it can be helpful to think of the outcomes that
will be incorporated in the outcomes chain as the reverse image of the problem or some reduction in the problem, its causes, and consequences. So:

- If the problem is a high rate of long-term unemployment for older
  workers relative to younger workers, then the intended outcome
  might be, "Reduction in the gap between the employment prospects
  of older and younger long-term unemployed people."

- If one of the causes of long-term unemployment of older workers
  is lack of relevant skills, then an intended outcome for inclusion in
  the outcomes chain might be, "Improvements in relevant skills of
  program participants."

- If some of the known consequences of long-term unemployment are
  financial dependence on government welfare, poverty, and ill health,
  then some intended outcomes for inclusion in the outcomes chain
  would be about improved financial independence, increased income,
  and improved health of older workers who participate in the program.

If no systematic and formally documented situation analysis has been
conducted, other approaches can be used to develop an outcomes chain.
For example, Cole (1999) discusses the use of problem etiology to construct
diagrams of causes of various problems and gives examples relating to causes
of homicides involving handguns, and causes of smoking among teenagers.
These problem etiology diagrams can become the basis for choice of which
determinants the program will work on and the construction of a chain of

intended outcomes for a program. This could be done as part of designing a program or as a point of reference for evaluating the adequacy of design and outcomes of an existing program.

Sources of information that can be used to develop a program theory were discussed in Chapter Six and are summarized here for convenience:

- Information about what gave rise to the program as shown in documentation and identified in discussion with staff and other stakeholders.

- Documented program objectives if they are expressed as outcomes.

- Program activities, observed and documented.

- Performance measures currently in use, but with a check to determine that they are about important, and not just easily measurable, aspects of the program.

- Program staff and management through interviews and group processes: the outcomes that are important to them and whether they foresee any positive or negative unintended outcomes.

- Program clients and other members of the community through interviews, group processes, market research, and formal and informal feedback collected by the program and others. Look for outcomes that are important for program clients and unintended outcomes, winners and losers.

- Other program stakeholders (funding agents, partner organizations) by talking to them and reviewing their correspondence to determine what outcomes are important to them and whether they foresee any positive or negative unintended outcomes.

- Data and program research that underpin the program.

- General research, literature, and evaluator expertise concerning similar programs.

- Relevant theories of change and program archetype outcomes chains (these are explained in Chapters Eleven and Twelve).

Whether working in a workshop situation or even when working alone to develop an initial version, it is helpful to put each outcome on a separate index card, sticky note, or something similar so that they can be easily sorted and arranged later. They should include all important outcomes, not just those that are measurable, any unintended outcomes (positive and negative) that can be reasonably predicted, and outcomes that might be relevant to different stakeholders. As well, they should include outcomes that are the focus of the program and those that need to be delivered by others even though they may be largely beyond the focus of the program. Outcomes that are important steps along the way but not necessarily easily influenced by the program itself are also important to include. If someone does not address them, they may become roadblocks.

Step Two: Cluster the Outcomes   Typically the first listing of outcomes will produce a larger number than is really necessary to give an overview of what the program is trying to achieve. Clustering the outcomes can be a useful way to manage the identified outcomes. Several different stated outcomes may be saying the same thing in different ways, or some may provide more specific detail than others. For example, there may be several different statements about desired changes in participant knowledge that could be clustered. It is then helpful to apply a working title to the cluster—something simple that can be included in a logic model diagram, for example, "Job motivation of program participants improves." It should not try to include all possible information about the outcome. However, it is a good idea to store all the statements that relate to each cluster of outcomes because at a later stage, the information might be useful for defining the outcome in more detail, identifying success criteria and other features of the program theory.

Step Three: Arrange the Working Label Outcomes in a Chain of *If-Then* Statements or Multiple Parallel *If-Then* Chains   At this stage, using the *if-then* approach is simply a means of ensuring logical thinking. It does not connote that the *if-then* relationships are water-tight or exclusive. A useful approach is to identify the highest and lowest levels first as anchor points for the remainder of the chain. A rule of thumb for determining what should be the lowest level in the outcomes chain is to identify the first point at which the program has some sort of effect on the target group—for example, getting members of the target group to participate in the program, especially if participation is voluntary. In this case, the lowest-level outcome is not really about directly addressing the causes of the problem but getting the target group to engage (respond to offers of assistance), a precondition if the program is able to work with target group members to address those causes.

Then gradually build up the middle sections, looking for any obvious gaps as you go, and insert additional outcomes as needed to make the logic relatively complete and plausible. Thinking about the types of outcomes that have been sought by programs similar to this one can also be a useful way to identify additional overlooked outcomes. Evaluator experience can be a valuable source of such insights. Some of these insights have been codified in outcomes chains for program archetypes, discussed in Chapter Twelve.

Sometimes it can be helpful to divide a large number of levels of outcomes (say, nine or ten) into immediate, intermediate, and ultimate outcomes or impacts if chunking them in this way helps with developing, communicating, or using the outcomes chain. You can use dividing lines in the outcomes chain to break the outcomes into immediate (lowest level and first point of impact of the program on target audience), intermediate (middle level, generally relating to identified causes of the problem), and ultimate and impact-level outcomes (generally relating to reducing the problem and its consequences). However, sometimes this process can generate unhelpful debate about the meaning of the words *immediate, intermediate,* and *ultimate outcomes* and exactly where to draw the lines between them. This can waste time that would be better spent on other aspects of program theory. Also for some programs, work will be occurring simultaneously on lower-level outcomes, such as achieving enabling policy reforms and building capacity, and higher-level outcomes, such as achieving ultimate health outcomes with pilot or sentinel communities for a program. Both types of outcomes are in some sense immediate.

Depending on the purpose for which the outcomes chain is being developed, additional outcomes that are necessary for the chain to make logical sense can be inserted even if program management and staff have never considered them. Attention needs to be drawn to the fact that the outcomes do not occur in the description of the program and to the possibility that their omission represents a program design flaw—a gap in the logic.

Some approaches to outcomes chains place outputs (which we define as the tangible products or services delivered by the program) at the bottom rung of the ladder. In general, we advise against this to avoid giving the misleading impression that outputs feed only into the outcome directly above them in the outcomes chain. In fact different outputs (and the different activities and resources that produced them) feed into different levels of outcomes in the chain. To illustrate, the mature workers employment program outputs that would be relevant to developing job-seeking skills (for example, résumés prepared) would be quite different from those that would be relevant to helping clients retain jobs that they had obtained through the program (for example, support visits made to employees in their workplace).

Nevertheless, on occasion, it can be helpful to include some outputs at the bottom of the outcomes chain. In the start-up stages of a program, before implementation takes place, some value may accrue to having at the bottom of the chain some milestones for completion of particular key activities. For example, when a new museum is being established, having the building completed by a particular time would undoubtedly be an important output and could be regarded as the outcome of the construction phase of the program. Including it in the outcomes chain is a means of recognizing its importance and demonstrating that significant progress is being made.

The choice of the way in which to incorporate these out-of-scope outcomes and whether to include them in an outcomes chain can be affected by various considerations. For example, if the program theory is being used to develop a common conceptual framework to coordinate the efforts or programs of several different partner organizations, it might be useful to include the main outcomes of all programs in the outcomes chain. It is also helpful to show who or what program or organization is primarily responsible for what. If simplicity of presentation and use for communication is important, then simplicity may require fewer rather than more outcomes to be shown in the outcomes chain. Outcomes to be achieved by others could instead be included in a category labeled "factors affecting outcomes." Chapter Eight discusses the importance of identifying these factors as a key feature of an effective theory of action.

Step Four: Identify Any Feedback Loops   Sometimes stakeholders debate the order in which outcomes occur. If this is because of the operation of feedback processes, it can be helpful to identify feedback loops in a logic model.

We advise against using so many feedback loops that the logic becomes meaningless. When feedback loops are incorporated, a balance needs to be struck between including all of them (because everything is related to everything else) and capturing some important ones. Showing that everything leads to everything else can make an outcomes chain very difficult to understand—what we call a spaghetti junction model. Nevertheless, some feedback loops can be critical to the success of a program and should be included. From a logic modeling perspective, we have found it helpful to use solid lines for directions of main effect and dotted lines for important feedback loops, as in the outcomes chain for the mature workers employment program.

Examples of programs to which feedback loops apply are those that are affected by supply and demand cycles and where level of use of the program (a lower-order outcome) is affected by target group knowledge of the success of the program in achieving impacts and outcomes at higher levels of the outcomes chain. So, for example, the more successful a program is in achieving high-level outcomes that are valued by potential clients, the more likely it may be to attract more clients.

Interactive effects among levels of outcome need to be identified and monitored. For programs whose effectiveness is affected by supply and demand, increasing demands on the program as a result of its success may in fact result in a degradation of outcomes unless the program is set up with sufficient resources to cope with the changing levels of demand.

When thinking about the relationships among the various levels of the outcomes chain, it can be helpful to apply systems theory principles or mental models such as Senge's (1990) systems archetypes. Some examples are those that relate to responding to delayed feedback (for example, overcompensating when it is too late), limits to growth (for example, when a market has been saturated), or shifting the burden (for example, by applying a short-term corrective solution that has long-term impacts in terms of capacity to deliver the fundamentals of the program and may have other unintended impacts).

Step Five: Validate the Outcomes Chain   The outcomes chain needs to provide the framework for telling a plausible *if-then* story around which various conditions and assumptions can be portrayed. Putting the outcomes chain to the plausibility test can help to reduce the likelihood of large gaps in the theory. (Chapter Ten provides suggestions about how to review the plausibility of a program theory, including its outcomes chain.) Using the stakeholder mental model approach directly incorporates validation as part of the development process. When deductive or inductive approaches are used to develop an outcomes chain, seeking additional validation is important. It may be useful to have several different causal chains relating to the different interests of different stakeholders. A program to provide financial assistance to those having difficulty paying their electricity bills had three outcomes chains: one for those who were having trouble with their bills, one for the electricity companies (for example, improving revenue and reducing costs of debt collection), and one for the nongovernment organizations that were distributing vouchers to pay the bills (for example, reducing long-term financial dependence of clients on them). While the needs of the program clients were central, the program also needed to meet some important needs of the two types of service providers. To be sustainable, the program needed to be a win-win-win one, so the relationships among the three chains were also important.

As we noted in Chapter Six, if a purpose of developing a program theory is to develop understanding, ownership, and maximum use, it is preferable to develop outcomes chains (and all other parts of program theory) in collaboration with program staff and management. The collaborative approach also allows questions about the logic of the program to be raised as its development progresses and can result in readier acceptance by staff of the possible need to change.

To engender ownership and improve program implementation, it is helpful to identify which stakeholders have particular responsibilities for each outcome in the outcomes chain. This information can be included in a logic model diagram. For example, within an organization, the marketing staff may be primarily responsible for achieving an immediate and low-level outcome that relates to whether enough of the right types of participants are attracted to the program. Program delivery staff may be primarily responsible for achieving various intermediate outcomes once the right people have been attracted.

Program design staff may be responsible for collecting information about higher-level outcomes and feeding this back into program design processes.

Some Key Points About Outcomes Chains

We address some key points to remember about outcomes chains here and summarize them in Exhibit 7.3:

EXHIBIT 7.3 SOME KEY POINTS ABOUT OUTCOMES CHAINS

- An outcomes chain is about the outcomes of program activities; it is not a flow diagram of activities and processes. If activities appear in an outcomes chain, they should be converted into the implied intended outcomes.

- The number of outcomes in the chain varies from one program to another.

- Although outcomes chains appear to be a unidirectional series of steps, feedback loops are common, and it is not always necessary or desirable to work through the steps in sequence.

- Nested outcomes chains can be used to supplement a main outcomes chain in order to keep the main outcomes chain simple for communication.

- The distance between the steps will vary, but large gaps with no plausible linkages or large differences in distances between steps can be a problem.

- It is useful to differentiate between outcomes in the chain that are the direct focus of program efforts, those that are within scope but not the focus and those that are out of scope.

- Outcomes chains can have branches to represent parallel outcomes chains contributing to achievement of the same ultimate outcomes. Links and interdependencies between parallel outcomes chains can be shown.

- An outcomes chain is only a model. It needs to be questioned during the process.

• *An outcomes chain is not a flow diagram of program processes and activities.* Rather it represents the intended outcomes or results of the various activities and puts them in order in a way that shows the contingencies among the various outcomes. Often program objectives are expressed in terms of activities to be undertaken. Evaluators can play a valuable role by translating those activities into assumed outcomes or, better still, helping staff and management to do so. Figure 7.2 contains a simple illustration: activities for a program to change the environmentally unfriendly practices of farmers. We have displayed the diagram in horizontal form because that is the way that flow diagrams are generally displayed. However, the diagram could also be displayed vertically as outcomes chains typically are.

• *The number of outcomes in the chain varies from one program to another.* Usually outcomes chains have several levels of outcomes. It is possible to have very short outcomes chains (for example, for a program whose sole purpose is to deliver a simple product to a target group). However, if an outcomes chain is very short (say, two or three levels of intermediate outcomes), the gaps between the levels may be too big, or perhaps too many assumptions are being made about what will happen between the levels (large "missing middles"). The more usual temptation of managers is to build in many steps in the chain, especially at the lower levels, to represent the complexities of their program. However, as a general rule, keeping the outcomes chain relatively simple is helpful. With practice, managers will find that many outcomes can be chunked into categories of outcomes (for example, improved skills) and that much of what they initially wanted to incorporate in the outcomes chain such as program activities and resources can be built into other parts of their theory of action (this is discussed in Chapter Eight).

• *Although outcomes chains appear to be a unidirectional series of steps, feedback loops are common, and it is not always necessary or desirable to work through the steps in sequence.* A program may be working at several levels concurrently, and achievements may begin to become apparent at higher levels before work has been completed at lower levels. Even when relatively unidirectional, it is rarely the case that an outcome in the outcomes chain would need to be fully achieved before moving onto the next outcome in the

Figure 7.2 **Conversion of a Flow Diagram of Activities to a Chain of Outcomes**

Note: Each activity has a corresponding outcome below.

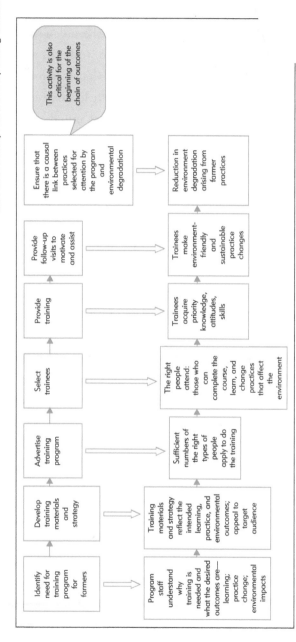

chain. Progress is not completely lockstep; it is often organic. In programs with complex aspects, there may be several alternative routes that run quite contrary to any notion of a single necessary chain of outcomes and feedback loops may make the beginning and the end of the chain somewhat arbitrary. A program manager may choose to work first at higher-level outcomes for various reasons. She may do so deliberately in order to obtain practical information about what happens in the field (for example, for an environmental education program) that can feed back into the refinement of approaches the program is making to achieve lower-level outcomes (such as building the capacity of environmental groups). Working at higher levels may also be a change management strategy for engendering interest from and engagement of key stakeholders by getting some small wins in terms of higher-level valued outcomes. In other cases, starting work at higher levels may be premature and may reveal that more groundwork is needed to be done at lower levels. Even this is a form of valuable organizational learning. We have seen all of these approaches in operation in programs such as community capacity-building programs and the evolution of environmental education programs in schools.

- *Nested outcomes chains can be used to supplement a main outcomes chain.* Nested outcomes chains can be used to keep the main outcomes chain simple for communication purposes. They are particularly useful for programs with complicated aspects and involving several partners (for example, whole-of-government strategies that involve several departments working on different aspects of an issue, such as reducing the use of drugs and mitigating their impacts). Nested chains can be of various types—for example:

  - Those that use the main outcomes chain as a template and then tailor it to various subprograms to reflect their more specific intended outcomes—for example, with different target groups or for local adaptations of a program (Funnell, 2006).

  - Those that unpack further detail within a particular outcome or between two outcomes by breaking down the *if-then* relationships into smaller steps. For the program to combat volatile substance abuse by young people, our main outcomes chain included very broad outcomes for each major component of the overall interdepartmental program—for example, the communication component that was about increasing community awareness and interest, the components that were about reducing supply of and access to drugs, and the components that were about strengthening community capacity. The combined outcomes chain with various strands showed how the main outcomes of each of these program components related to each other and, in particular, their interdependencies. More detailed nested outcome chains were developed for the various components, which would be more useful for the design, planning, monitoring, and evaluation of each component while still anchoring that component to the overall program.

- *The distance between the steps in the chain will vary.* Typically differences are greater toward the top level of the chain (concerning ultimate impacts), where program managers have less detailed knowledge. Conversely, distances between levels tend to be smaller at the lower end of the chain, where managers often have a great deal of knowledge and more direct influence over the achievement of outcomes. Managers and evaluators need to be alert to very large differences between levels of outcome, especially at the top of the chain, because they may suggest that they are making massive leaps of faith about the relationship between addressing a particular problem and how doing so has the potential to contribute to ultimate intended outcomes. Too great a space between the steps of the ladder is a factor that could cause the climber to fall off. An example would be an outcomes chain that jumped straight from an outcome of improved knowledge of the target audience to alleviation of the social issue that gave rise to the program. Too great a leap of faith would be involved and insufficient follow-up action might be undertaken unless the chain of intended outcomes also included other intermediate steps that would need to occur—for example, behavior change and the immediate impacts of behavior change.

- *It is useful to differentiate between outcomes in the chain that are the direct focus of program effort, those that are within scope but not the focus, and those that are out of scope.* This can be done in several ways. Color coding or various shapes of text boxes can be used to differentiate outcomes that are in focus, in scope, and directly achievable or may be able to be influenced from those that are out-of-scope, overlooked, or neglected. Out-of-scope or out-of-focus outcomes could be scattered across the outcomes chain or across several branches of an outcomes chain rather than being consistently above or below a particular line of attribution. This scattered effect will be especially noticeable when many other partners are contributing to the whole outcomes chain or when there are gaps in the program theory: various intermediate outcomes may have been identified as critical, but there are as yet no corresponding program activities for those outcomes. Another approach is to draw a line of attribution or accountability in the outcomes chain: above that line it would be unreasonable to expect that the program would make a significant direct contribution even though it might have significantly affected enabling conditions that can more substantially contribute to impacts. Sometimes we have found it useful to label out-of-scope outcomes as assumptions about conditions rather than outcomes to avoid confusion with program outcomes. When it comes to selecting measures of program performance, these distinctions can be useful for discerning which measures are those of a program's relatively direct effects, those that are measures of downstream outcomes that are partially a consequence of what the program has done and those that are external conditions that may affect how well outcomes are achieved.

- *Outcomes chains can have several branches.* Any given program may consist of several interlinking outcomes chains whose collective influence is required to bring about the desired outcome. Sometimes, as we noted in our discussion of complex programs in Chapter Five, the outcomes chains will need to be shown as alternative pathways. On other occasions, they will be parallel but interlocking or spliced, as for complicated programs. When parallel outcomes chains are developed, it is important not to lose sight of the fact that a program that consists of several outcomes chains typically depends for its success on the interaction of the different strategies represented in the different outcomes chains.

An overall pictorial portrayal showing the interrelationships among various outcomes chains can be very useful, as it was for the interagency program to combat volatile substance abuse among young people. For that project, we conducted a risk analysis that explored the risk and impact of breakdown in relationships among the components and how these risks might be averted or managed. It can also be important to determine the relative importance of each of the various chains or alternative pathways in achieving the outcomes that are desired ultimately. These judgments of relative importance can be made in part on the basis of research evidence concerning the causes of the need or problem that gave rise to the program. Weiss (2000) suggests a number of criteria for making choices about which theory to study (or which outcomes chain to focus on). One criterion concerns the centrality of the assumptions to the program. If a program revolves primarily around one theory (or outcomes chain), then it makes sense to make this the primary focus of evaluation efforts. However, if a multipronged approach involving several different outcomes chains is genuinely needed, it would be foolish to concentrate on a single chain. Moreover, the relative importance of the different chains may change over time. Programs that use a combination of carrots, sticks, and sermon strategies (see Chapter Twelve) are an example of types of programs that may need multiple outcomes chains in operation either concurrently or sequentially.

• *An outcomes chain is only a model that needs to be tested and in any case applies at a particular point in time.* The purpose of developing an outcomes chain is rarely to make statements about absolute truth. Rather, it is to make the working assumptions of program staff and others explicit and more readily testable. An outcomes chain should therefore be thought of as a model of relationships among various types of outcomes for a given program in a certain set of circumstances at a particular time. It is, above all, a heuristic device whose validity must be regularly questioned logically and empirically. Moreover, an outcomes chain that is valid at one time may no longer be valid, as the causes of program need change.

SUMMARY

In this chapter, we have discussed three components of a theory of change whose inclusion can improve a program theory: a robust situation analysis, scoping that both establishes the boundaries of a program and sets it in a wider context, and the development of an inclusive outcomes chain that mirrors the situation analysis and change mechanisms selected for the program. We have shown how the inclusion of these components can assist with avoiding some common traps encountered when developing and using a program theory. We have described each of the three components of a good theory of change and shown how each can be developed. We have given particular attention to outcomes chains since we see these as the centerpiece of a program theory, to which the other important aspects of a program theory such as factors that will affect success, program resources, activities, and outputs should be connected. The outcomes chain is also the main link between a program's theory of change and its theory of action.

EXERCISES

Exercise 1

Following is a brief situation analysis for a program to increase participation in preschool of rural children with disabilities.

The Situation That Gave Rise to the Program

Problem: Children with disabilities. especially in rural areas, have low rates of participation in preschool.

Its consequences: Nonparticipation is contributing to reduced developmental opportunities for the children, which in turn contributes to low levels of readiness to start kindergarten and reduced workforce participation for parents.

Its known causes: Participation is low for these reasons:

- Inadequate opportunities for families to access preschools, including the fact that the preschools in the area are not large enough to admit all children who are entitled to attend

- Parental working conditions

- Parental ignorance of or lack of confidence in what services are available; concern for their children

- Difficulty in attracting suitably qualified teachers to rural areas

- Preferences of some parents to stay home with their children

- Distance from preschools and transport difficulties

Scoping   *Causes that are to be addressed by this program:* Lack of access to child care opportunities, parental ignorance of what preschools are available, and concerns about how their children will be treated in a mainstream context. The program is not directly addressing parental working conditions, preferences to stay at home, transport, or wages and other conditions that are barriers to entice qualified teachers to move to rural areas.

How the program proposes to address these causes: Provide funding for new facilities appropriate to the needs of children with disabilities; recruit staff to improve staff-to-child ratios; train staff to work with children with disabilities; provide relevant child care standards; consult and communicate with parents to develop their confidence.

From this analysis of the situation:

1. Identify what an outcomes chain for this program might look like.

2. Are there likely to be any feedback loops?

3. What assumptions does the chain you have developed make about the causal relationships among the various outcomes?

4. What effects are the out-of-scope causes likely to have on the achievability of the program's intended outcomes? Is there anything the program could do to manage the risks associated with these out-of-scope causes? Work with boundary partners?

Exercise 2

1. Develop a statement for your program that shows:

   a. The situation that gave rise to the program:

      - The problem

      - Its consequences

      - Its causes

b. Scope of the problem

- Which of those causes the program addresses or does not address

- How, in broad terms, it addresses those causes (what strategies, such as infrastructure and communication, and what resources are used)

2. Develop an outcomes chain that reflects your analysis of the situation and identify which outcomes in the chain are in and out of scope of your program.

8

Developing a Theory of Action

IN THIS CHAPTER we identify some features of a program's theory of action that can improve the contribution of program theory to program design, planning, monitoring, and evaluation and show how these features can avoid common traps. Although these features are not included in all approaches to program theory, we have found them to be useful for improving the quality of a theory of action.

As we noted in Chapter Seven, although we present a sequential process for identifying the six key elements of a program theory (Table 8.1), the process is in fact an iterative one, and in practice, the development of the two aspects of a program theory often takes place as a single exercise.

In simple terms, the theory of action is about what the program does or expects to do in order to activate the change theory. It includes the choices that are made and priorities that are set in relation to a range of possible features of each outcome in the outcomes chain and the choices that are made about what will be done to achieve those outcomes and their selected

Table 8.1   **Components of a Program Theory**

Theory of Change (Chapter Seven)			*Theory of Action (This Chapter)*		
Situation analysis: Identification of the problem, causes, opportunities, and consequences	Focusing and scoping: Setting the boundaries of the program, linking to partners	Outcomes chain: the centerpiece of the program theory, linking the theory of change and theory of action	Desired attributes of intended outcomes; attention to unintended outcomes	Program features and external factors that affect outcomes	What the program does to address key program and external factors

features. So if the change theory for a particular program is designed to bring about changes in behavior and achieve community benefits by changing social norms and stimulating peer pressure, what will the program do to activate that change theory? Will it use community-based activities? Educational programs? Peer mentoring? Changes in legislation? Incentives, rewards, and sanctions to change norms? Public communication campaigns? Some combination of all of those? Who will it target, and why? What resources will it devote to bringing about change?

Importantly, the theory of action includes the rationale for these choices. It clarifies what the program would look like in terms of outcomes and operations if it were working well. It also clarifies the assumptions and information about what other factors will affect the achievement of outcomes and might need to be considered when managed as part of program implementation or when monitoring and evaluating program effectiveness. Exhibit 8.1 summarizes the components of the theory of action discussed in this chapter.

This chapter also presents a program theory matrix as a tool to assist with identifying and portraying the elements of a well-rounded program theory. The matrix is used in conjunction with and complements the type of logic diagram that is built around an outcomes chain as described in Chapter Seven. It includes

---

EXHIBIT 8.1 USEFUL FEATURES FOR INCLUSION IN THE THEORY OF ACTION

- *A detailed statement about each of the broadly stated outcomes in the outcomes chain.* This statement identifies the choices that have been made about success criteria for each outcome. For example, if improved knowledge of safer sex practices is a broadly stated intended outcome in the outcomes chain of a program to combat HIV/AIDS, who in particular is to become more knowledgeable: priority target audiences in terms of demographics, needs, and eligibility for assistance, or another audience? What choices have been made about the types of safer sex practices about which target audiences are to become more knowledgeable?

- *Assumptions about features of the way the program needs to operate* (such as adequacy of staffing to meet client demand, quality of service delivery). We refer to these as program factors.

- *Assumptions about other external factors* (such as the economy, partners, other programs) that could affect whether the outcomes are achieved. We refer to these as nonprogram factors.

- *What the program has chosen to do to address these factors:* its resources, activities and management strategies, outputs, and throughputs. For example, has the program chosen to use mass media? Teaching? Mentoring and advising? Face-to-face methods? Telephone counseling? What has it done to address external factors and external actors that will affect success?

---

reference to the three elements of theory of change discussed in Chapter Seven and the three elements of the theory of action discussed in this chapter.

We show how effective attention to the features of the theory of action that we discuss in this chapter can avoid common traps such as those we identified in Chapter Three. Table 8.2 provides a summary of those traps to which the various features of theory of action are particularly relevant.

Table 8.2 Theory of Action Features That Can Avoid Common Traps

Program Theory Inadequacy	Features to Avoid this Trap		
	Identify Desired Features of Intended Outcomes: Success Criteria, Including Any That Relate to Unintended Outcomes	Identify Assumptions About Program and External Contextual Factors That Will Affect Outcomes and Their Desired Features	Identify What the Program Does to Address Those Outcome Features and Program and Other Factors That Will Affect Outcomes
Trap 1: No actual theory Failure to show how program activities will contribute to particular outcomes or features of outcomes, including outcomes that are further along the chain of outcomes	✓	✓	✓
Trap 2: Poor theory, implausible theory Failure to show the expected mechanisms for change and what the program will actually do to achieve its intended outcomes		✓	✓
Trap 3: Poorly specifying intended results Failure to specify important features of intended outcomes Excessive focus on what is measurable and exclusion from the program theory	✓		

of any outcomes that are too difficult to measure, which can lead to goal displacement

Trap 4: Ignoring unintended outcomes
Success criteria for outcomes that do not attend to known or suspected potential unintended outcomes ✓

Trap 5: Oversimplifying when there are important aspects that are complicated or complex or vice versa ✓
Failure to recognize that programs rarely function as closed systems ✓
Insensitivity to the inherent unpredictability of the contexts within which programs operate
Failure to acknowledge the influence of external factors and the fact that these factors may differ depending on choice of success criteria
Overclaiming or underclaiming observed outcomes

PREPARING THE THEORY OF ACTION

The best time to prepare a theory of action is at the time of designing the program, and ideally it should be done by program designers, possibly with assistance from program theorists or evaluators. In practice, program theorists and evaluators are often called on at the time of developing a monitoring and evaluation plan to retrofit the theory of action based on what the program does. This may involve recognizing where there may be no particular underpinning theory, as when choices were made primarily for reasons other than likely effectiveness in achieving outcomes, or in the absence of good information about what would affect outcomes. It is important when doing this to clearly differentiate between the priorities that have been set and the choices made by or for the program from what program designers and staff recognize should be in place and the choices they would have liked to have made given unlimited resources and freedom from political, practical, and other constraints.

SUCCESS CRITERIA FOR A THEORY OF CHANGE

Whereas the statement about each outcome in a chain of outcomes can be quite general, a theory of action is more specifically defined, including clear statements about what each successful outcome would look like and what would constitute effective program performance.

Success criteria identify the desired features of each of the outcomes in the outcomes chain. They define more precisely the nature of the outcomes sought and are the link between the stated outcome and the performance measures for that outcome. All programs need to make choices about which aspects of outcomes they consider more important and less important—for example, who will be the priority target audiences for the program and why. *Success criteria* for outcomes differ from success *factors*: success criteria are about the basis for judging the worth of an outcome, whereas success factors are about what affects the achievement of the outcome.

Ideally a success criterion for an outcome has two elements: attributes and comparisons. *Attributes* relate to specific features of the outcome; for example, if changed farming practices are a desired outcome, what specific changes are sought? Is there a range of alternative acceptable behaviors? Typically attributes relate to quality, quantity, timeliness, cost, and priority target groups with which the outcome should be achieved. It may also be possible to identify any expected trade-offs that the program may need to make among these attributes—for example, the relative importance of quantity versus quality. Attributes may also relate to the interests of various stakeholders that need to be protected even if those stakeholders are not the target group of the program. In this way, attributes can give deliberate attention to known or suspected unintended outcomes. *Comparisons* make it possible to tell whether the attribute is being achieved to a desired level, moving in the right direction, and so on. Standards, norms, targets, change in the desired direction over time, comparisons with baselines, and redistribution of outcomes among target groups are all examples of comparisons.

A simple example illustrates these two components of a success criterion. Suppose that a program that trains potential astronauts has as a lower-level outcome, "The right potential trainees apply for the positions." What does *right* mean? At a simple level, an *attribute* of right may relate to height (in the same sense that physical attributes have sometimes been specified for the armed forces). The *comparison* to be applied to this attribute might be that the height of all trainees should fall between five feet six inches and six feet three inches.

We are following the mature workers employment program introduced in Chapter Seven throughout this chapter too. Recall that the outcomes chain for this program included a simply stated outcome: "Older workers who obtained a position (through the program) retain their jobs." This low level of specificity about the outcome was fine for an outcomes chain but needed to be explained in more detail for purposes of program design, implementation, monitoring, and evaluation. The success criteria were identified from the literature on employment programs and long-term chances of success, program financial break-even information, policy statements concerning priorities for different subgroups, interviews with experts in employment programs, and data used for monitoring and evaluating those programs. Some information was also available for estimating break-even levels and for setting targets based on comparison with general employment statistics for this demographic group and with other labor market programs.

The following example shows a selection of success criteria for one outcome in the chain of outcomes for the mature workers program:

SUCCESS CRITERIA FOR ONE LEVEL OF OUTCOME IN THE MATURE WORKERS PROGRAM

Outcome: Older workers who obtain a job through the program retain their jobs.

	Relevant Attributes	**Some Possible Comparisons**
About older workers	Older workers are those who are forty-five years of age or older. [They are the target group for the program, so this attribute also applies to other outcomes in the outcomes chain.]	Comparisons among several subcategories based on research about the relative ease of getting a job following long-term unemployment—for example, forty-five to fifty-four years, fifty-five to sixty-four years, sixty-five years and older.
	Who is retained? Demographic features include prior length of unemployment, gender, ethnicity, and type of employment (industry classification). [These attributes are also related to the other outcomes in the outcomes chain.]	Outcomes for all of the above to be benchmarked in relation to similar subgroups in the wider community. Results for participants should exceed what would be expected from community statistics.
About the jobs in which they are retained	A job includes paid jobs that could be full time, part time, or casual. [This attribute also applied to the outcome that was about getting a job: one level below the "job retention" outcome in the chain.]	Full-time level preferred to part time, which was preferred to casual. Retention should be at a level equal to or better than the level obtained through the program.

	Job satisfaction—important in its own right but also as a predictor of likely future retention.	Eighty percent or more participants to report moderate to high levels of job satisfaction at six- and twelve-month follow-up
About the meaning of retention	Duration of retention: Consider three categories: zero to six months, six to twelve months, and more than twelve months.	Desired standard was twelve months or more. Less desirable or acceptable was six to twelve months. Least desirable was zero to six months.
	Retained: Could be in same or different job.	Equally acceptable but change of more than two jobs during the twelve-month period may be of concern for future employment prospects. Also any changes affect total number of months employed within a given period.
	Continuity of retention is of interest, not just whether they happened to be in a job at the exact time of the follow-up.	Desirable that 80 percent or more of the calendar period of employment should have been in employment rather than between jobs.

Some types of comparisons are more readily available for some programs (for example, employment programs) than for many other social programs. It can also be easier to set reasonable and informed nonarbitrary targets for some programs and some problems than for others. Historical and benchmarking data can assist with setting targets.

In many cases, deciding on the desired attributes even without identifying comparisons can add value to the program theory. Giving deliberate consideration to possible comparisons as part of the definition of a desired outcome can alert program managers and evaluators to various possibilities.

However, we are reluctant to advise programs to set targets for the sake of setting targets if doing so can have perverse effects such as goal displacement.

Note that the mature workers program included among its success criteria levels of performance that were ideal or preferred, as well as those that were less than the ideal: employed up to six months rather than the ideal of twelve months or more and employed noncontinuously rather than the ideal of fully continuous employment. In this way, it is possible to use the step of identifying success criteria to set up a scale from least desirable to most desirable performance against each outcome. When it comes to monitoring and evaluation, this scale will provide a useful basis for judging success.

In this chapter, we focus on attributes of success. We revisit comparisons in Chapter Fourteen when we discuss the use of program theory for monitoring and evaluation.

The Importance of Identifying Desired Attributes

Typically the outcomes in a chain are expressed in brief summary terms for purposes of presentation, communication, and ease of memorization. As such they are generally insufficient for designing or evaluating a program.

Success criteria can be seen as the essence of evaluation because they identify the values or criteria by which performance should be judged. Even more important, success criteria provide direction for sound program design and implementation. In Chapter Six we explained how desire to avoid conflict may contribute to stakeholder reluctance to be clear about various aspects of program theory. This reluctance is perhaps most likely to emerge when we ask stakeholders to be clear about their success criteria within a program theory especially if lack of clarity serves a purpose. Yet some degree of clarification is needed for purposes of program design, performance monitoring, and evaluation, even if this clarification is only to the degree of setting boundaries around what is acceptable and desirable and what is unacceptable. A program theory and a program as implemented should focus on the desired outcomes and desired features or attributes of outcomes, not just the measurable features or attributes. Carefully selected important success criteria can help to immunize a program against goal displacement. So it is important to identify all important success criteria, not just those that are easy to measure. By making criteria explicit and matching these to the selection of performance measures, discrepancies between important aspects of performance and the aspects of performance that are actually measured can be more easily identified.

USE OF SUCCESS CRITERIA TO AVOID GOAL DISPLACEMENT

A program that provided income support to veterans included among its intended outcomes that "veterans receive their *correct* entitlements in a *timely, consistent, and courteous* manner." Success criteria were therefore those relating to correctness, timeliness, consistency, and courtesy. A review of the program's performance measures showed that the measures related to the more easily monitored attributes of correctness, timeliness, and consistency but not courtesy.

Sometimes services that are driven by a concern for timeliness of response do so at the expense of courtesy and related service quality attributes. Hence, to avoid goal displacement, it would be useful to obtain some measures of courtesy of service, if not as part of routine monitoring, then as part of occasional evaluation studies.

Inclusion in the theory of action of the more difficult-to-measure but important attributes serves as a reminder of the possible need for occasional data about those attributes, as well as the need to give attention to those attributes as part of program design—for example, when training staff.

Knowledge of important attributes of success—for example, quality, quantity, timeliness, equity, different target audiences and other features, comparisons—that are desired for each of the outcomes in the outcomes chain can also provide a framework for undertaking the next step in describing the theory of action: identifying factors that will affect successful achievement of outcomes. Different factors will be relevant to achieving different attributes. For example, achieving an outcome with many participants (an attribute of success that relates to quantity) will require consideration of different factors from those that would need to be considered for achieving an outcome to a particular intensity or in a sustainable manner (a success criterion relating to quality). The measurement and evaluation differences with respect to relative emphasis on quality versus quantity are well known.

To some extent, the importance of stipulating attributes in detail varies according to the type of program—whether it is largely simple or has complicated or complex aspects. Specification will be more important and useful for simple and complicated programs than for complex ones. In general, it is also relatively easy to identify success criteria for simple programs, such as direct service delivery of universal services. Programs with complicated aspects typically require coordination among several components, and common and clearly understood success criteria can help to ensure that those delivering the components are all in agreement.

For programs with complex aspects and that are organic and evolving, the outcomes are emergent and not easily specifiable in advance. In this situation, it can be more helpful to have a menu of choices of attributes that fit the overall intent of the program rather than a one-size-fits-all approach. So, for example, when we were developing the theory of action for a complex program to strengthen families and communities, we had many different types of initiatives and projects working with many different types of individuals and groups. So for each outcome in the outcomes chain (expressed in general terms such as greater understanding, skills, and capacity for the initiative of participating communities as an organizing concept for the whole program), we identified quite a long list of examples that we considered relevant to or illustrative of each outcome.

As another example, in a program that aimed to improve the quality of life of people with a mental illness, the specific quality-of-life improvements would vary enormously across participants according to need and entering position, aspirations, circumstances, and the way these evolve. We would neither expect nor desire the same outcomes for all. For some, the outcomes may be improved social relationships or social participation; for others, improved income; for others, being able to retain stable housing. It would be quite inappropriate to set expectations about, say, improved housing or improved employment for all participants. However, that does not mean that consideration of success criteria is irrelevant. For example, a case management plan might set individualized and a000daptable objectives, and the desired attributes might be realization of personal short-term objectives or, more broadly, identifiable progress toward overall personal goals, accepting that the short-term objectives may change. The program might also set some boundaries around the range of attributes that it would work to achieve.

How to Identify Desired Attributes

Desired attributes can be identified in many different ways, including various combinations of these approaches:

- Defining the terms in an outcomes statement

- Defining what, when, where, how, why, and who for each outcome

- Consulting with stakeholders and reviewing program documentation concerning expressed stakeholder needs

- Identifying social justice considerations that might otherwise be overlooked

- Reviewing literature that identifies the types of attributes that are important for each type of outcome

- Identifying what is realistically achievable given what is known about the nature of the problem being tackled, needs of different target group members, resources available to the program, and other constraints and opportunities

Define the Terms   This is the simplest approach. It takes each of the key words in the outcomes statement and expands on them. This was the approach that we showed for the mature workers employment program. The attributes included what was meant by "older workers," by "job," and by "retention." Another example comes from an outcomes chain that we developed and used to evaluate an organization's program of program evaluations that had been conducted over several years. We were evaluating this program with respect to its utilization (Figure 8.1).

The first cut at attributes of success for Figure 8.1 was based on the evaluation utilization literature available at the time and was used to design data collection tools. Further insights during the evaluation resulted in some additional attributes and helped to shape our understanding of the many and varied ways in which evaluation processes and products can be used.

Define What, When, Where, How, Why, and Who for Each Outcome   One approach to identifying the attributes that might be included in success

Figure 8.1   **Utilization of Evaluations: Outcomes Chain and Definitions of Successful Outcomes**

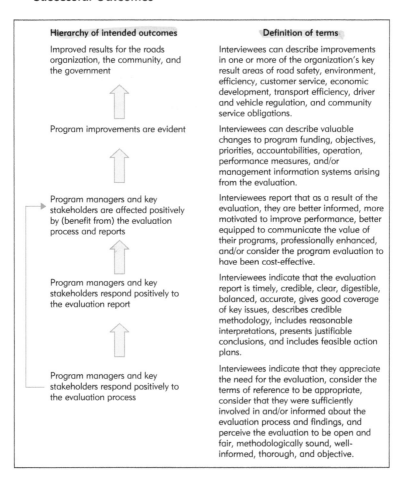

Hierarchy of intended outcomes	Definition of terms
Improved results for the roads organization, the community, and the government	Interviewees can describe improvements in one or more of the organization's key result areas of road safety, environment, efficiency, customer service, economic development, transport efficiency, driver and vehicle regulation, and community service obligations.
Program improvements are evident	Interviewees can describe valuable changes to program funding, objectives, priorities, accountabilities, operation, performance measures, and/or management information systems arising from the evaluation.
Program managers and key stakeholders are affected positively by (benefit from) the evaluation process and reports	Interviewees report that as a result of the evaluation, they are better informed, more motivated to improve performance, better equipped to communicate the value of their programs, professionally enhanced, and/or consider the program evaluation to have been cost-effective.
Program managers and key stakeholders respond positively to the evaluation report	Interviewees indicate that the evaluation report is timely, credible, clear, digestible, balanced, accurate, gives good coverage of key issues, describes credible methodology, includes reasonable interpretations, presents justifiable conclusions, and includes feasible action plans.
Program managers and key stakeholders respond positively to the evaluation process	Interviewees indicate that they appreciate the need for the evaluation, consider the terms of reference to be appropriate, consider that they were sufficiently involved in and/or informed about the evaluation process and findings, and perceive the evaluation to be open and fair, methodologically sound, well-informed, thorough, and objective.

criteria is to use Rudyard Kipling's (1902) time-honored "I Keep Six Honest Serving Men":

> I keep six honest serving men
> (they taught me all I knew);
> Their names are What and Why and When
> and How and Where and Who.

Not all of these will be relevant on every occasion for every outcome, but it is useful to consider whether each would be relevant before dismissing them or overlooking their possible importance. Also we would advise considering *what* and *who* first and then *where, when,* and *how. Why* can be used as a ruler to run over all of the identified possible success criteria to assess their importance.

Here are some examples of ways to use these to identify attributes of success:

What features the outcome should have

- Description of the nature of the outcome—for example, what skills are to be acquired or behaviors to change

- The quality—for example, reliable, accurate, to the client's satisfaction, comprehensive, equitable

- The quantity—for example, how much, what proportion, whether to increase or decrease

With whom each outcome should be achieved and who has an interest

- Numbers and types of clients

- Target groups identified by types of needs

- Attributes important to different stakeholders

Where the results should be achieved

- Intended geographical location of results

- Particular sites

When the outcome should be achieved

- Timeliness of response—for example, of services to client requests for assistance

- Time targets or milestones for achievement of each outcome to a certain level

How results should be achieved, including addressing side effects, unintended outcomes both positive and negative, and other agendas (but not the process for achieving the outcome)

- At a cost of $X

- With minimal disruption to other programs and other stakeholders

- With equitable distribution of results

- In compliance with statutory requirements

- Preserving clients' dignity and self-determination

Why the criteria are important

- Stakeholder interests and concerns

- Link to overall effectiveness of the program (research base)

- Necessary to achieve higher levels in the outcomes chain—features required of lower-level outcomes, such as percentage of target population reached, that are a prerequisite for achieving higher levels of outcomes

- Mandated by professional standards.

Consult with Stakeholders and Review Program Documentation   Stakeholders may be not only those who are the recipients, developers, deliverers of the program (including policymakers), experts, and other partner programs but also those who may be in some way affected by the program or denied access to it. These people will want assurance that their interests will not be traded off against those of the direct beneficiaries of the program and will want certain safeguards to be built into the program's success criteria. Market research and social and environmental impact analyses can feed into the development of success criteria. As far as possible, obtain information directly from stakeholders, and presume as little as possible about how other stakeholders might define success.

It is not that the views of all stakeholders are equally weighted or that they are the final determinant of success criteria. But they certainly should be factored into the equation. Almost inevitably stakeholders will have differences of opinion about the relative importance of various criteria, and some may appear to be mutually exclusive.

Identify Social Justice Considerations That Might Otherwise Be Overlooked   This is really a subset of stakeholder interests, but we discuss it separately because it is often overlooked. The discussion here focuses on addressing the needs of subsets of the client group whose interests might particularly need to be considered. For example, in relation to the mature workers employment program, the very long-term unemployed would be a particularly disadvantaged subset. It would be important to ensure that the program was not creaming off the relatively short-term unemployed at the expense of the long-term unemployed for greater surety of outcomes and meeting targets. Among social justice considerations are these:

• *Ensure the overall relevance of the program and of its specific objectives to the needs of disadvantaged groups* if these groups are among the target group, rather than assuming that one size fits all. In some cases, this may mean not only that the attributes of success for a particular outcome will need to be different for different subgroups (for example, for the mature workers program, time within which it would be reasonable to expect members of different subgroups to get a job) but that the outcomes themselves might need to be different (say, a different set of outcomes for those who cannot get a job but whose social, emotional, health, and financial needs still need to be met in some way through the program or by referral from the program to another more suitable program).

• *Have success criteria that draw attention to the distribution of program outcomes to identify the winners and the losers and not just average results.* Averages can conceal the fact that some groups are particularly advantaged and others disadvantaged. Breaking the target group up into subgroups can help, as shown in the mature workers program.

• *Ensure that targets do not prejudice service deliverers against giving service to the most needy,* as in the example of creaming in relation to the mature workers program.

• *Ensure that equity of access considers not only what is done to achieve equity but also actual levels of take-up.* What is done to improve access might include such strategies as child care services, interpreter services, hours of service delivery, and mode of service delivery according to need and sensitivities. Ultimately programs may be able to improve opportunities only for access rather than ensure improved uptake. However, it makes sense to include among the attributes of success some that relate to actual levels of uptake. Are the members of the target audience accessing the program? Are some doing so less than others? If so, why? Which of the strategies seem to be working and which not working? Can the program do more? Is a different type of program needed for some subgroups, and should their needs be highlighted with policymakers?

• *Ensure that outcomes are achieved and services delivered in a way that preserves the self-respect and dignity of disadvantaged groups.* One of the examples we have discussed is a program that provides vouchers to pay for the electricity accounts of people in crisis. A key factor affecting whether people would apply for this assistance was whether they believed they would be publicly identified and stigmatized when they used their vouchers to pay. Accordingly, the success criteria relating to use of vouchers included reference to the fact that when clients used the vouchers, they should feel they were treated with respect by the electricity company, like any other customer. This had program design implications for who delivered the service and training given to staff.

Review Relevant Literature   A review of relevant literature can be used to identify the types of features that might be important. The mature workers program example drew on literature about employment programs with the age group that the program targeted. The outcomes chain and success criteria for the utilization of program evaluations case example reflected some common understandings in evaluation literature at that time (the early 1990s) concerning utilization. Some of these definitions may well be different in the light of what has been learned about utilization since then.

Identify What Is Realistically Achievable   Taking account of what can realistically be achieved can affect the choice of success criteria. Ideally one would do this in the light of the end results that are to be achieved. However, aspirations need to be tempered by the realities of the wide range of factors that may affect whether outcomes are achieved. These include the realities of program resourcing and implementation, the nature and size of the problems to be addressed, and the issues for different target group members. Making allowances for these constraints is not, however, an open invitation to set targets low without regard for the effects on overall achievement of outcomes for the program.

SETTING REALISTIC SUCCESS CRITERIA

The service providers of a program for disadvantaged families questioned the short time frames that had been set for achieving the outcomes. They did this on the basis of their knowledge of the different intermediate outcomes that would need to be achieved with various families. They also stressed that the time frames would need to be flexible for each family to enable each to move at its own pace. For some families, a great deal of work needed to be done to establish rapport and trust and develop individual family plans. The differences among families and the need to adopt an adaptive case management approach made it difficult to set universal criteria relating to the time by which outcomes should be achieved. The success criteria needed to be flexible, reflecting the objectives of particular families.

ASSUMPTIONS ABOUT FACTORS THAT AFFECT SUCCESSFUL ACHIEVEMENT OF OUTCOMES

In order for us to expect that intended outcomes will be achieved, we have to make various assumptions about what will occur to facilitate their achievement and about the absence or management of factors that may impede achievement. It is helpful to identify these assumptions in a systematic way and incorporate them in the program theory.

Factors affecting outcomes fall into two broad classes: program factors and nonprogram factors. These factors should be considered in relation to the program as a whole and to each outcome in the outcomes chain since different factors will have greater impact on some outcomes than on others. For example, we can assume that the relevance of the messages and timing of marketing strategies may be primarily important for attracting the right people to participate in the program (usually a relatively low-level outcome in the outcomes chain). Thereafter, we can assume that the effectiveness of program delivery will be a factor that will play a more important role in achieving higher-level outcomes with clients.

Assumptions About Program Factors

Program factors are those that are largely within the control of or can be largely influenced by program funders, program management, and staff.

A theory of action needs to include assumptions—testable assumptions—about how well these various program factors will function—for example:

Program resources (controlled or influenced by program funders) that are needed to deliver the program so that it will achieve outcomes

- Adequacy of resourcing—financial and human

- Certainty and dependability of resourcing

- Diversity of resource base

- Flexibility permitted for use of resources (for some programs)

All aspects of program delivery controlled or influenced by program staff that are needed to achieve outcomes

- Quality and quantity of activities and service delivery

- Quality, quantity, and timeliness of services delivered

- Quality, quantity, and timeliness of outputs and throughputs

Management activities and processes controlled or influenced by program management that are needed to support program design, delivery, sustainability, and so forth to achieve outcomes

- Effective, efficient, and economical use of available resources

- Appropriateness of staff selection, training, and other aspects of human resource management

- Program leadership that is appropriate to the type of program: complex, complicated, or simple (refer to the discussion in Chapter Five)

- Management strategies, accountabilities, governance arrangements, incentives and systems for rewards and sanctions, and so on, that are conducive to the particular program's success and send the right messages

- The effectiveness of strategic planning and program design processes, including the soundness of data available for doing so

- The effectiveness of monitoring, evaluation, and program improvement processes

- The effectiveness of relationship management, including stakeholder management, partnerships, and communication strategies

- The effectiveness of strategies to influence external factors that may impinge on the program's success or to manage the risks associated with those factors.

Program theory approaches often provide descriptions of what the program does or purports to do rather than identifying the quality, value, or adequacy of what it does as key factors that underpin the theory about what is needed to make the program work. Our contention is that the inclusion in program theory of not just what a program does but also the assumptions about how well and to what extent it must be done to achieve results adds some valuable dimensions to the program theory.

Assumptions About Nonprogram External Factors

Nonprogram external factors are those that lie beyond the direct control of program management and staff but nevertheless have a significant impact on outcomes. Typically these are about the operating environment of the program. The program will be able to influence some of these external factors (for example, influence other policies through their partners) and may be able to monitor and partially manage the risks associated with others that it cannot influence (for example, risks for clients associated with economic downturn, media).

Examples of external factors follow, with some clearly being better able to be influenced by or catered to through program design (for example, catering to past experiences, expectations and demands of participants) and others requiring risk management strategies (for example, changes in the economy):

- Economic climate

- Political climate

- Industrial climate

- The weather, climate, and natural disasters—both patterns and erratic events

- Crises and emergency situations

- The changing nature and scale of the problems addressed by the program

The following external factors are often closer to the program:

- Government policies, regulations, and priorities
- Changing policies and priorities within the organization
- The activities of partners and other key actors
- Competing, parallel, or interdependent programs
- The influence of public opinion groups and the media; program critics with their own agenda

Characteristics of clients, referred to in some program theories as inputs, are to some degree factors that the program may need to address but may not be able to influence:

- Demographic features of clients such as gender, age, ethnicity, socioeconomic status, health status, and educational background
- Past experiences of clients and what they bring to the program
- Client demands, expectations, and specific needs that emerge over time

The nonprogram factors could be on a grand scale or a relatively local scale, such as a particular community crisis or other unplanned event that either diverts community and program efforts or rallies a community around a program, giving it greater profile, importance, resources, and so on.

The drivers-of-change approach used in some international development work provides some systematic ways for thinking about contextual factors that can affect whether change occurs and what might need to be addressed if change is to occur. This approach identifies three broad categories of drivers that apply in the development context but many would also apply to programs in the developed world:

"In this model the three elements are defined, in ascending order of flexibility and relative speed of change, as follows:

- *Structural features*—the history of the state; natural and human resources; economic and social structures; demographic changes; regional issues;

- *Institutions*—the informal and formal rules that determine the realm of possible behaviour by agents. Examples are political and public administration processes.

- *Agents*—individuals and organisations pursuing particular interests. Examples are the political elite; civil servants; political parties; local government; the judiciary; the military; faith groups; trade unions; civil society groups; the media; the private sector; academics; and donors" [Warrener, 2004, p. 8].

The effects of nonprogram factors may be direct (for example, a change in the economic environment may directly influence the probability of an unemployed person obtaining a job) or indirect through influencing the extent to which the program can be implemented in the manner intended (for example, competing priorities, loss of staff, introduction of new technology). In both cases, direct and indirect, nonprogram factors may be:

- Positive, in which case they may improve the results achieved

- Negative, in which case they may dampen the impact of an otherwise appropriate and effective program

- Interactive, by improving results for some clients but detracting from results for other clients

We provide some examples of program and nonprogram factors for one level of outcome in the mature workers employment program. These lists of factors came from an initial discussion with staff of this program about program and nonprogram factors that would affect whether those they had placed in positions would retain their jobs. Further discussion led to the conclusion that staff may actually be able to influence some of the nonprogram factors that appeared to be outside their control, such as employer expectations. Rather than complaining about unrealistic employer expectations beyond their control, staff could redefine them as program factors. When this is done, some additional program activities or ways of working with employers and employees could be incorporated in the program and perhaps some resources dedicated to those processes. The program might also consider adopting some strategies to prepare the workplace for the older worker. Given the boundaries and limited resources of the program, this would occur only if the receptiveness of other staff to older workers proved to be a major barrier to retention and would probably need to occur on an employer-by-employer basis rather than as a universal program strategy.

ASSUMPTIONS ABOUT PROGRAM AND NONPROGRAM FACTORS

Factors that will affect this outcome: "Older people who are clients retain jobs"

SOME PROGRAM FACTORS

- Adequacy of program staff understanding of client and employer needs

- Appropriateness of match of person and placement

- Adequacy of follow-up support after placement

- Whether performance targets encourage quantity and/or quality of placements

SOME NONPROGRAM FACTORS

- Change in employee circumstances, for example, health, relocation, grandchildren to care for

- Employer expectations of employee and whether they are realistic

- Business factors that require layoffs or may encourage greater use of mature age workers

- Receptiveness of other staff toward older workers

- Employer attitudes to mature-age workers

Reasons to Identify Assumptions About Factors That Will Affect Achievement of Outcomes

Identifying factors that we believe will affect how well outcomes are achieved is a means of developing an understanding of the mechanisms that will need to be in place if the program is to work. Once those factors are identified, the next step is to find out what activities and resources are used (or planned) to ensure that the mechanisms are in place. So if credibility of advisors is identified as an important program factor or mechanism in influencing changes in client knowledge, what does the program actually do to ensure that its advisors are credible? If effective communication strategies are identified as a key program factor for ensuring that the right people participate in a program, what communication strategies does the program have in place? If a deteriorating economic situation (a nonprogram factor) is adversely affecting the employment prospects of mature-age workers, what does the program do to cushion or mitigate the impacts of that situation and the other effects of unemployment on their health, self-esteem, and so on?

Incorporating assumptions about program and nonprogram factors in a program theory is useful in terms of:

- Program design and program improvement by attending to factors that are, should, or could be within the control of the program and choosing activities to address the most important of these factors

- Risk management of factors that are beyond the direct control or influence of the program and might affect it adversely

- More effective use of opportunities outside the program, including activities of partners that could be conscripted to contribute to its success

- Causal contribution by comparing the relative impact of program and nonprogram factors on the outcome when evaluating a program

Identification of program and nonprogram factors and their likely relative impacts can assist with making judgments about which outcomes in the chain should be measured and possibly attributed to the program and which should be measured for other purposes, such as monitoring the nature and extent of the problem and the changing nature of the context within which the program operates. It can also assist with identifying what else other than outcomes need to be measured for purposes of interpreting findings about so-called outcomes and for causal attribution.

We sometimes find it helpful to show a line of attribution or line of accountability in the outcomes chain. This is the point above which the non-program factors so outweigh the program factors that it would make little sense to substantially attribute the outcomes to the program, even though it may still be possible on occasion to identify some contribution of the program to the outcomes above that line. However, the line of attribution will not work if a program's efforts are scattered across an outcomes chain, with some outcomes being addressed primarily by others or not at all.

In our experience, program management and staff have a tendency to see rather more factors as being outside their control or influence than do program funders and policymakers. The daily encounters that program staff have with the factors that complicate implementation and impacts of the program often give staff a certain sense of powerlessness or at least humility with respect to their own contribution. Sometimes funders and policymakers see the cautiousness of program staff as an excuse for not being held account-able for results, and undoubtedly their perceptions are sometimes correct.

One consequence of the different perceptions of nonprogram factors is that program staff sometimes want to measure their success to a lower point in the chain of outcomes and avoid measuring the higher-level outcomes that they see as beyond their control. Program funders and policymakers whose sights are set on achieving higher-level impacts and addressing needs would like to have the program's success measured in terms of these higher levels in the chain; the lower levels are of little interest to them. Of course, the opposite can also occur: program staff may be eager to measure and claim higher-level outcomes even though attribution is difficult. Policymakers on the other hand, aware that a wide range of programs, initiatives, and other factors contributes to outcomes, may take a more conservative approach. Either way, the identification of program and nonprogram factors can be a fruitful starting point for discussing and resolving some of the differences in expectations, typically to achieve a middle and realistically achievable ground.

When using program theory for program design and management, if a nonprogram factor is identified as a barrier to success, various approaches can be taken—for example:

- Expand the scope of the program to address that factor head-on.
- Focus efforts on some aspect of the barrier to manage the risks in a particular situation.
- Work around the barrier rather than try to overcome it.

These three general options are illustrated in the example that follows.

MANAGING AN IDENTIFIED NONPROGRAM FACTOR: MATURE WORKERS PROGRAM

Changing employer attitudes on a wide scale is not a major focus of the mature workers program even though employer attitudes are recognized as a nonprogram factor that sometimes poses an impediment to employment of older people and therefore directly affects the outcome in the outcomes chain labeled "clients obtain appropriate jobs." In the event that these employer attitudes become a serious impediment to the success of the main in-scope activities in securing employment for mature-age people, the program could consider one or more of a number of paths:

1. Expand the program to incorporate an active and comprehensive strategy to change the attitudes of employers, for example, by significantly expanding the scope and resources of the program to include different policy tools. Examples are a mass communication campaign, working with government to change regulations around age discrimination, and offering tax or other incentives to employers to hire older people.

2. Work more intensively to break down the attitudinal barriers of select groups of employers who might otherwise be well positioned to offer jobs to mature-age workers, for example, Rotary clubs and other community groups. This approach requires adjustment at the margins with respect to scope.

3. Work around the employer attitude problem by focusing all energy on sympathetic employers. This approach requires no change of scope.

Although the program does not include wide-ranging strategies to change employer attitudes and it is beyond its mandate and its resources to do so (option 1), it does include some activities to work with specific employers and groups of employers where it identifies potential to generate job opportunities and interest in mature workers (option 2). It also fosters ongoing positive relationships with friendly employers by ensuring that the program staff refer only appropriate employees to them and give any postplacement support that may be needed. This is a means of working around the barriers of attitudes to mature-age workers in the short term rather than trying to overcome them (option 3).

HOW TO IDENTIFY FACTORS THAT ARE LIKELY TO AFFECT OUTCOMES

There are many different ways to identify factors that affect success at each level of the outcomes chain. Some are program specific and some are more general:

Program-Specific Research and Investigation Techniques

Sources of information

- Market research, discussions with clients, focus groups, formal and informal client feedback

- Program staff experience and expertise through interviews and feedback from routine debriefings and critical reflection by staff

- Past evaluations and program monitoring—one of the first ports of call when developing a program theory

Analytical techniques

- Analysis of helpers and hinderers of the program in consultation with staff and others

- Use of particular success criteria for each outcome as a starting point for asking about factors that will affect program achievements in relation to those success criteria. Examples: For timeliness as a success criterion: What factors will affect the time by which the outcomes will be achieved? For equity as a success criterion: What factors will affect whether the outcome is distributed equitably across the target group?

- Analysis of strengths, weaknesses, opportunities, and threats

- Critical incident analysis and best-case and worst-case analyses to identify causes

- Total quality management methodologies—for example, analysis of sources of statistical variation in program performance monitoring data; factors that affect larger portions of the variance (the 80:20 rule) may require more attention than others

Program Theory and General Research

- Literature review and discussion with key informants

- Review of quality assurance, professional, and other management standards in related areas

- Factors that seem to commonly affect the success of different types of services (for example, human services, physical products services)

- Activities and standards as part of program templates that have been developed for particular types of programs (Scheirer, 1996)

- Factors associated with the success of theories of change (such as stages of change theory; see Chapter Eleven) and archetypal programs (advisory, regulatory, service; see Chapter Twelve)

- Evaluator experience of what affects the success of other programs that share some features in common

It is best to use several techniques and sources—clients, staff, experts, and so on—since all will have different perspectives on factors that affect outcomes. For example, clients may not be able to accurately identify factors that influence their own behaviors, and program staff may draw inappropriate inferences. Examples of the application of a few of these techniques follow:

- *Discussions with potential users* of a crisis counseling service for problem gamblers identified the types of factors that would affect whether they would or would not use the service: whether face to face or by telephone, group or individual, hours of operation, whether monitored for service quality, appeal of different approaches to people of different ethnic backgrounds, gender, and others.

- The program theory *drew on the experience of the staff of a program* to deliver vouchers to people who could not afford to pay their electricity bills. Staff identified the development of trust between worker and client, worker understanding of the components of a successful plan to assist with future budgeting, and distancing of the program from the government as some of the factors that would affect outcomes.

• A program was established to encourage tenants of public housing to participate in democratic decision-making processes concerning the design features of public housing to address specific community needs, such as design for single-parent families and for particular ethnic groups to reflect such preferences as communal eating areas. The program theory development *drew on the experience of the program manager.* He recognized that many factors would affect whether the whole target group would participate. Among other things, these included whether child care was available so that single parents could attend meetings; whether there were language barriers to participation; and whether there were cultural barriers to participation, such as lack of experience of some ethnic groups in participating in democratic meetings and in using deliberative democratic processes rather than, say, attempted bribery of officials or resolving differences through physical means. Having identified these factors, he was able to put in place appropriate strategies such as provision of child care, provision of interpreters and facilitators for meetings of groups of people who did not speak English, and provision of educational programs to develop skills in running and participating in meetings and working with officials.

• Scheirer (1996) comments that *templates,* in their original use for formative evaluation, "summarized recommended components for teachers' professional development. . . . These templates provided guidance and external standards concerning program activities in order to address the key question: Does the specific program include all components recommended for best practice in the field?" (p. 1). Templates reflecting best practice are drawn from research.

• The evaluation of the mature workers program *drew on research literature and published standards.* It included among program factors whether best practices that had been identified by the Commission of the European Community for government-funded employment programs were in place for this program. The evaluation compared various practices (funding, setting targets, monitoring) of the government agency running the program with the commission's best practices.

• Another program management factor common to many programs is the *relative expenditure of funds on program delivery* compared with program administration. In response to this general program factor, some government agencies set expectations or ideals concerning the distribution of total funds between service delivery (what is available for direct use with the target group) and administration costs (for example, no more than 5 percent of total funds should be spent on administration).

- Use checklists for thinking about the types of factors that are likely to be relevant. Examples are *sets of principles that have been developed for the operation of various types of services or programs* (for example, the principles for early intervention in early childhood identified in Rogers, Edgecombe, and Kimberley, 2004). Another example of checklists is that relating to desirable features of different types of services. Table 8.3 shows some of these factors for three different types of services.

IDENTIFYING WHAT THE PROGRAM DOES

An important part of the theory of action is clearly specifying what the program is expected to do (for program theory developed in the planning stage) or what it does.

What Is Included in What the Program Does

What the program does includes everything that the program consumes, does, and produces to achieve its desired outcomes. Resources, activities,

Table 8.3 **Factors That Affect the Success of Different Types of Services**

Human Services	Information and Professional services	Physical Products and Services
Factors that affect the quality of the relationship	Factors that affect the usefulness of the service	Factors that affect the value of the product to the consumer
Responsiveness	Accuracy	Reliability
Empathy	Credibility	Fit for purpose
Courtesy	Clarity	Price
Communication	Relevance to need	Convenience
Trust	Honesty	Availability

and outputs are what the program does to trigger the change mechanisms encapsulated in outcomes chain and the factors that will affect successful achievement of outcomes. As for all other parts of the program theory, the links among resources, activities, and outputs, and between what the program does, the program and nonprogram factors, and the intended outcomes need to be made clear. Resources, activities, and outputs relate to program factors identified in the previous section:

- Allocation and expenditure or use of resources, financial and human, from within the program and from outside it

- Service delivery, including principles that apply to service delivery

- Program management and support, including what the program does to influence and risk-manage nonprogram factors.

Note that program activities are also about managing or influencing nonprogram factors.

For programs with complicated and complex aspects that involve several partners, it is also useful to identify who is responsible for what. For example, as part of the overall theory of change developed for a multiagency program to combat drug abuse among youth in an ethnic community, we identified which agency had primary responsibility for each outcome (or suboutcome) in the outcomes chain. This helped to identify the various points for key interfaces between particular agencies (typically, but not always, those in closest proximity to them on the chain) to ensure that the theory of change worked as a whole.

Why It Is Important to Identify Resources, Activities, and Outputs

Identifying what the program actually does is critical to drawing conclusions about the role the program plays in producing the outcomes. This applies no matter what research methodology is used when conducting empirical evaluations of program outcomes. Even randomized control trials require clear description (and control) of what the program delivers in order to draw conclusions about program impact. We have to know what the program actually does.

The program theory identifies planned resources and activities of the program. Plans are theory rather than practice. The theory provides one focus for identifying what information should be sought during the evaluation about how the program operates in practice and whether the operation is consistent with the theory. Identifying the connection (or lack thereof) between particular outcomes and what the program does (or fails to do) is useful for program improvement purposes.

To ensure the link between the activities and the outcomes, the activities should be identified for each outcome rather than simply listed for the program as a whole. Of course, some activities will appear against several different outcomes. For example, some teaching strategies may be used concurrently to develop knowledge and self-confidence and to directly foster behavior change. There will also be some outcomes that have no particular activity devoted to them, as in situations where a higher-level outcome flows primarily from lower-level outcomes, with little or no further effort required of the program.

Identifying planned and actual resources, activities, and outputs is a precursor to identifying whether they match or in some way address the program and nonprogram factors that have been identified as critical to program success. This is a key step for reviewing a program theory (see Chapter Ten) and for designing, evaluating. and improving a program.

Low-cost techniques such as interviews with a small number of key informants can be used to build up and assess a program theory on paper. An empirical evaluation would find out whether, in this case, the new communication strategy is actually effective.

How to Identify Resources, Activities, and Outputs

With a checklist of items such as those already listed in program and nonprogram factors, program staff and management are usually able to identify relatively quickly a program's resources, activities, management strategies, and so on—or at least what they believe to be the case. Reality testing through site observations and interviews with service delivery staff and with clients may prove otherwise and is an important function of most evaluations.

Because program activities and resources are usually relatively easy to identify (at least on paper), we have focused more on how to identify the

COMPARING PROGRAM ACTIVITIES WITH PROGRAM FACTORS
ASSUMED TO AFFECT OUTCOMES

One of the lower-level intended outcomes of a program that provided crisis financial assistance to people having difficulty paying their electricity bills was that members of the target group should be aware of the availability of assistance and how to obtain it and that they should feel comfortable about applying for it. Therefore, one program factor that emerged as important for this program was the extent to which an effective strategy was in place to communicate with all members of the target group concerning their entitlements.

A listing of activities showed that in one region in the state, the electricity company, although it had identified effective communication as a key factor that would affect success, was taking little action to inform people who were having difficulty paying their accounts of the availability of the assistance and how to obtain it. The communication efforts of community-based welfare organizations that dispensed the vouchers would miss some pockets of the target group, especially older people, who tended to be outside the social welfare network. They valued self-sufficiency and often believed that there was always someone worse off than themselves. They were also more likely to feel stigmatized if they sought assistance. The company prepared a new communications strategy to address the acknowledged gaps in communication activities to ensure awareness of the service, appreciation of entitlement, and reduction in stigma for this subgroup.

other important components of a program theory that are often underplayed or overlooked: situation analysis, scoping, outcomes chains, success criteria, and factors that affect the success of the program. Here we give a relatively brief description of resources, activities, and management strategies while recognizing that a huge literature abounds on specific techniques such as activity-based costing that cannot be covered in this book.

We do draw attention to two features of resources and activities that are sometimes overlooked. First, in relation to resources, it is important to include resources other than program funding that may well be critical to the delivery and sustainability of the program. These other resources are often overlooked in program theories. Dependence on the resources of partners and others outside the program are especially deserving of consideration. Changes in their priorities or circumstances and in their commitment of resources (financial, in kind, and so on) can sink a program. Second, among the activities, we include the principles of operation that underpin program implementation. Although these are important, they are sometimes overlooked in descriptions of activities. What program practices are within and off limits? What are acceptable and unacceptable approaches to working with clients (for example, to meet certain standards or principles, without disadvantaging particular groups, to maximize self-determination and empowerment)?

AN EXAMPLE OF RESOURCES, ACTIVITIES, AND OUTPUTS: THE MATURE WORKERS EMPLOYMENT PROGRAM

RESOURCES

Funding (the amount would be specified) is provided to twenty-six community-based organizations to apply a case management approach (as the central policy tool) to assist mature-age people who are looking for work.

These organizations contribute in kind to varying degrees, and this contribution, including its social capital in the local communities, is critical to their success.

MAIN ACTIVITIES

- Case assessment and planning with clients, resulting in an agreed case management plan

- Personal counseling (and, as needed, referrals to specialist services) to address motivational and self-esteem issues, family relationships, and other consequences of long-term unemployment

- Reinstating work habits for long-term unemployed clients by helping them to participate in voluntary work before returning to the open work market

- Exploration of career change options

- One-on-one advice, written materials, and educational activities to assist with job search, job application (résumés), and job interview skills

- Training to develop or update skills, such as technology training

- Assistance for finding suitable positions

- Facilitating placements with employers by bringing employers and prospective employees together

- Providing postplacement support to increase the probability of retention

- Arranging work experience and community volunteer work as a transition to paid employment for some and as a desired alternative to paid work for others

- A limited amount of ongoing personal support for those who are unable to obtain a job

EXAMPLES OF OUTPUTS AND THROUGHPUTS OF THESE ACTIVITIES

- Number of case management plans produced

- Templates for preparing résumés and number of résumés produced

- Number of computer skills training programs run and number of people who completed them

- Number of job interviews obtained

- Number of postsupport placement visits made to employees in their place of employment.

EXAMPLES OF PROGRAM SUPPORT AND MANAGEMENT

- The community organizations, as well as the program as a whole, have a marketing strategy to attract potential target group members. Information kept about the success of the program contributes to this marketing.

- Community organizations work with the funding agency to ensure that it understands the context and sets targets and other parameters that will reduce the likelihood of cherry-picking and other counterproductive approaches.

- Different service providers use their own systems for recording case management information. However, the program has common monitoring and reporting templates that apply to all service providers.

OTHER ACTIVITIES TO MANAGE THE RISKS ARISING FROM UNSUPPORTIVE EMPLOYERS

To lay the groundwork and increase the general receptiveness of employers to employing mature-age people: personal contact individually, group events such as Rotary clubs, and through working with local media. The program also places great effort on keeping friendly employers friendly by going to considerable effort to understand their needs and ensuring that employee placements match those needs.

PRINCIPLES FOR SERVICE DELIVERY

Among the principles that guide program practice are those that relate to case management standards, including such practices as empowering the people who contact the service to make their own decisions about what types of career paths they wish to follow and the speed at which they move through various stages of improved job motivation, job readiness and skills, and job seeking. These principles sometimes require trade-offs against performance measures (required by the funding agency) that relate to maximizing the number of people who secure paid jobs within a minimum amount of time. Different community organizations have different approaches for translating the principles into practice.

The program as a whole is also committed to applying social justice and equitable access standards through such practices as taking onto its books anyone within the target audience who approaches it for help, not just those who are easier to place.

PULLING THE THEORY OF CHANGE AND THE THEORY OF ACTION TOGETHER IN A MATRIX

In the previous chapter and this one, we have identified some important features of a theory of change and of a theory of action. The features within and between these two subtheories are interrelated, and so we need a means of recording not just each set of features for a given program but also the interdependencies among them. A matrix (Funnell 1997, 2000) is a tool to assist with the process of identifying the various features and recording them. The outcomes chain is the linchpin between the theory of change and the theory of action, and the matrix is constructed in relation to the outcomes chain. Each row in the matrix is dedicated to an outcome from the outcomes chain.

The matrix has these columns:

1. *Outcomes extracted from the outcomes chain (which in turn derives from the situation analysis).* These are placed in approximate ascending order. The order is approximate because an outcomes chain may have several strands or paths with several outcomes at similar levels.

2. *Success criteria* for the outcome, including attributes and, as appropriate, standards and other comparisons.

3. *Assumptions about program factors that affect the outcome*—largely within the control of the program.

4. *Assumptions about nonprogram factors that affect the outcome*—those largely outside of control but some which may be influenced or managed.

5. *Outputs and throughputs* produced by the program.

6. *Activities* (otherwise often referred to as processes) undertaken and used by the program.

7. *Resources* (otherwise often referred to as inputs)—financial and human, including resources from outside the program.

On particular occasions for particular purposes, we may choose to focus more on some parts of the matrix than on others. For example, we may focus on program and nonprogram factors to tease out the assumptions that underpin the program. We have found that the format shown in Table 8.4 is a convenient way to include outcomes, their desired attributes, and program and nonprogram factors that affect outcomes.

Table 8.4 An Alternative Approach for Focusing on Outcomes, Their Desired Attributes, and Factors That Affect Success

Assumptions About Nonprogram Factors That Will Affect This Outcome and Its Particular Attributes	Outcomes and Attributes	Assumptions About Program Factors That Will Affect This Outcome and Its Particular Attributes
Factor 1: Factor 2: Factor 3:	**Outcome 1:** Attribute 1: Attribute 2:	Factor 1: Factor 2:
Factor 1: Factor 2:	**Outcome 2:** Attribute 1: Attribute 2: Attribute 3:	Factor 1:

The matrix is a thinking tool rather than an end in itself. Applying the principles discussed in this chapter and incorporated in the matrix is far more important than filling in the boxes of the matrix. Moreover, the relationships among the boxes are as important as what is in the boxes themselves. Tracing the connections horizontally across the matrix and vertically up the matrix can help to clearly demonstrate the logical or expected relationships amongst the components. With practice, application of the principles becomes a natural part of the way of thinking about programs, and the matrix can become redundant.

SUMMARY

In this chapter, we have discussed three components of a theory of action whose inclusion can improve a program theory by avoiding common traps in the development and use of a program theory. One component is the identification of important success criteria (including both attributes and comparisons) for each of the intended outcomes. Another component is the identification of the types of program features (such as level of staff competencies and effectiveness of communication strategies) and external contextual factors (such as past experience of participants, economic climate, other programs) that are expected to affect how well outcomes are achieved. A third component is what the program actually does in terms of using resources and delivering activities and outputs to ensure that desirable program features are in place and external factors are in some way addressed to achieve the intended outcomes. We have also included a matrix that can assist with recording the components as they are developed and pulls together the theory of change and the theory of action as a complete program theory.

Table 8.5 Application of the Program Theory Matrix to One Outcome from the Outcomes Chain for the Mature Workers Program

Outcomes Chain	Success Criteria	Program Factors	Nonprogram Factors	Outputs and Throughputs	Activities, Processes, Principles	Resources and Inputs—Financial; Human
Improved health, financial well-being of program participants; benefits to community						
Older people who are clients of the program retain jobs	Age: 45–54, 55–64, 65 or more years Demographic features such as gender, ethnicity, type of industry Paid or unpaid Level of job satisfaction Employment for up to six months, seven to twelve months, and twelve months or more Same versus different job	Whether: Person and job are well matched Client expectations are well managed There is adequate follow-up support after placement Performance targets encourage sustainable placements	Whether: Employee circumstances change Employer has realistic expectations of employee Business factors require layoffs Other staff are receptive	Postsupport placement visits made to employees in their place of employment (outputs) Clients receiving postsupport placement visits (throughputs)	Meet with potential employers to understand their needs Activities with client and employer to develop realistic expectations, such as workplace visits and trial periods Offer and provide postplacement support to all clients Performance indicators support sustainable jobs by including adjustments for clients who may be difficult to place	Program delivery funds for postplacement support Employer time to assist with the adjustment of a long-term unemployed person

Clients of the program obtain appropriate jobs					
Clients acquire job-seeking skills					
Clients acquire job skills					
Clients acquire job motivation, confidence, work habits, life skills					
Clients sign up to appropriate case management plans					
Older unemployed people know about and register with the program					

EXERCISE

Choose a level of outcome in the outcomes chain for the mature workers employment program, other than the level of outcome for which an example has been provided on page 238, and develop the matrix for that outcome, applying the concepts described in this chapter for:

- Success criteria

- Assumptions about program and nonprogram factors that affect successx

- Program resources, activities, and outputs

Make sure that you can show the links across the matrix—that the resources, activities, and outputs that you choose can be shown to be linked to addressing the assumptions that you have identified as relevant to that specific outcome.

What were the challenges that you experienced in applying this process?

9

Representing Program Theory

THE REPRESENTATION OF program theory must be useful for those who have created it and comprehensible and engaging for others who will use it. It must effectively communicate the main messages. And it must not be cluttered yet include all necessary details or provide a framework within which to elaborate on those details. This chapter sets out different ways of representing program theory, including ways of addressing complicated and complex aspects of interventions and situations. We provide guidelines for assessing the quality of logic models and show some logic model makeovers.

OPTIONS FOR REPRESENTATION

How we think about program theory influences how we represent it, and how we represent program theory influences how we think about it. Rick Davies (2004), an evaluator who has explored ways of linking program theory and network analysis, has described clearly the importance of representation:

> How can ... organizations represent the complex processes of
> change that they are engaged with, at local, national and interna-
> tional levels, along with a host of other actors, many of whom do
> not share the same objectives? Without adequate representation,
> it is much more difficult for an organization to propose, test, and
> improve its "theory of change," and to communicate this refined
> knowledge to others, enabling its wider use and impact [p. 102].

It is therefore important to consider carefully the form of representation that will be used. It is also important to be aware of the different ways that program theory can be represented. If you are introducing program theory to an organization, you can choose the version that will be most useful for you. If you are using a particular version that has been standardized across your organization, it is still important to understand that your partners in other organizations might be using quite a different format. Lack of awareness of the different possible forms of logic models can lead to confusion and difficulty when working across organizational boundaries.

There are four broad approaches to representing program theory:

- *Outcomes chain logic models,* which show a sequence of results leading to the ultimate outcomes or impacts of interest, were used in
  the first examples of program theory (Kirkpatrick, 1959; Suchman,
  1967; Weiss, 1972).

- *Pipeline logic models,* which represent an intervention as a linear process, where inputs go in one end and impacts come out the other end,
  with activities and outputs in between, have been widely popularized
  through publications such as the United Way guide (1996) and the
  W. K. Kellogg *Foundation Logic Model Development Guide* (2004).

- *Realist matrices* represent program theory in the form of a table that
  shows the particular context (the implementation environment or
  participant characteristics) in which causal mechanisms operate to
  generate the outcomes of interest. This approach was developed in
  the 1990s when British sociologists Ray Pawson and Nick Tilley

(1997) set out their approach to realist evaluation, which focuses on how interventions trigger particular causal mechanisms only in favorable circumstances.

- *Narratives* set out the logical argument for a program in the form of a series of propositions that explain the rationale for the program—why it is needed and how it operates. The narrative can tell the story of how inputs produce a series of outcomes or how participants move through the program to achieve the intended results. Most other types of diagrammatic representation need to be accompanied by a narrative that explains them. Diagrams alone are rarely sufficient.

Later in the chapter, we discuss how logic models can also incorporate other types of diagrams, such as systems dynamics diagrams.

Pipeline models have become a common way of representing program theory, but there is often value in using one or more of the other types of representation in addition to or instead of a pipeline logic model. We can see what these different options look like by applying them to a specific intervention: a project that provides a summer school in information technology and follow-up support to students from disadvantaged schools to improve their educational and occupational outcomes. These examples are based on an actual project, with some details changed for the purposes of this exploration.

A Pipeline Logic Model of a Computer Project

A four-box version of a pipeline logic model (Figure 9.1) might represent this project in terms of inputs, activities, outputs, and outcomes. (In Chapter Thirteen we present different types of pipeline logic models).

The main problem with this logic model is that it does not explain what it is about the computer courses that will produce (or contribute to) the outcomes of interest. For example, although it refers to skills, it is not clear whether the program aims to develop skills in using specific software, such as commonly used word processing and spreadsheet packages, or skills in learning how to learn to use software that can be applied to quite different software in the future. It is also not clear what it is about the various components of the project that make the program work.

Figure 9.1   **A Pipeline Logic Model of a Computer Skills Project**

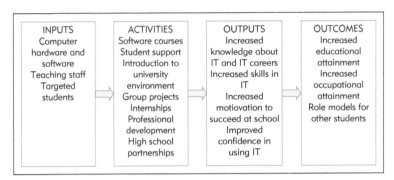

Considerable revision would be needed to more specifically define the expected outputs in ways that clarify how the project is understood to work. As we discussed in Chapter Seven, development of a program theory should begin with an analysis of the situation in terms of needs, problems, strengths, and opportunities. A program theory is an explanation of how an intervention is expected to solve the identified problems, meet the identified needs, and draw on the identified strengths and opportunities. In this case, a situation analysis might identify the following problems: the reason that many students at disadvantaged schools are excluded from postsecondary IT courses is that they lack basic skills and self-confidence in using computers because their schools, homes, and communities do not provide access to using computers. In addition, their teachers have few computer skills. They are also from communities where educational and employment prospects are very poor, and they have no role models of local people achieving educational or occupational success.

Drawing on descriptions of similar programs and observations of this specific program, we might identify three important mechanisms: skill development, reimagining the future, and forming strong social networks for students.

The example also demonstrates three other common problems when using pipeline logic models to represent interventions with complicated or complex aspects. First, the activities are all shown near the start of a pipeline logic model, even though they might occur at quite different times. In this project, internships occur after the computer courses, once students have developed some basic skills, but the logic model shows these activities together because all the activities are at the beginning of the model. Second, pipeline logic models like this do not clearly show the different causal strands that relate to the students and to the high schools but jumble them together. For example, the activity of professional development relates to improving teachers' skills so that they can then teach the students, whereas some of the other activities relate directly to students. Finally, the logic model does not show a very important feedback loop, where success with an initially small group of students might lead to success with a larger group of students in subsequent years (a reinforcing loop).

There are ways of adapting pipeline logic models to better represent complicated and complex aspects of the project, as we see later in this chapter. If the activities are not all at the beginning of the causal chain, a stacked pipeline can be drawn showing several sequences of input-activities-outputs. If there are separate causal pathways, a logic model can be drawn with distinct causal strands. An arrow doubling back to the beginning can be added to show a reinforcing feedback loop.

However, an outcomes chain logic model is often more effective in explaining how an intervention is understood to work and identifying variables that should be included in an empirical evaluation.

An Outcomes Chain Logic Model of a Computer Project

The distinguishing feature of an outcomes chain logic model is that every box is a result: an initial result, a subsequent result, or a final result of interest (Figure 9.2). The model might represent this project in terms of a causal chain that goes from left to right (or top to bottom or bottom to top). It shows the chain of results but not the activities that produce or contribute to them. Activities are instead shown in the detailed description of the theory of action (discussed in Chapter Eight).

This logic model in Figure 9.2 explains how the project is understood to bring about the intended educational and occupational objectives. Students who engage in the summer courses develop human capital

Figure 9.2 **An Outcomes Chain Logic Model of the Computer Project**

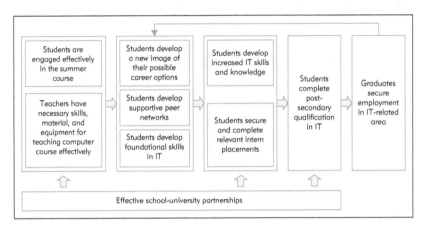

(skills and knowledge), social capital (supportive peer networks), and a new vision of their possible future. The skills and knowledge they develop in the form of basic skills in particular software make it possible for them to undertake intern placements. These placements, and the support they gain from supportive peer networks, help them to continue to increase their information technology skills and knowledge. This makes it more likely that they will complete postsecondary qualifications in information technology and go on to work in a related area. These successful participants then become role models helping new students to imagine this as a possibility for themselves.

Drawing this as an outcomes chain also makes it easier to show how the university-school partnerships are understood to contribute to the success of the program throughout the process: by developing teachers' skills to deliver the summer course, helping to set up internships, and helping them to succeed in their college applications.

Comparing Pipeline Models and Outcomes Chain Models

While outcomes chain models are usually more successful in articulating both the theory of change and theory of action, there are some instances in which a pipeline model might be appropriate.

For some organizations, developing even a simple logic model could be a major step forward. Some projects and programs have continued for years without explanation of what it is they are supposed to achieve. Their monitoring and evaluation systems focused only on counting activities have reinforced the focus on what is being done rather than on why. When the United Way introduced its guide to logic models in the 1990s, it was in a context in which many services had been delivered for years by volunteers without articulation of how these were intended to contribute to longer-term and broader impacts. They found that introducing a simple pipeline form of a logic model helped staff, long-term volunteers, and new volunteers move to a focus on results (Hendricks, Plantz, and Pritchard, 2008).

However, after two decades of corporate management, results-based management, activity-based costing, and new public management, many organizations are not in this situation of undertaking activities without having to explain why. They have already articulated their mission, goals, and objectives. Repackaging these into the four or five boxes of a pipeline model with slightly different terminology often adds little value and can crowd out a more useful way of representing program theory.

Another possible advantage of using a pipeline logic model is that its fixed format makes it easier to produce logic models that can be readily understood across an organization. Without care, outcomes chain logic models can become so-called spaghetti models of boxes and lines that are incomprehensible to anyone not involved in creating them.

Pipeline models are more likely to be appropriate for interventions where the causal mechanisms are well understood and all of the activities occur at the start of the process. Then a series of outcomes can be confidently assumed to flow from that.

The absence of explicit activities in outcomes chains can become a disadvantage. If those developing the program theory become too absorbed by what they would like the program to be achieving and the ideal chain of outcomes, they may forget to pay adequate attention to what the program actually does. In this case, a disconnect between the theory of change and the theory of action may occur.

Realist Matrices

Realist matrices are not an alternative to pipeline or outcomes chain logic models but a complementary approach that focuses on one or more causal mechanisms in a program theory and explores what it is about the program that makes this causal mechanism work.

This is not simply a matter of identifying moderators (contextual variables that affect the result). A realist approach to causal mechanisms includes attention to both agency (explanations that focus on the individual's decisions and actions) and structure (explanations that focus on the social and political situation). Programs are understood to provide resources or to change contexts to which participants respond; it is the interaction between resources and response that creates outcomes. Outcomes arise from a combination of "the stakeholders' choices (reasoning) and their capacity (resources) to put these into practice" (Pawson and Tilley, 1997, p. 66). A realist analysis also pays attention to identifying contexts within which these mechanisms will operate and those where they will not, addressing complicated aspects of interventions in terms of, "What works for whom in what situations?"

The computer project has a number of different intermediate outcomes where it might be useful to understand the mechanisms and the contexts within which they operate. Examples are how teachers become sufficiently competent in delivering the summer program, how the program succeeds in engaging appropriate students, and how students achieve the level of expertise needed to obtain internships. We now look at the last one of these.

A realist matrix (Table 9.1) would show the causal mechanisms understood to be involved in this process and the contexts within which they are likely to operate. For example, not all students will be able to develop IT skills in a summer program. The program will work only for conscientious students who attend regularly and engage actively with the program and who have strong literacy and numeracy skills. They have no prior computer experience but will quickly learn IT skills. So even within the one school, there will be different contexts—in this case, the characteristics of

Table 9.1   **A Realist Matrix Logic Model of the Computer Project**

Context	Mechanism	Outcome
Conscientious students with strong literacy and numeracy skills but no prior computer experience	Skill development	Achieve a threshold level of computer skills that makes it possible to start an internship
Students with literacy and numeracy problems and poor behavior	Skill development does not occur.	Do not achieve a threshold level of computer skills
Students with existing computer experience, skills, and confidence	Skill development does not occur.	No change to level of skills

the students. A realist matrix shows the contexts in which these mechanisms work and the contexts in which they do not.

Narratives

Narrative representations of program theory can be extremely important. Not everyone likes diagrams, and for these people, a narrative account will be preferable to any of the types of logic models. Narrative representations can be particularly appropriate for communicating with people who have not been involved in developing the program theory and do not find a diagram engaging. Sometimes people like developing diagrams to help clarify their own thinking but do not find others' diagrams particularly useful and prefer to follow a logical argument set out in narrative terms.

In many cases, a narrative can be a useful complement to a diagram. It is easy to miss key parts of a program theory if it is communicated only through diagrams, as there is not usually enough detail in a diagram for someone who is not familiar with that type of intervention. Moreover, it can be difficult to understand all the meanings and implications of the connecting arrows without further explanation. Nevertheless, a diagram can provide a structure that makes a narrative more coherent. (In Chapter Sixteen, we discuss how to use a narrative representation of the program theory, together with evidence from the evaluation, to report findings.)

Ideally the narrative will not only describe a sequence of events but communicate the essential features of the theory of change and the theory of action. A narrative that simply describes the sequence for the computer project, identifying intermediate outcomes, but not how the project brings these about, might go something like this:

> The project provides disadvantaged students with access to courses in software through a computer summer school, together with follow-up support, to improve their skills, knowledge, motivation, and confidence in using IT in order to improve their long-term educational and occupational attainment. Successful students will also provide role models for subsequent students.

A narrative that explained the program theory, what it was about the program that made it work, and in what contexts it was likely to work might go something like this:

> The computer project works with conscientious students from disadvantaged backgrounds who do not have access to computers. As students develop foundational IT skills, opening up opportunities to undertake internships, they develop a new image of their possible future. At the same time, they are creating a supportive community of students who make a safe space to learn and to make, and implement, ambitious plans about their future education and career.

Developing More Than One Logic Model

It may well be that different approaches to representation are needed at different times, for different purposes, and with different people. For example, you might need to develop a pipeline logic model version using a particular format to report to partner organizations and funders, even if you use an outcomes chain internally to provide more detail about how the intervention works or complement a diagram with a narrative. It can be helpful to have a simple logic model that fits easily on one page for the executive summary and a more detailed one with supporting evidence in a more detailed evaluation report. Having both is usually best, but this depends on resources, the availability of evidence, and the history of development of the program theory.

You might also develop different versions over time. For example, you might develop a rough version as part of the process of articulating people's mental models and then streamline it for subsequent use, or you develop a broad version for general use and then a more detailed version for frontline service deliverers. An intervention with important complicated aspects might usefully be represented by an overall diagram, together with close-ups of particular levels, components, or contexts. Some software programs (which we discuss in Chapter Thirteen) can easily provide this option. For an intervention with important complex aspects, it might be useful to begin with a broad diagram and then fill in the details during implementation, both supporting and documenting the learning.

Options for Formatting Logic Models

Opinions differ about whether logic models should flow left to right, top to bottom, bottom to right, inside out or outside in, and whether there are advantages to producing them in color. The options are set out in Table 9.2.

REPRESENTING COMPLICATED PROGRAM THEORY

It might be important to adequately represent complicated aspects of program theory that relate to the intervention or to the situation. Logic models can be adapted to show interventions that involve multiple organizations working together in parallel or in sequence, involve multiple causal strands, or work in conjunction with particular contexts. We provide examples from real evaluations of each of these.

Table 9.2 Options for Formatting Logic Models

Option	Advantages	Disadvantages
Left to right	Follows the pattern of normal reading from left to right (in European languages) Emphasizes the production line metaphor of simple logic models (where this is an advantage)	Constrains how many levels of results can be shown unless a landscape page format is used, printing sideways across the page; therefore best suits a four- or five-box pipeline model or a variation on this
Top to bottom	Follows the pattern of normal reading from top to bottom (in European languages) More space than a left-to-right model to show multiple levels of results and connections between causal strands	Shows the final result at the bottom, which can focus attention primarily on the early results rather than the ultimate result
Bottom to top	Emphasizes the ultimate result Consistent with corporate planning representations and organizational charts More space than a left-to-right model to show multiple levels of results and connections between causal strands	Can be confusing if it is not familiar because it does not follow the pattern of normal reading from top to bottom (in European languages)

Inside to outside	In a circular diagram, can be used to emphasize how the one intervention contributes to a range of final results Can emphasize the holistic approach of an intervention	Can be confusing because it is not familiar
Outside to inside	In a circular diagram, can be used to emphasize how multiple interventions contribute to a shared final result Can emphasize the holistic approach of an intervention	Makes it hard to show contributions from other interventions and external factors Can be confusing because it is not familiar
Use color	Can be used to distinguish different stages of a pipeline logic model Can highlight different strands of a results chain, including showing how two different elements need to combine to produce the result (shown by combining yellow and blue to make green, for example) Can be used to highlight additional aspects of the logic model, such as sources of evidence	Makes it hard to show contributions from other interventions and external factors Loses information if it is photocopied in black and white Can add significantly to the cost of printing and distributing the logic model Adds to the amount of information readers have to take in; to be used only if it adds meaning, not just decoration Not essential to show stages, as these can usually be shown through use of columns

Multiple Organizations Working in Parallel

It can be very useful to show how different organizations contribute to producing impacts, and not necessarily as part of the same intervention. The example from Health Scotland in Figure 9.3 has a pipeline logic model for each agency and shows how these are expected to combine to produce the final outcomes. The multiple results chains are presented as a means of communicating the idea that different partners make different contributions to common or shared outcomes and can be used to represent the strategic priorities of an organization.

The contributions of different partners can instead be shown in terms of a vertically integrated outcomes chain. In Figure 9.4 the right hand side shows an outcomes chain for an interagency program to combat graffiti. The left hand side shows which agencies contribute most directly to each outcome. As the activities and their outcomes may be occurring concurrently rather than sequentially, it would not be appropriate to classify the outcomes as short, medium, or long term according to their position in the chain. Each of the outcomes might occur only to a small degree in the short term and then strengthen in the long term, especially as reinforcing feedback loops come into play.

The two examples shown in Figures 9.3 and 9.4 have complicated aspects. However, interdependencies among the outcomes being delivered by various partners may produce somewhat unpredictable results and then the programs will also have complex aspects.

Multiple Interventions Working Together Sequentially

Many interventions can achieve their objectives only if another intervention follows them. Showing this other intervention can be important for managing the transition for identifying appropriate indicators for evaluation. As in a relay race, the outcome depends on the performance of all the partners in the process, and if the baton is passed clumsily or dropped, there is often dispute about the allocation of blame.

One way to represent this type of intervention is to use a stacked pipeline logic model. The logic model in Figure 9.5, for a research and technology and deployment program developed for the U.S. Department of Energy (McLaughlin and Jordan, 1999), shows four linked pipeline logic models. The

Figure 9.3   **Complicated Logic Model Showing Multiple Organizations**

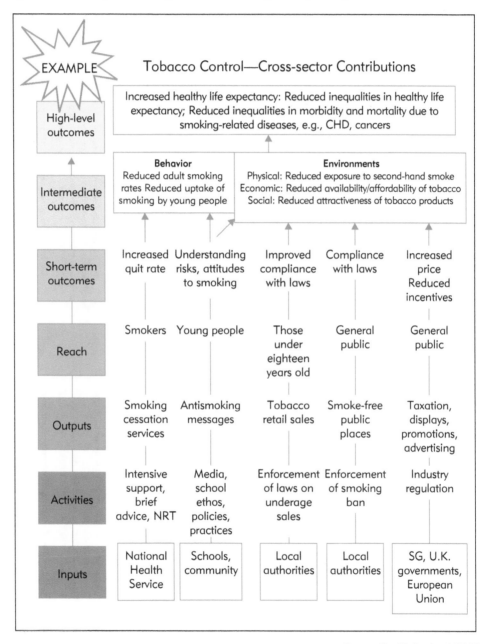

Source: Health Scotland (2009)
Note: CHD: coronary heart disease; NHS: National Health Service; SG: Scottish government.

Figure 9.4 Logic Model for a Program Involving Multiple Partner Organizations

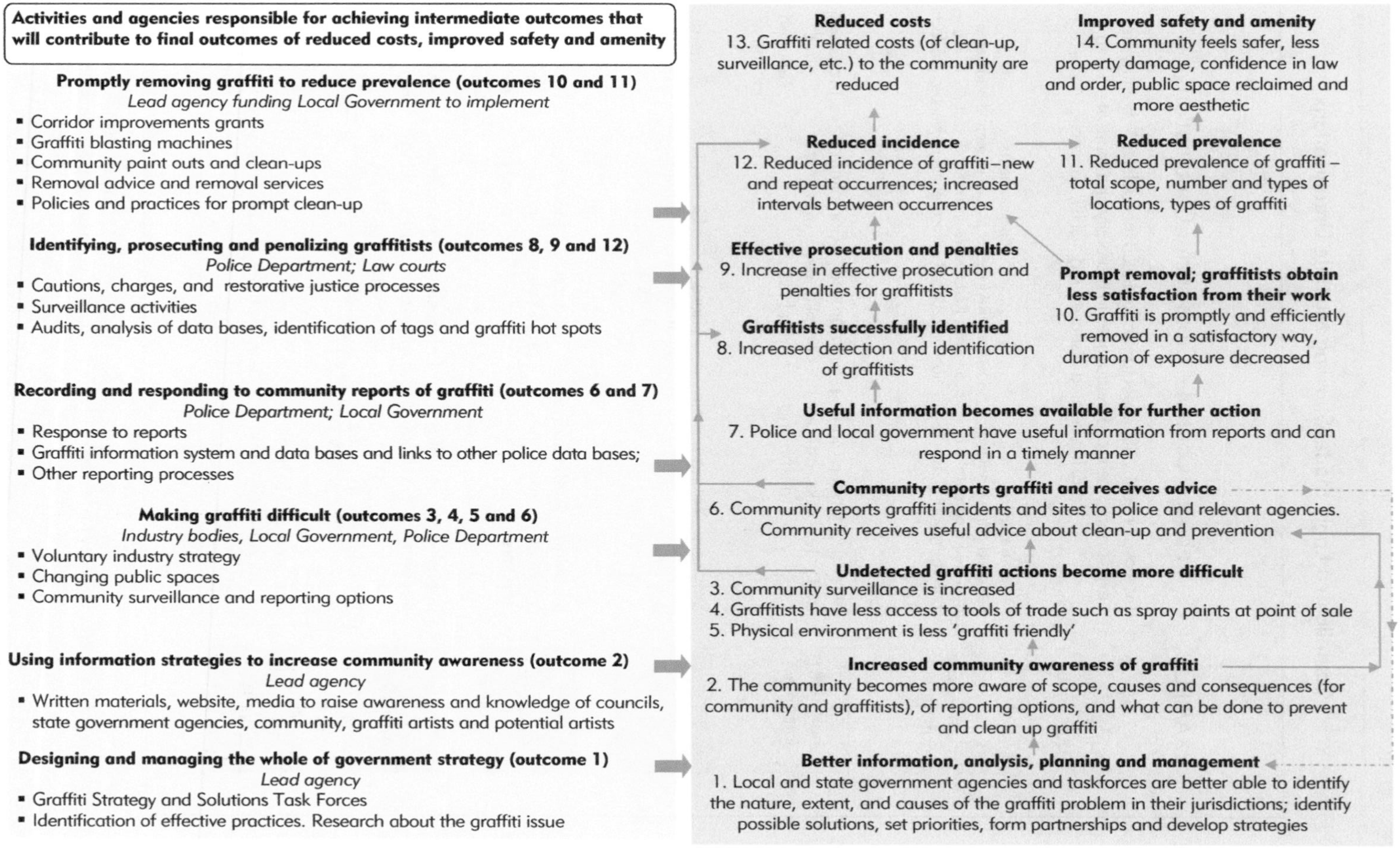

Activities and agencies responsible for achieving intermediate outcomes that will contribute to final outcomes of reduced costs, improved safety and amenity

Promptly removing graffiti to reduce prevalence (outcomes 10 and 11)
Lead agency funding Local Government to implement
- Corridor improvements grants
- Graffiti blasting machines
- Community paint outs and clean-ups
- Removal advice and removal services
- Policies and practices for prompt clean-up

Identifying, prosecuting and penalizing graffitists (outcomes 8, 9 and 12)
Police Department; Law courts
- Cautions, charges, and restorative justice processes
- Surveillance activities
- Audits, analysis of data bases, identification of tags and graffiti hot spots

Recording and responding to community reports of graffiti (outcomes 6 and 7)
Police Department; Local Government
- Response to reports
- Graffiti information system and data bases and links to other police data bases;
- Other reporting processes

Making graffiti difficult (outcomes 3, 4, 5 and 6)
Industry bodies, Local Government, Police Department
- Voluntary industry strategy
- Changing public spaces
- Community surveillance and reporting options

Using information strategies to increase community awareness (outcome 2)
Lead agency
- Written materials, website, media to raise awareness and knowledge of councils, state government agencies, community, graffiti artists and potential artists

Designing and managing the whole of government strategy (outcome 1)
Lead agency
- Graffiti Strategy and Solutions Task Forces
- Identification of effective practices. Research about the graffiti issue

Improved safety and amenity
14. Community feels safer, less property damage, confidence in law and order, public space reclaimed and more aesthetic

Reduced prevalence
11. Reduced prevalence of graffiti – total scope, number and types of locations, types of graffiti

Reduced costs
13. Graffiti related costs (of clean-up, surveillance, etc.) to the community are reduced

Reduced incidence
12. Reduced incidence of graffiti–new and repeat occurrences; increased intervals between occurrences

Prompt removal; graffitists obtain less satisfaction from their work
10. Graffiti is promptly and efficiently removed in a satisfactory way, duration of exposure decreased

Effective prosecution and penalties
9. Increase in effective prosecution and penalties for graffitists

Graffitists successfully identified
8. Increased detection and identification of graffitists

Useful information becomes available for further action
7. Police and local government have useful information from reports and can respond in a timely manner

Community reports graffiti and receives advice
6. Community reports graffiti incidents and sites to police and relevant agencies. Community receives useful advice about clean-up and prevention

Undetected graffiti actions become more difficult
3. Community surveillance is increased
4. Graffitists have less access to tools of trade such as spray paints at point of sale
5. Physical environment is less 'graffiti friendly'

Increased community awareness of graffiti
2. The community becomes more aware of scope, causes and consequences (for community and graffitists), of reporting options, and what can be done to prevent and clean up graffiti

Better information, analysis, planning and management
1. Local and state government agencies and taskforces are better able to identify the nature, extent, and causes of the graffiti problem in their jurisdictions; identify possible solutions, set priorities, form partnerships and develop strategies

Figure 9.5   **Stacked Logic Model**

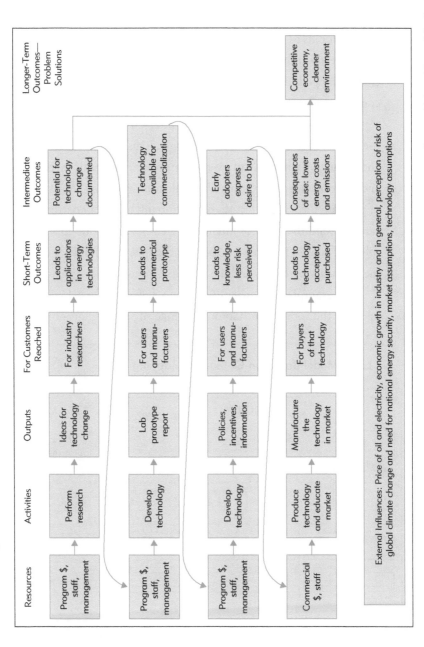

Source: McLaughlin and Jordan (1999)

first layer relates to basic research that identifies the potential for technological change. The outcomes from this become inputs to the second layer, which relates to the development of technology. The outcomes from this second layer become inputs to the third layer of commercialization of the technology, which in turn leads to the actual adoption of the new technology and consequential energy and pollution savings. This stacked logic model clearly shows the interconnections among the programs, as well as their separate accountabilities.

Another way to represent multiple interventions working sequentially is to use the concepts from outcome mapping. This is an approach to monitoring and evaluation consisting of twelve steps covering intentional design, outcome and performance monitoring, and evaluation planning. At its heart is a form of program theory that focuses on identifying boundary partners who need to carry forward the chain of intended outcomes and distinguishes between the sphere of control (one's own activities), the sphere of influence (the activities of the boundary partners), and the sphere of interest (the consequential results).

The example in Figure 9.6 shows a template for a logic model developed by the Indonesian Country Office of the Belgian nongovernmental

Figure 9.6   **Logic Model Based on Outcomes Mapping**

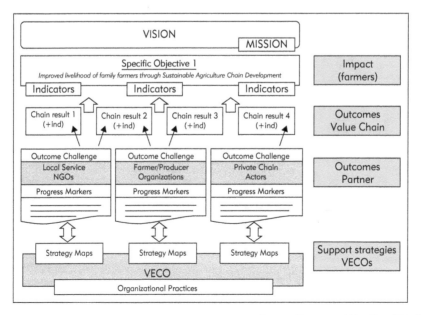

Source: Deprez and Van Steenkiste (2010)

Vredeseilanden (VECO Indonesia). Its program of sustainable agricultural chain development supports farmers producing healthy rice, cacao, coffee, cashew, and groundnuts in eastern Indonesia, Bali, and central Java. This template was then used across different objectives and outcomes value chains (different crops). This logic model shows the sphere of control (the support strategies VECO conducts), the sphere of influence (the outcome challenges of their outcomes partners), and the sphere of interest (the outcomes in terms of each value chain and the impact for farmers in terms of improved livelihoods).

Multiple Causal Strands Within an Intervention

Interventions often have multiple and potentially conflicting objectives, and it is important to represent these adequately in a logic model to ensure they are both addressed. The Danish evaluator Peter Dahler-Larsen (2001) has discussed how this might be done with reference to a program for unemployed people. The central theory of change is that increasing people's formal qualifications will improve their employability, which will lead to successful applications for jobs. However, a second causal strand is also operating: participants need to have sufficient self-confidence for their improved qualifications to transform into job-seeking behavior. We might represent this complicated logic model as shown in Figure 9.7.

The program will be successful only if both causal paths are successfully completed. This can be achieved by accepting only participants who already meet this requirement or by including strategies within the intervention to

Figure 9.7 **Complicated Logic Model Showing Multiple Causal Strands**

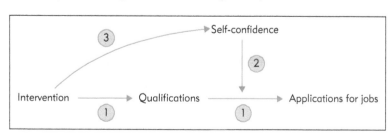

Source: Dahler-Larsen (2001, p. 341)

improve participants' self-confidence. If, however, a simple program theory that focuses only on qualifications is used, the risk is that the program will not improve participants' self-confidence and may inadvertently decrease it. For example, a strict, mandatory program focused on participants' deficits rather than their strengths might convey the message that participants have few capacities and are not competent to manage their own lives, undermining the self-confidence that is needed. (This is an example of a negative program theory, as discussed in Chapter Six, which explains how an intervention might contribute to negative outcomes.) Dahler-Larsen (2001) refers to these as "tragic" programs, that is, they unknowingly inhibit the change they are seeking to achieve. Conversely, programs that affect a moderator so that it helps enact an idea in the program theory, acting as positive, self-fulfilling prophecies, can be labeled as "magic": they change the situation so that a causal path that was not possible becomes possible.

In Chapters Seven and Eight, we discussed an alternative theory of change for an employment program. In this example, the theory of change was that participants needed to have enough self-esteem in general terms and motivation to become employed in order for them to commit to developing job-related skills and job-seeking skills. For some unemployed people, there would be a need to develop their sense of self-worth around issues that were initially not directly related to employment. In this example, different issues might arise. To what extent would they need to develop self-confidence and job motivation in order for them to be responsive to opportunities to develop job skills? To what extent does actively working on development of job skills even where there is no job motivation help with the development of that motivation?

Multiple causal strands might relate to different parts of an intervention. The logic model developed for the Recognised Seasonal Employer (RSE) Policy of the New Zealand government demonstrates the value of identifying multiple, interacting causal paths that need to be achieved for success (Figure 9.8). The RSE Policy allows the temporary entry of offshore workers, especially from the Pacific Islands, to work in the New Zealand horticulture and viticulture industries. The systems logic model developed for the policy was designed to show that while the needs of New Zealand industries and employers were the driving force behind the policy, its success was dependent

Figure 9.8   Logic Model Showing Two Different Sets of Intended Outcomes

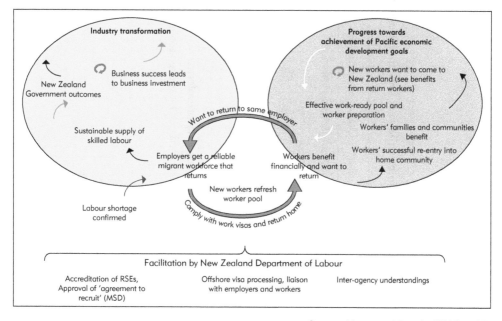

Industry transformation

Business success leads
to business investment

New Zealand
Government outcomes

Sustainable supply of
skilled labour

Want to return to same employer

Employers get a reliable
migrant workforce that
returns

New workers refresh
worker pool

Labour shortage
confirmed

Comply with work visas and return home

Progress towards
achievement of Pacific economic
development goals

New workers want to come to
New Zealand (see benefits
from return workers)

Effective work-ready pool and
worker preparation

Workers' families and communities
benefit

Workers benefit
financially and want to
return

Workers' successful re-entry into
home community

Facilitation by New Zealand Department of Labour

Accreditation of RSEs,
Approval of 'agreement to
recruit' (MSD)

Offshore visa processing, liaison
with employers and workers

Inter-agency understandings

Source: Nunns and Roorda (2009, p. 29)

on its also meeting the needs of the Pacific workers (Department of Labour, 2010; Nunns and Roorda, 2009).

If the interactions between these different sets of intended outcomes cannot readily be predicted (for example, where an imbalance between the two systems precipitates other unexpected changes in the other), it might be useful to also consider these as complex aspects and watch carefully for unexpected results.

Interventions That Work Only in Particular Contexts

Very few representations of program theory show clearly the difference that context can make to outcomes, although some include a box where different types of relevant contexts are listed. Realist matrices specifically focus on this aspect of complication. The example in Table 9.3 represents the Priority Estates Program (PEP), implemented in London and Hull in the United Kingdom with an experimental and a comparison housing

Table 9.3 A Realist Summary of Foster and Hope's Study of the Effects of PEP on Crime in the Hull Experimental Estate

Context	+	New Mechanisms	=	Outcome Pattern
Area A: Poor-quality, hard-to-let housing; traditional housing department; lack of tenant involvement in estate management	+	Improved housing and increased involvement in management create increased commitment to the estate, more stability, and opportunities and motivation for social control and collective responsibility	=	Reduced burglary prevalence
Area B: Poor-quality, hard-to-let housing; traditional housing department; lack of tenant involvement in estate management	+	Increased involvement in management creates conditions for increased territoriality, but unattractive housing maintains high turnover, retaining vulnerable families	=	Stable burglary prevalence concentrated on the more vulnerable

Area C:
Three tower blocks, occupied mainly by the elderly; traditional housing department; lack of tenant involvement in estate management

+ Concentration of elderly tenants into smaller blocks and natural wastage creates vacancies taken up by young, formerly homeless single people inexperienced in independent living. They become the dominant group. They have little capacity or inclination for informal social control and are attracted to a hospitable estate subterranean subculture in the housing

= Increased burglary prevalence concentrated among the more vulnerable; high levels of vandalism and incivility

Overall
Traditionally managed, low-quality housing in hard-to-rent public housing with a criminal subculture and some difficult families

+ Variations in changes wrought by PEP and tenant allocation trigger varying mechanisms in differing parts of the estate, leading to more social control in some areas, creating or reinforcing criminality in others

= Mixed patterns of increasing burglary; stable burglary and decreasing burglary according to area

Source: From Pawson and Tilley (1997, p. 101)

estate (public housing) in each city. The program involved changing the ways in which services on housing estates were delivered and managed, including increased local involvement. The impact evaluation of the Hull experimental estate found that the program had led to reduced crime in one area, stable rates in another area, and increased crime in another (Foster and Hope, 1993).

Pawson and Tilley (1997) drew on this evaluation to show how a realist matrix could summarize the multiple causal mechanisms and contexts involved in these results. Rather than showing a sequence of outcomes, linked by specific causal mechanisms, the table presents a number of mechanisms triggered by the program in the specific context to explain the observed differences in outcomes.

REPRESENTING COMPLEX PROGRAM THEORY

Complexity presents considerable challenges for most approaches to program theory that are based on the assumption that specific plans can be made to achieve an agreed goal. For example, Jonny Morell (2009), an experienced evaluator and editor of the journal *Evaluation and Program Planning*, which has published many papers on using program theory and logic models, has described the view that logic models are not appropriate for programs that are deliberately poorly specified (where they are being developed through rapid prototyping) or where they are unstable, because models imply stability. We think otherwise. Clearly, when program theory is used for complex, evolving interventions, it cannot be a static blueprint for action and accountabilities, but we have found that it can still be a useful guide for planning and evaluation. For interventions with important complex aspects, logic models, and the program theories they represent, need to be used as organizing heuristics rather than as a formula for implementation.

Five Options for Addressing Complexity

There are five quite different ways of using and representing program theory for interventions with significant complex aspects (Table 9.4). One way

Table 9.4 **Options for Representing Interventions with Important Complex Aspects**

Option	Description
1. Fixed ultimate outcome and emergent program theory	Iteratively developed and revised during implementation
2. Structured processes for developing emergent program theory	Program theory developed using structured processes such as U-process or Appreciative Inquiry
3. Generic theory of change with emergent theory of action	Program theory initially focuses on articulating the theory of change, and the theory of action is emergent
4. Vertical integration of outcome chains from different agencies	Program theory represents how different agencies will work together to support adaptive implementation
5. Diagrams and concepts from systems approaches	Diagrams such as sociograms and systems dynamics diagrams and simple rules from complex adaptive systems

is to set sail with a clear end point in mind but with no defined plan of how to get there and to develop, use, and revise program theory along the way to create this plan during implementation. This is what South African development practitioner Doug Reeler (2007) has called "a theory of emergent change," where "we make our path by walking it." An essential component of this approach is to have effective processes for regular review and revision of plans and program theories. A slightly different approach is to develop emergent program theory using structured and prespecified processes. A third way is to have a broad program theory and iteratively develop a theory of action. For interventions that are both complicated (with many contributing agencies) and complex (with unpredictable synergies and interactions), a logic model that shows vertical integration can be used as a heuristic to support adaptive implementation. Finally, diagrams and concepts from systems approaches can be used, such as sociograms (from network modeling), systems dynamics diagrams, and simple rules from complex adaptive systems.

Fixed Ultimate Outcome, Emergent Intermediate Program Theories

In these situations, the intended impact of an intervention is fixed, but the means to achieve it are open. This draws on the notion of tight ends and loose means outlined by Peters and Waterman (1992) and the subsequent mantra of "let the managers manage," addressing accountability in terms of incentives for achieving objectives rather than compliance with specific procedures or detailed plans. It is also congruent with agile project management (Cockburn, 2001), an approach to software development that does not assume that specific requirements can be identified adequately at the start and then remain stable, or that software development is then primarily about compliance with specifications. Instead, agile project management explicitly uses notions of complex adaptive systems, focusing on "setting the direction, establishing the simple, generative rules of the system, and encouraging constant feedback, adaptation, and collaboration" (CC Pace 2003–2009).

An example of this type of use of program theory comes from an intergovernmental initiative to improve justice-related outcomes for Aboriginal people in Western Australia (Williams, 2008). The higher levels of social disadvantage and imprisonment among Indigenous people is an example of a wicked problem, where there is dispute about the causes of the problem, many different organizations need to collaborate to solve the problem but there is no clear way to do this, and the risk of unintended negative outcomes is high. In such situations, an ongoing process of developing, using, and revising program theory might be the best way forward, without an expectation of producing an enduring blueprint. A logic model was developed that had clear ultimate outcomes and a framework for the intermediate outcomes (in terms of activities creating enabling outcomes), but with the intention of developing the details in the middle through ongoing consultation between the different stakeholders involved and fast cycles of iterative planning, implementation, and evaluation.

This approach to dealing with complexity is most likely to be successful for small interventions with a small number of stakeholders who can be engaged in the journey of iteratively developing and revising program theory. It will be more difficult to use this approach when dealing with a large number of bureaucracies with a need for more clarity about direction and more lag time in agreeing to emergent changes.

Structured Processes for Developing an Emergent Program Theory

A number of different procedures can be used to structure the process of developing an emergent theory of change, including Appreciative Inquiry and the U-process.

Appreciative Inquiry (Cooperrider and others, 1995; Cooperrider, Whitney, and Stravos, 2002) is an approach to organizational development that has also been used for evaluation (Preskill and Catsambas, 2006; Preskill and Coghlan, 2004). Rather than focusing on the problems and gaps in an organization, Appreciative Inquiry works through a process of identifying strengths and then developing plans to build on these. The approach is often worked out in practice by using the 4-D model:

> *Discover*—people talk to one another, often via structured interviews, to discover the times when the organization is at its best. These stories are told as richly as possible.
> *Dream*—the dream phase is often run as a large group conference where people are encouraged to envision the organization as if the peak moments discovered in the discover phase were the norm rather than exceptional.
> *Design*—a small team is empowered to go away and design ways of creating the organisation dreamed in the conference(s).
> *Destiny*—the final phase is to implement the changes [Seel, 2008, pp. 7–8].

An elaboration of this, the Seven D approach to community capacity development (Figure 9.9), systematically works through iterative cycles involving seven components: developing relationships, discovery, dream, directions, design, delivery, and documentation:

Figure 9.9   **The Seven D Approach to Developing and Using an Emergent
Program Theory**

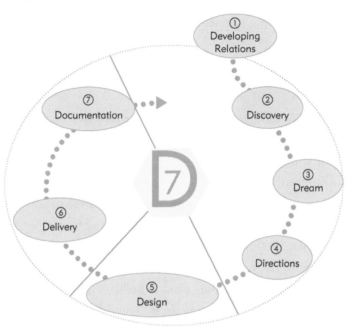

Source: Dhamorathan (2009)

The first steps focus on developing trustful, respectful relationships, and encouraging community members to imagine a desired
future. This motivates them to initiate necessary actions. The next
steps are intended to strengthen their ability to analyse their potential and the challenges they face, to reach consensual decisions on
collective action and reflect on the outcomes. Throughout the
process, behavioural change is envisaged as an incremental, collective, continuous effort by everyone involved by creating a space
in which desired behavioural patterns can emerge [Dhamorathan,
2010, p. 7].

The U process is another structured approach to developing emergent
program theory, although one with a diametrically different approach to
Appreciative Inquiry (Figure 9.10). It begins by convening leaders from a
poorly performing system who are committed to changing it. They then

Figure 9.10   **The U Process Theory of Change**

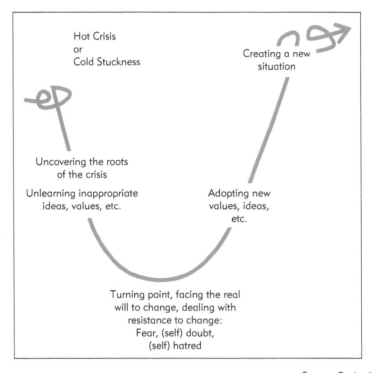

Source: Reeler (2007, p. 12)

work through three stages. In co-sensing, they develop a deep shared understanding of the current reality. In co-presencing, they develop clarity and commitment about what they need to do to create a new reality. In co-realizing, they build prototypes and plots that begin to change the system (Generon Consulting, 2005).

Theory of Change, with Emergent and Iterative Theory of Action

The third alternative is to have a broad theory of how change will come about, which is operationalized quite differently in specific situations in response to emerging needs and opportunities—that is, to have a clear theory of change with an emergent theory of action.

The example in Figure 9.11 comes from an agricultural research program that involves clear theories of horizontal scaling up (other villages use new

Figure 9.11   **A Broad Theory of Change with Emergent Theory of Action: Dealing with Striga**

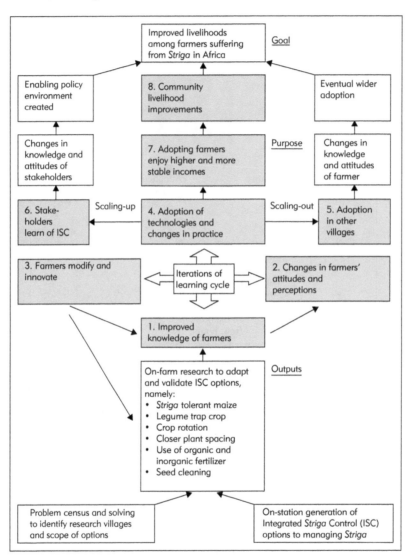

Source: Douthwaite, Kuby, van de Fliert, and Schulz (2003)

agricultural methods) and vertical scaling up (involving different levels of government and other organizations). Iterative cycles of learning and adaptation in the middle of the process deal with the uncertainty around the wicked problem of a parasitic weed, striga.

The example in Figure 9.12 is a logic model that was used to represent the program theory of evaluation capacity development efforts. These were iterative and opportunistic but within a conceptual framework that clearly builds on the community capacity-building archetype that we discuss in Chapter Twelve. This generic logic model could be customized to include details of the specific activities undertaken, such as developing templates and materials and providing training, and could be elaborated to articulate the ways by which evaluation contributes to improved programs, such as informing resource allocation, identifying problems, and demonstrating unmet needs.

Being creative about the shape and structure of the logic model can make it easier to show cyclic, iterative, and evolving processes. Andrea Johnston, from Neyaashiinigmiing, Ontario, Canada, developed the Waawiyeyaa Evaluation Tool for the Tending the Fire program, which works with Aboriginal men with multiple needs, including issues to do with substance abuse, involvement in the criminal justice system, housing, employment,

Figure 9.12   **A Theory of Change with Emergent Theory of Action: Evaluation Capacity Building**

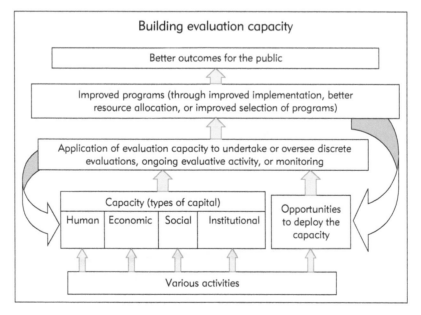

Source: From Rogers (2004)

and emotional healing (Figure 9.13). *Waawiyeyaa* is an Ojibway word that refers to a circular process that can lead to rebirth and transformation. The evaluation tool represents the program as five circles, from crisis through awareness, ownership, releasing/letting go, building on strengths, and standing tall. Each of these involves "four stages of growth [that] include recognizing needs, planning and willingness to move forward, putting plans into action, and outcomes of the actions" (Johnston, 2010). It is intended as a self-evaluation tool that can be used to gather, make sense of, and report

Figure 9.13   A Theory of Change with Cyclic Learning at Each Stage

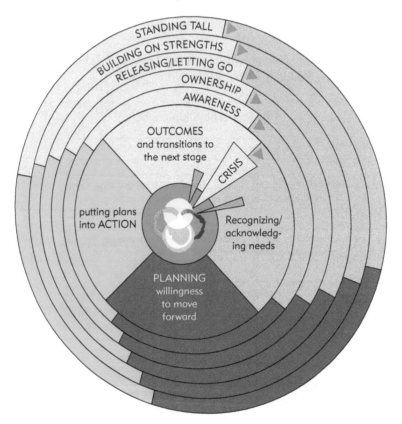

TENDING THE FIRE

Program Model

Source: From Johnston Research, Inc. (2010)

stories and other evidence about processes and outcomes, in a way that is accessible to program participants.

An adaptive approach can be taken to an individual service or a responsive, individually tailored set of services. For example, the logic model in Figure 9.14 represents the program theory of the Kalamazoo Wraps initiative, one of the more than 140 Substance Abuse and Mental Health Services Administration systems of care. This is a coordinated network of community-based services and supports organized to meet the challenges of children and youth with serious mental health needs and their families. The initiative involved responding to needs and opportunities within a clearly stated framework of principles and core values that required services to be family driven, youth guided, culturally and linguistically competent, individualized and community based, and evidence based. The logic model was developed by a large group of community stakeholders—including young mental health consumers and their parents—led by Peter Dams; evaluators Carolyn Sullins and Becca Sanders then used the logic model to guide stakeholders in developing evaluation questions and methods for answering them.

Vertical Integration of Outcome Chains from Different Agencies

Interagency collaborative programs might be essentially complicated if the relationships are predictable and formalized. They have complex aspects when these relationships are adaptive and emergent, or when the outcome chains from the different agencies interact in ways that produce somewhat unpredictable results. The vertical integration style of logic model we showed earlier to represent the interactions between different organizations can be used as a heuristic to support adaptive implementation.

Diagrams and Concepts from Systems Theory

As systems thinking becomes more widely used in evaluation (Williams and Imam, 2007; Williams and Hummelbrunner, 2010), many of the diagrams and concepts are being used in representing program theory. Here we discuss three of these: network theory, systems dynamics, and simple rules.

Network Theory   Network theory (discussed in Chapter Eleven) focuses on relationships among people, organizations, global networks, and so on. These

Figure 9.14 A Theory of Change with Emergent Theory of Action: Mental Health

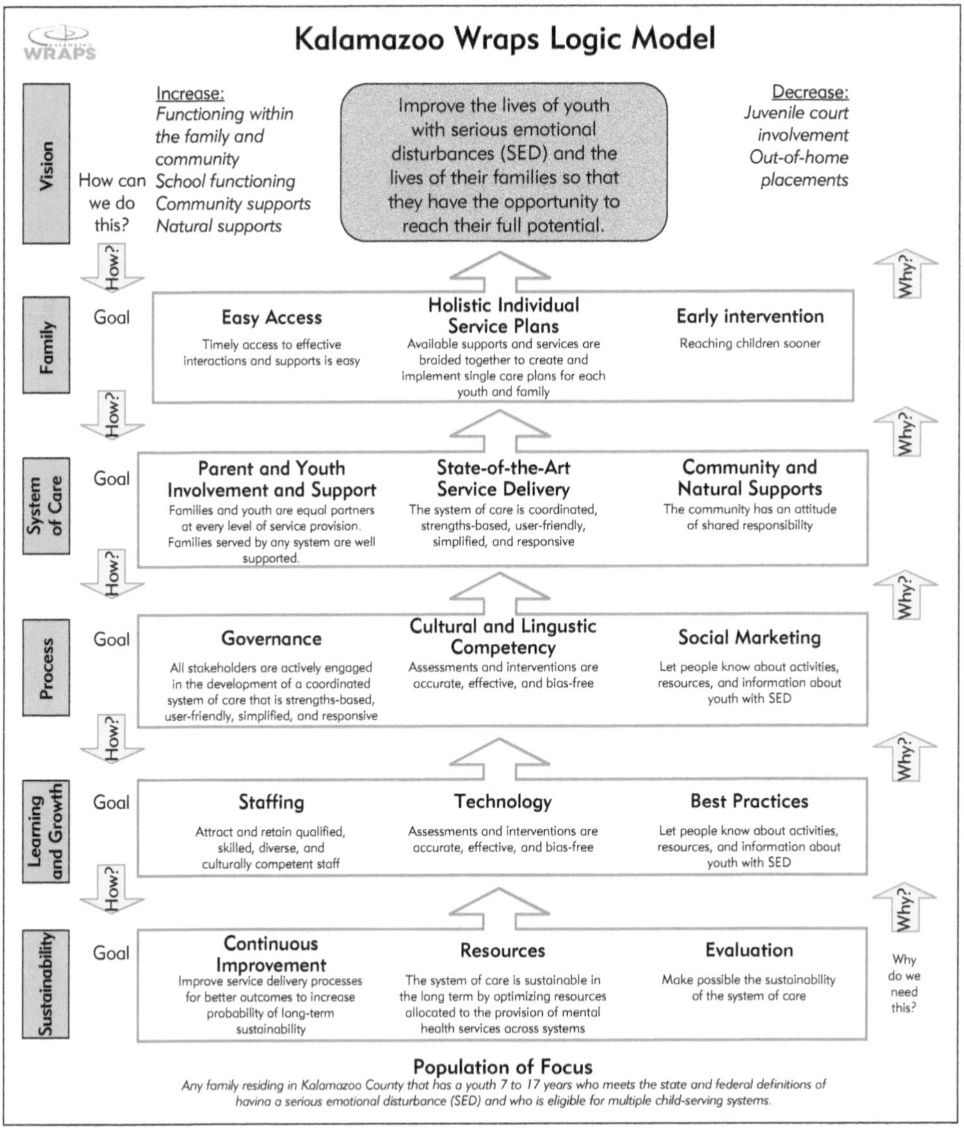

Kalamazoo Wraps Logic Model

Vision	How can we do this?	**Increase:** *Functioning within the family and community* School functioning Community supports Natural supports	**Improve the lives of youth with serious emotional disturbances (SED) and the lives of their families so that they have the opportunity to reach their full potential.**	**Decrease:** *Juvenile court involvement Out-of-home placements*

Family	Goal	**Easy Access** Timely access to effective interactions and supports is easy	**Holistic Individual Service Plans** Available supports and services are braided together to create and implement single care plans for each youth and family	**Early intervention** Reaching children sooner
System of Care	Goal	**Parent and Youth Involvement and Support** Families and youth are equal partners at every level of service provision. Families served by any system are well supported.	**State-of-the-Art Service Delivery** The system of care is coordinated, strengths-based, user-friendly, simplified, and responsive	**Community and Natural Supports** The community has an attitude of shared responsibility
Process	Goal	**Governance** All stakeholders are actively engaged in the development of a coordinated system of care that is strengths-based, user-friendly, simplified, and responsive	**Cultural and Linguistic Competency** Assessments and interventions are accurate, effective, and bias-free	**Social Marketing** Let people know about activities, resources, and information about youth with SED
Learning and Growth	Goal	**Staffing** Attract and retain qualified, skilled, diverse, and culturally competent staff	**Technology** Assessments and interventions are accurate, effective, and bias-free	**Best Practices** Let people know about activities, resources, and information about youth with SED
Sustainability	Goal	**Continuous Improvement** Improve service delivery processes for better outcomes to increase probability of long-term sustainability	**Resources** The system of care is sustainable in the long term by optimizing resources allocated to the provision of mental health services across systems	**Evaluation** Make possible the sustainability of the system of care

Population of Focus
Any family residing in Kalamazoo County that has a youth 7 to 17 years who meets the state and federal definitions of having a serious emotional disturbance (SED) and who is eligible for multiple child-serving systems.

Source: From Kalamazoo Wraps (2010)

relationships can be powerful drivers of change. Also, interventions can be designed to draw on or influence the relationships. Interventions that are expected to work through changing the relationships between organizations (for example, establishing or strengthening collaborative networks) may seek to use network diagrams to represent the existing and intended relationship structures they are seeking. The most common means of portraying networks has been through graphs, matrices, and sociograms (Durland, 2005). Participatory impact pathways analysis (Douthwaite and others, 2007; Douthwaite, Alvarez, Theile, and Mackay, 2008) explicitly includes a step where participants draw both the current status of relationships with other organizations and the intended future status. An example is increasing the number and strength of direct relationships between organizations rather than their being mediated through a central hub organization.

Systems Dynamics   Traditional logic models do a poor job of representing iterative relationships. Systems dynamics diagrams (Figure 9.15) are particularly useful in these cases:

- Representing stocks and flows, that is, accumulations and their rates of change)

- Reinforcing loops, where an initial effect is amplified over time—for example, initial success leads to greater effort which increases success, and so on

- Balancing loops, where an initial effect weakens over time as compensatory processes take effect—for example, initial success leads to complacency and less effort, which leads to subsequent poor performance, which leads to increased attention and effort and hence improved performance, and so on

The purpose of these models is not to make a simple prediction of the future situation but to engage a stakeholder group in discussions that are informed by an understanding of the possibly counterintuitive results of interacting iterative causal loops (Richardson, 2007; Dyehouse and others, 2009).

Simple Rules   Another option is not to represent program theory in a diagram at all but in the form of simple rules or principles within which program theory will emerge. Flocking birds are often used as an example to show how the application of simple rules addressing separation, alignment, and cohesion (avoid

Figure 9.15   A Systems Dynamics Model of a Tobacco Control Initiative

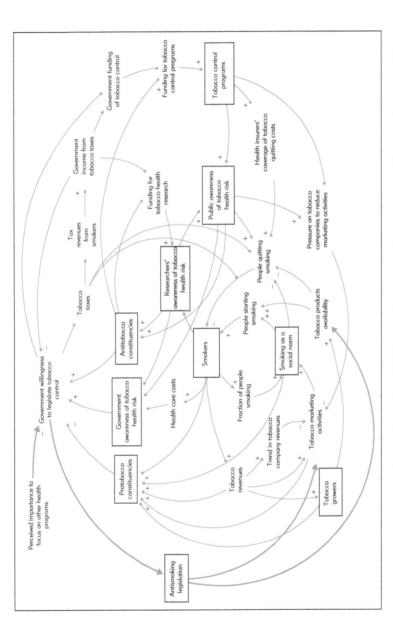

Source: Richardson (2007). An earlier version of this figure was published as Figure 1 in A. Best et al. 2006. Systemic transformational change in tobacco control: An overview of the Initiative on the Study and Implementation of Systems (ISIS), in *Innovations in Health Care: A Reality Check*, eds. A. Casebeer, A. Harrison, and A.L. Mark. Houndmills, UK: Palgrave Macmillan, Reproduced with permission of Palgrave Macmillan.

neighbors, steer toward the average heading of neighbors, and steer toward the average position of neighbors) produces complex, adaptive behavior.

The Harlem Children's Zone (2009) is a community-based organization that provides education, social services, and community-building programs to children and families living in an area in Harlem that spans over one hundred city blocks. It provides an example of a highly adaptive program that operates in terms of a set of five principles:

1. Serve an entire neighborhood comprehensively and at scale.

2. Create a pipeline of support.

3. Build community among residents, institutions, and stakeholders who help to create the environment necessary for children's healthy development.

4. Evaluate program outcomes and create a feedback loop that cycles data back to management for use in improving and refining program offerings.

5. Cultivate a culture of success rooted in passion, accountability, leadership, and team work.

WHAT MAKES A GOOD REPRESENTATION OF PROGRAM THEORY

It can sometimes be hard for those who have developed a program theory representation to view it critically. As we discussed earlier, this might be due to the IKEA effect, where people develop an irrational attachment to something they have put together. In addition, a logic model that is incomprehensible or misleading for outsiders might be seen as valuable and insightful by those who created it, who can fill in its gaps and correctly interpret ambiguous terms. However, there are risks in settling for a logic model that is agreed on by its developers without taking a more critical eye. Inevitably the model will be shown to, and used by, others, some of whom will read it literally and perhaps skeptically. Reviewing the logic model should be done throughout the process of developing it and formally as part of reviewing the program theory (which we discuss in Chapter Ten).

A logic model has three essential features: it is a coherent causal model (a comprehensible explanation of the causal processes that are understood to lead to the outcomes), it is logical (subsequent outcomes are plausibly consequential), and it communicates clearly. These three simple ideas provide a powerful filter for assessing logic models. When we were researching this book, we reviewed many logic models that were publicly available in published reports and articles and on Web pages. A surprising number of them failed to meet this simple standard. Many do not explain how the intervention works, which seriously limits the usefulness of the program theory. The logic model is essentially, "We get stuff, we do stuff, and stuff happens." Many of the diagrams are quite illogical, describing outcomes in terms that are in fact the reverse of what is meant or using arrows that cannot possibly be meaningful. Many do not communicate clearly; instead they present a logic model that does not provide a clear overview of the intervention. In Exhibit 9.1, we set out ten guidelines for drawing logic models that present a coherent causal model, are logical, and communicate the main messages clearly.

In the following sections we discuss each of these in more detail, show some poor examples (based on real logic models we have seen), and demonstrate some logic model makeovers. The examples all relate to a Meals on Wheels program. The program theory might be represented in narrative form as follows:

Meals on Wheels provides nourishing and culturally appropriate meals to elderly people who are unable to provide meals for themselves. The meals, together with daily contact by caring volunteers and early notice of accidents and dangerous situations, allow frail, homebound persons to avoid or delay moving to institutional residential accommodation and to remain in their own homes, where they want to be.

Guideline 1: Explain How the Intervention Contributes to the Results

A logic model needs to communicate a coherent message about how the activities undertaken contribute to the results of interest. One of the common

EXHIBIT 9.1 GUIDELINES FOR DEVELOPING EFFECTIVE
LOGIC MODELS

PRODUCE A COHERENT CAUSAL MODEL

1. Explain how the intervention contributes to the results.

2. Avoid dead ends.

BE LOGICAL

3. Make every arrow meaningful.

4. Indicate the direction of expected change.

5. Clearly show sequential and consequential progression.

COMMUNICATE CLEARLY

6. Focus on the key elements.

7. Avoid too many arrows and feedback loops.

8. Remove anything that does not add meaning.

9. Ensure readability.

10. Avoid trigger words or mysterious acronyms.

mistakes in drawing logic models is to draw them as a flowchart of activities rather than a causal model of what produces the impacts.

The logic model in Figure 9.16 shows the activities (preparing and delivering meals) and the end results (people maintain good health and therefore avoid moving to a nursing home) but does not communicate what it is about the service that contributes to these results. Is it the nutrition in the food, improved mental health through regular social contact, the opportunity for early warning of an accident, a combination of these, or different combinations for different types of clients?

The diagram is also pointlessly messy, with boxes not aligned and arrows askew. All of these peripheral details distract from communicating the main messages (see guideline 6).

Figure 9.16   **A Poor Logic Model in Terms of Focusing on the Main Elements and Explaining How It Works**

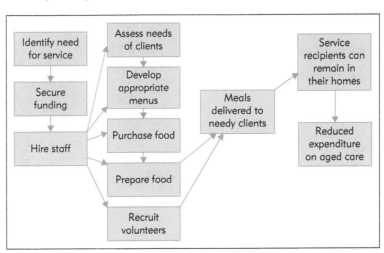

Guideline 2: Avoid Dead Ends

A logic model cannot include all of the activities and consequences of an intervention. Even if attention is paid to explicit and implicit goals, intended and unintended impacts, and the triple bottom line of economic, social, and environmental impacts, a single logic model cannot capture the consequences of every ripple it causes.

Therefore, every causal strand in a logic model should lead to an identified longer-term impact of interest (good or bad, intended or unintended). If it is important enough to include the strand in the logic model, then it must be because it will contribute to some important final impact, and this should be shown in the logic model.

In Figure 9.17, a causal strand relates to the impact of the Meals on Wheels program on the volunteers who participate, but this is a dead-end causal strand because the impacts shown in the logic model relate only to the benefits to service recipients.

There are two options for resolving a dead-end causal strand. One is to add another outcome of interest to the final level, but this can be tricky. In one project we worked on, some of the stakeholders wanted to put

Figure 9.17   **A Poor Logic Model in Terms of a Dead End**

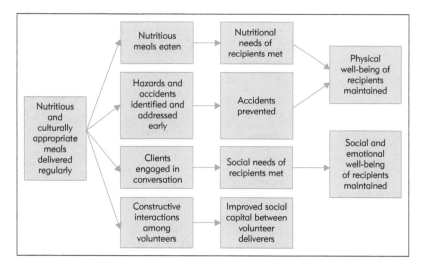

"increased status for the profession" as a final outcome from a program that was intended to be improving services for the community. Putting it as an ultimate outcome might have been interpreted as showing that the program had been designed for the benefit of staff rather than the community.

Another option is to consider ways in which it might plausibly be connected to the outcomes of interest. For example, the sustainability of the service might be increased if the volunteers find it satisfying and therefore support each other to do it well and continue their involvement. Figure 9.18 shows how this might look.

Guideline 3: Make Every Arrow Meaningful

Poorly drawn logic models put arrows between all the boxes, even when there is no plausible causal connection. In well-drawn logic models, every arrow is meaningful and two boxes linked by an arrow can be read either as, "If A, then B" (for simple attribution) or, "When A is achieved it helps to bring about B" (complicated contribution).

Logic models with arrows connecting every box to every box may sometimes result when path analysis has been used to develop the program theory,

Figure 9.18   **Improved Logic Model in Terms of Linking a Dead End to an Ultimate Impact**

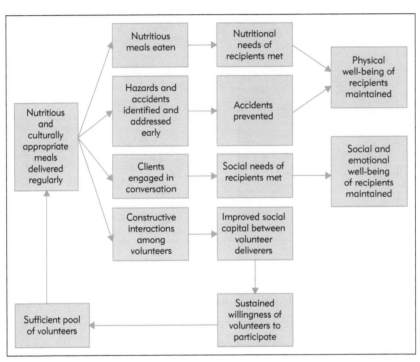

and a correlation has been calculated for every possible combination of variables. Although there might be a small correlation between two variables, it does not necessarily mean there is a causal relationship between them; it may well be spurious and due to a third variable that has not been measured and is therefore not included in the path analysis. Even where there is a causal relationship between two variables, if it is small, it might be more helpful not to include it and instead to focus attention on the main causal relationships.

Figure 9.19 shows a spider web that links every intermediate outcome with every final outcome. Doing this formulaically when drawing logic models will mean that some of the arrows are meaningless, and the important arrows are lost in the maze of lines.

When we read this diagram narratively, some of the parts work well. The top strand is, "Nutritious and culturally appropriate meals delivered regularly," which leads to "Nutritional needs of recipients met," which contributes to, "Physical well-being of recipients maintained." But others

Figure 9.19   **Poor Logic Model in Terms of Meaningful Arrows**

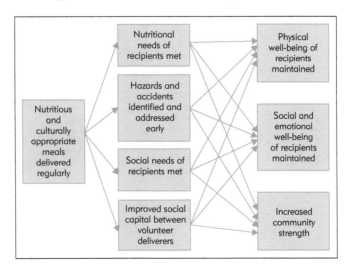

are less credible. One of the bottom strands is, "Nutritious and culturally appropriate meals delivered regularly," which leads to "Improved social
capital between volunteer deliverers," which contributes to "Physical well-
being of recipients maintained."

As it stands, it is hard to imagine a credible, direct link between the
deliverers' social capital and the well-being of recipients. The link should
be either elaborated or removed. It is possible that improved social capital
between deliverers might increase the amount of informal communication,
leading to better exchange of information about recipients, increased awareness of early warning signs of problems, and, hence, earlier intervention.
However, if this possible link was investigated and found to be not credible, not supported by any evidence, or not important, it would be better
to remove it.

We might begin by removing the meaningless arrows (Figure 9.20). We
can further improve the diagram by simplifying it as shown in Figure 9.21.

Guideline 4: Indicate the Direction of Expected Change

A good logic model can be read as a coherent and plausible story of cause and
effect. This requires attention to how the components are labeled. Sloppy
shorthand references to intermediate results might be acceptable during the

Figure 9.20   **Improved Logic Model in Terms of Meaningful Arrows**

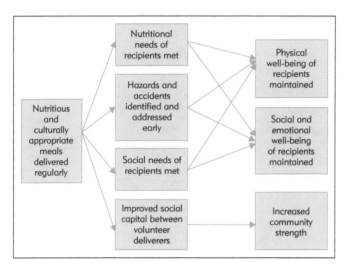

Figure 9.21   **More Clearly Drawn Logic Model**

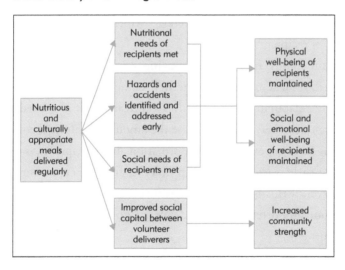

process of developing a program theory but need to be clearly specified when drawing up a logic model for wider use. Figure 9.22 shows outcomes that are poorly specified and therefore not logical.

In Figure 9.22, "Nutritious and culturally appropriate meals delivered regularly" leads to "Social isolation" and "Accidents," which contribute to

Figure 9.22   **Poor Logic Model in Terms of Representation of Direction of Intended Change**

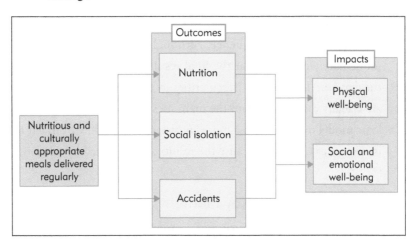

"Physical well-being" and "Social and emotional well-being." Meals on Wheels is intended to reduce social isolation, not to produce it. (It is possible that this is not the case, and that delivering meals actually does increase social isolation by reducing clients' need to go out shopping, but if the logic model were intended to show this alternative program theory, the stated impacts would be different.) Meals on Wheels is not expected to lead to accidents; instead it might be expected to reduce the incidence of accidents (cooking-related accidents) or provide early notification that they have occurred. Figure 9.22 is actually a confused blending of a problem analysis diagram and a results chain. "Accidents" and "social isolation" would appear in a problem analysis, whereas something like "reduced accidents" and "improved social participation" would appear as their mirror image in a chain of intended outcomes for an intervention.

Guideline 5: Clearly Show the Sequential and Consequential Progression

Poorly drawn logic models fail to show the sequence and consequences of activities and intermediate results. One of the advantages of using pipeline logic models is that this is built into the format. Results chain logic models, which are more free form, are at risk of not making this clear.

Figure 9.23 shows what can happen when someone uses a free-form results chain logic model format and does not pay attention to clearly showing sequence and consequence. Although the arrows make sense, it is not immediately obvious what the final result is because it is not at the end of the diagram. In addition, irregularities in the arrows and alignment of boxes are visually distracting from the main messages. Figure 9.24 shows the same logic model redrawn with a more consistent logical flow from left to right. It is also easier to follow because the results are organized in columns.

Guideline 6: Focus on the Key Elements

A logic model is meant to support communication, so decide on the main message, and draw it in a way that clearly gets it across. Use symmetry and alignment to reduce distraction and support your main meaning. If there are

---

Figure 9.23   **Poor Logic Model in Terms of Representation of Sequence and Consequence**

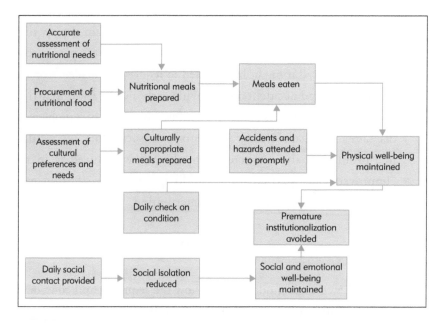

Figure 9.24   **Improved Logic Model in Terms of Sequence and Consequence**

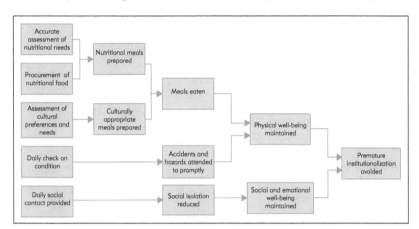

multiple strands, try to lay them out symmetrically. For example, if the intervention has three main stages that are fundamentally different, show this. You might put each of the three stages in a large box so the message stands out (make sure you align the boxes to reduce distracting extraneous detail; see guideline 9). If the intervention needs two simultaneous causal strands to work, show this. You might show these as parallel boxes of equal size, even if the amount of text is different, to emphasize that they are equally important, or use different size boxes if you want to represent one as more important than the other. If you have lots of small boxes, which make it hard to see an overall picture, try to group some together and put a labeled box around them.

Figure 9.25 comes from the Public Service Commission of South Africa, a central agency of the national government charged with the responsibility of evaluating the performance of national and provincial government. Their logic model for its program of monitoring government performance was designed to emphasize a two-stage process, involving reporting and follow-up, and several causal paths, not only identifying and correcting deficiencies.

Guideline 7: Avoid Too Many Arrows and Feedback Loops

Although reality is often complicated and it can be important to recognize this in program theory, logic models should be as simple as possible.

Figure 9.25 **Logic Model Emphasizing Four Causal Strands, Each with Two Stages**

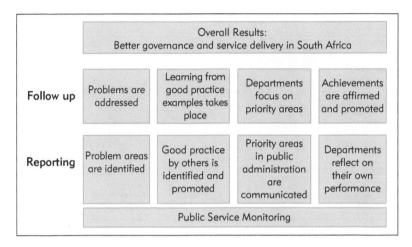

Source: Office of the Public Service Commission of South Africa (2006)

Although in some ways, everything might be connected to everything else, that does not mean it is going to be useful to show this.

It is worth remembering an important idea from systems theory, first stated by Polish American scientist and philosopher Alfred Korzybski: "The map is not the territory." A logic model cannot show every causal factor, every causal connection, and every feedback loop and at the same time be coherent and useful. Figure 9.26 shows what can happen when feedback arrows are used indiscriminately.

In Chapter Seven, we discussed better ways of using feedback loops, including prioritizing the ones that are included in the representation, and possibly using dotted lines to show feedback loops.

Guideline 8: Remove Anything That Does Not Add Meaning

Resist the temptation to add extra borders, flourishes to arrows, bullet points in a list, or anything that does not actually add meaning. These will clutter the diagram and distract from the main messages. For example, in Figure 9.27, the boxes for individual components produce an overload of visual information without adding meaning to the diagram.

Figure 9.26   **Poor Logic Model in Terms of Too Many Indiscriminate Feedback Lines**

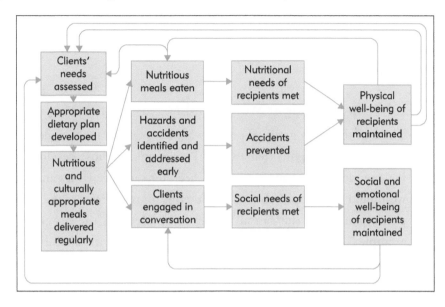

Figure 9.27   **Poor Logic Model in Terms of Visual Elements That Do Not Add Meaning**

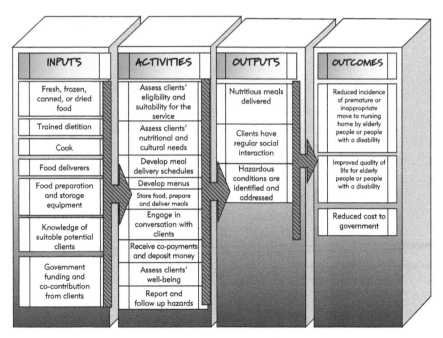

Guideline 9: Ensure Readability

Small type sizes make it difficult to read logic models. If a logic model is very detailed, it may be better to do an overview of the main elements, and then unpack these on separate pages rather than trying to squash it all onto one page.

Guideline 10: Avoid Trigger Words and Mysterious Acronyms

Think of the wider group of people who may read the logic model. Check carefully for acronyms that are well known among those developing the logic model but not more widely. Check for sensitive issues and careful wording. For example, choose which terms to use to refer to the program participants and the causal processes involved in the program. For example "disabled people" is shorter than "people with disabilities" but less acceptable.

SHOULD LOGIC MODELS INCLUDE SMART MEASURES?

SMART measures are specific, measurable, achievable, realistic, and time-bound. Opinions differ about the desirability of including specific numbers of outputs and outcomes that are expected if a logic model is implemented. For example, the Hecksher Foundation, which promotes the welfare of children in New York and elsewhere throughout the United States, requires its applicants to include specific, measurable outcomes in logic models—for example, "At least 90% of 9th graders will graduate from high school on time" (Hecksher Foundation, 2009).

Obviously a monitoring and evaluation plan needs to address the issues of what will be considered successful performance and what data will be collected on actual performance, and in Chapters Eight and Fourteen, we outline ways to do this. However, unless you are required to do so by a funder or central agency, we advise you to be cautious about including SMART measures in your actual logic model. These can distract attention from the actual program theory. It is not coincidental that most logic models that include SMART measures lack an actual theory of change and theory of action.

We have found it is better to focus on clarifying and negotiating an agreed theory of change and theory of action before moving to negotiate evaluative criteria and sources of performance data.

Deciding on performance indicators and measures at the time of developing the logic model can also reduce the attention that can be given to developing credible and appropriate measures. There is a higher risk of defining performance measures in terms of what is easy to count rather than what will be useful and meaningful and on measures or indicators of quantity without reference to quality. Logic models with SMART measures often focus on completed activities (such as number of classes held or number of students enrolled) rather than achievements (such as number of students who learned from the course).

There are, of course, situations in which it is highly appropriate to include a specific performance measure in a logic model. This is when a particular measure or target has been previously agreed to by a key organizer for the program. For example, a highly successful road safety campaign in a state in Australia in the 1970s was organized around the figure of 1,034, the number of people killed in traffic accidents in one state in one year. A community campaign advocated for a number of safety measures, including the introduction of compulsory seat belts, in order to reduce the annual road toll below 1,034. Although this was a very clear target, it was still open to change, and in the following years, the annual road toll has continued to decline. In 2009, less than a third of this number (290 people) were killed on roads in the state, despite a large increase in the total population.

SUMMARY

How to represent program theory is an important decision that should take account of the forms of logic models that are currently in use or familiar to stakeholders but also the strengths and limitations of the different options. It might be helpful to have different versions for different purposes or for different interventions or to insist on using a common template across an organization.

EXERCISES

1. Find a logic model for a type of intervention you know quite well by reviewing evaluation reports or articles or searching on the Web. To what extent does it represent a credible and relevant theory of change and a theory of action rather than a sequence of events?

2. Find a logic model for a type of intervention you are not highly familiar with by searching journals or the Web. What do you understand about the intervention from the logic model? What questions do you still have about how it works?

3. Assess a logic model in terms of the guidelines outlined in this chapter. Are there areas where it could be improved? Redraw the logic model to address these issues.

10

Critiquing Program Theory

THIS CHAPTER PRESENTS a range of criteria for undertaking a desk review or critique of a program theory as it appears on paper or as described, some considerations when involving stakeholders in reviewing a theory, and some types of recommendations that might arise from a review of a program theory. These criteria and processes can be used on their own or in conjunction with empirical testing of the theory. This chapter dispels the myth that the only way to test a program theory is by formal experimental methods. Nevertheless, empirical theory testing is useful and is one of the tasks that can be undertaken by a program evaluation.

When we conduct a critique or desk review of a program theory, our objective is to ensure that it will be useful rather than perfect. Also we are not reviewing all aspects of its validity for all time. As we have shown in previous chapters, program theories evolve over time in response to changing needs and circumstances, new knowledge about effective policy tools, and new evidence about what works for this particular program and others like it.

The theory to be reviewed could be an explicit one that has been with the program for some time, or it could be a new one. The review can be undertaken before, during, or after a program has begun. Sometimes a review is prompted by a change in circumstances, such as a change in relevant policy. However, reviews can also take place on a regular basis. Sitaker (2008) reports on the usefulness of undertaking a review of the program logic as an integral part of the annual process of applying for funding for the Washington State Heart Disease and Stroke Prevention Program (WaHDSPP) and on some of the changes that have been made to the logic as a result of that process:

> The WaHDSPP has used the annual occasion of preparing a continuing application as a time for collective reflection, not only on how to better refine existing activities or develop new ones but also as time to critically examine and improve the logic model and evaluation tools [p. 6].

Review of the theory could be undertaken by program staff or other users or target audiences of the program theory, including partners. Review could also be usefully undertaken by people with program theory expertise or expertise in the subject matter areas being addressed by the program. As for eliciting and documenting a program theory, a review of a program theory can be done as a collaborative exercise or as a desk job, depending on the purposes of the review process and the circumstances.

A program theory critique can guide several types of decisions. There may be decisions to make about whether the objectives of the program should be changed to something more realistic and achievable, whether the design of the program needs changing, whether more research is needed concerning the validity of the assumptions underpinning the program, and whether it makes sense to proceed with an empirical impact evaluation. The review of a theory can also be used to explore hypotheses about whether program failures are more likely to be attributable to a poor theory, poor implementation of the theory, or how the program is situated within its wider enabling (or disabling) context.

Review of a theory can be used to check whether it has fallen into any of the program theory traps. Here we express questions about these traps in positive terms:

- Does the theory demonstrate the logical and defensible relationships between what the program does and the outcomes it is trying to achieve?

- Does the theory present a plausible solution to a problem and/or tackle one or more causes of a problem?

- Does the theory adequately specify the intended outcomes so that it is possible to ascertain whether the chosen approach to implementation is relevant to those outcomes? Are the intended outcomes stated in a way that their achievement could be recognized, if not measured?

- Does the theory incorporate consideration to unintended outcomes where these can be anticipated or are known?

- If the program has complicated or complex aspects, does the theory incorporate elements that reflect the nature of the program—for example, coordination and links among various sets of intended outcomes and activities of a complicated program, open systems considerations relating to complex programs including issues around feedback loops, or emergent outcomes and fluid boundaries?

- Is the program theory such that it is potentially useful for evaluation? Can it be used to generate helpful evaluation questions or set priorities for evaluation? For example, is it constructed in a way that shows the most important outcomes to look at and the most important factors that are likely to affect outcomes so that they can be explored as part of an evaluation or a research project?

- Is the type of program theory that has been developed fit for the purpose (for example, with respect to presentation for particular audiences), and has it been adapted to reflect changing contexts and circumstances?

In addition to incorporating these questions in a critique of program theory, we suggest two sets of criteria: one that relates to internal validity and one concerning external validation. *Internal validity* is about:

- Clarity of description

- The outcomes chain as the central organizing principle for the program theory

- Demonstration of how desired outcomes relate to addressing the problem

- The strength and plausibility of the logical argument about the connections of all parts of the program theory, the gaps in the logic or breakdowns in temporal sequencing, and the likely adequacy of the program to produce the desired outcomes, given the program theory

- Articulation of mechanisms for change that underpin the choice of outcomes

External validation considers:

- The evidence base for the program theory, including the currency of relevant evidence

- Consistency with accepted theories

- Whether the program theory incorporates consideration of context and its impact on outcomes

- The ethical base for the program theory

- The intended use of the program theory and purposes for which it is being developed

- Whether it was worth the time, effort, and resources given the purposes

CRITERIA FOR ASSESSING INTERNAL VALIDITY

Internal validity is about whether the program theory hangs together in a way that makes sense and tells a clear, coherent, believable, and logical story about the outcomes the program is trying to achieve, why those outcomes are important, and how the program will contribute to the outcomes.

Clarity of Description

This is about whether the program theory is spelled out sufficiently clearly that there is enough detail to see how the program will work (without going into all the daily operational detail) and to be able to raise questions about the logic of the argument and the feasibility of implementation. At a minimum, the program theory should include the following key elements: description of the extent, nature, causes, and consequences of the problem or issue that the program will address; intended outcomes and how achieving those outcomes will contribute to addressing the problem or issue; planned key activities;

resources; and external and contextual factors that are likely to affect how well outcomes will be achieved and how these factors will be addressed or managed. Chapters Seven and Eight provide information about the types of details that should be included.

A clearly articulated program theory should differentiate between what the program is trying to achieve in terms of outcomes and how it is going to do so—the inputs, activities, and outputs that are being used to achieve those outcomes. For example, a training program might have an activity of running workshops. There could be several desired outcomes, such as improved understanding among participants of the topic at hand or improving networking and peer support. These need to be specified rather than implied.

The description of the program theory needs to appropriately differentiate among outcomes, activities, resources, and so on. Identifying whether program theory terms and concepts are used appropriately can be done only by looking at their use in context. The terms *input, activity, output,* and *outcome* are relative terms. What might be the outcome of one program could be the input for another. For example, noticing that a particular item has a dollar sign in front of it does not make that item an input for a given program.

INPUTS, OUTPUTS, AND OUTCOMES ARE RELATIVE

- An advocacy program to generate funds to address HIV/AIDS had the successful generation of funds as its main outcome. This in turn was an input to the operation of other programs whose direct role was to prevent HIV/AIDS and mitigate its impacts. All programs or subprograms had the same end goals, but each achieved different outcomes along the way.

- A partnership program incorporated many activities to develop trust among partners so that they would work harmoniously together to address environmental sustainability issues. The direct outcome of those partnership formation activities was that the partners began to work together in an integrated and constructive manner. On the face of it, working together may appear to be an activity, but if one were building a program theory for the team-building activities, then working together was actually an outcome that in turn would become an input to a successful environmental sustainability program.

The Outcomes Chain as the Central Organizing Principle for the Program Theory

The outcomes chain is the link between what the program is doing and the problems it is seeking to address. However, many program theories fail on the basic test of being outcome focused. They concentrate excessively on what the program will do rather than the outcomes it will achieve. Some may just have a box that lists outcomes in no particular order. This may make it difficult to link particular activities to particular outcomes and to understand the relationships (temporal, causal, and other) among outcomes. In contrast, a focus on outcomes will typically:

- Draw attention to the fact that a program will have several outcomes ranging from immediate, to intermediate, to ultimate outcomes that can be arranged in sequence (but not, as we showed in Chapter Seven, necessarily linear or unidirectional)

- Clarify what is to be achieved with whom, who are the beneficiaries, perhaps the timing of outcomes, and so on (see the success criteria in Chapter Eight)

- Draw attention to the fact that different activities and resources may be needed for different outcomes and possibly at different times

- Provide a point of reference for thinking about what else outside the program might affect the achievement of each outcome (see the discussion of nonprogram factors that affect outcomes in Chapter Eight)

If a program theory is expressed primarily in terms of activities, those activities need to be reformulated as intended outcomes. Figure 7.2 gave an example of such a reformulation.

Conversely, we have on occasion seen program theories that consist only of outcomes chains. These too are inadequate. The outcomes chain should be the backbone to which all other "bones" of the program theory (resources, types of activities, and outputs) and then "flesh" (detailed action plans and so on) should be attached.

Demonstration of How the Desired Outcomes Relate to Addressing the Problem

When reviewing a program theory, we should ask whether the intended outcomes can be linked to the needs that gave rise to the program. Do the intermediate outcomes mirror the selected causes of the problem that the program is to address? Say there is a lack of parental encouragement to children to read at home, and specialists have identified this as a factor contributing to the fact that some children do not read outside school. The program wishes to tackle this lack of parental encouragement with a view to increasing children's reading and improving literacy. Is there an intended outcome in the program theory that relates to parents' becoming more actively supportive of their children's reading?

Does the program theory also include outcomes (such as low-level outcomes) that are important enabling outcomes not directly related to the causes of the problem being addressed? For example, some immediate outcomes of activities to promote a program to its target audience would be that the right numbers and types of people are attracted to the program and choose to participate in it. Nonparticipation in the program is not a cause of the problem since the problem exists regardless of whether the program exists. However, participation in the program is clearly a necessary condition for the program to achieve outcomes that will address the problem.

The Strength and Plausibility of the Logical Argument

The strength of the logical argument includes considerations relating to coherence, sequencing, completeness versus glaring gaps in the logic, and temporal arrangement of outcomes and their relationship to the timing of program activities. The logic of the outcomes chain can be tested from both directions: starting at the end point (or ultimate intended outcome) or starting at the beginning point (the immediate outcomes).

Starting at the end point requires asking questions about how each outcome will be achieved: What prior outcomes and what additional activities will need to feed into that outcome? Then work down to the next level of outcome. For example, if we have a high-level outcome that relates to changing the behavior of individuals, we could ask what needs to occur that will prompt individuals to change their behavior. Answers might relate to changes in their motivation, including perceptions of benefits and costs or changes in their opportunity or capacity to act. We might then ask more about what would influence their capacity to act, being careful to elicit other prior outcomes rather than activities. Perhaps we would obtain answers around practical considerations, financial capacity, or improved knowledge and skills.

Starting at the first step in an outcomes chain—say, successfully getting the target audience to participate in the program—ask a series of questions about why each step in the program theory would be expected to lead to or be followed by the next step until you reach the ultimate outcome. Any large gaps in the logic should become apparent through applying these processes. The process is somewhat similar to that used when eliciting a program theory from stakeholders.

Sometimes the lack of logic may become apparent from the sequencing of outcomes or activities. Whether a temporal order is correct is not, however, always easy to establish, especially when strong feedback loops are in operation. Also, progressive strengthening of a single set of outcomes (increasing ability or increasing self-esteem, for example) might call for a logic model that portrays a spiraling of outcomes or looks something like a double helix.

Some of the process of reviewing the connections among the various parts of the program (outcomes, activities, resources) can be done with minimal collection of data. Gaps can often be identified by interviews with program staff about activities included in the program design (aside from its implementation) compared with what apparently needs to be done and therefore about gaps in the theory. For example, it may be apparent, through analyzing program factors that will affect success, that a communication strategy to attract members of the target group should be an important part of a program design and should be incorporated in its theory. Such a communication strategy may or may not have been planned. It is easy to ask staff whether such a strategy is part of the design. However, often the comparison between important program factors, on the one hand, and activities and resources, on the other hand, will not be as clear-cut as this. Questions are more likely to be raised about the adequacy and quality of the program activities required to address the identified program factor rather than simply whether an activity exists.

The inclusion of outcome objectives that are unrealistic given the role of the program and the resources it commits relative to the problem and to other resources beyond the program is a common deficiency of programs and their theories. So adequacy of resources to achieve the desired outcomes is a key consideration when reviewing a program theory. Questions about whether, in the program theory on paper, all intended outcomes are matched by program activities should also be raised. In some cases, the absence of any activities for an outcome could point to an omission in the logic. However, it is not essential that all outcomes have identifiable activities since achievement of prior outcomes may be the main means of leverage from one level in the outcomes chain to the next.

DECIDING WHETHER IT IS IMPORTANT THAT EACH OUTCOME BE MATCHED BY SPECIFIC ACTIVITIES

The activities of a program to change farmer practices in ways that are known to have positive environmental effects may stop at the point in the outcomes chain that relates to changing farmer practices. These practice changes may be the main outcome the program achieves in order to bring about environmental improvements, and it might not have any further specific activities to achieve environmental improvements. It would, however, be important to include those subsequent environmental improvements in the program theory and outcomes chain since they provide the reason for the program and affect which practices the program chooses to influence and which farmers it targets.

Program activities and resources that cannot be linked to one or more outcomes would also be a subject for discussion as part of the process of reviewing a program theory. Activities that are not linked to outcomes may be unnecessary. Or they may suggest other outcomes that have not been shown in the outcomes chain and should be included.

Tests for internal coherence can contribute to an evaluability assessment. Wholey (1983) identifies the evaluability process as having several steps, among them the development of a program logic and the conducting of plausibility analyses: "The plausibility analysis represents the evaluator's judgments on the likely success of the program—judgments based on available information and subject to later revision" (p. 49).

Nevertheless, demonstrated plausibility is no guarantee of success. As Weiss (2000) has pointed out, some of the most plausible models have fallen afoul of the evidence from empirical evaluations, and some programs appear to work in spite of their questionable logic. A role of the evaluator can be to elicit or construct a program theory that would or could account for the success of a program whose theory appeared to be implausible and, conversely, to identify why a program with a plausible theory did not work.

An initial review of the plausibility of a theory is only the first step. The theory itself should continue to be fair game as an evaluand rather than taken as a given. Among the issues that should continue to be considered in relation to a particular program theory, even when that theory is being used to design an evaluation, are the following:

- A plausible program theory does not rule out equally plausible competing theories, including those based on different value systems.

- What if your program has the desired effects but came about through some other mechanism? We evaluated a program for families affected by drug abuse that included art activities for mothers and their young children. One program theory might recommend that all programs have art activities. But was it the art activity that was a vehicle for strengthening mother-child communication and relationships by, for example, providing new, more detached, and perhaps less confronting ways for communicating with each other, or would any activity in which mothers and children spent enjoyable time together achieve the same effect? A critique of the program theory encouraged exploration of these alternatives.

- Unintended outcomes can turn out to be as important as or even more important than intended outcomes.

- A program with a sound theory may still fail because implementation did not adhere to the theory. So program theory can be used to provide a framework for monitoring the fidelity of program implementation.

- Failure to implement may also point to failures in the theory. The theory may have made incorrect assumptions about causes or what is feasible.

Articulation of Mechanisms for Change

A program theory should articulate its mechanisms for change, for example, reducing ignorance, creating peer pressure, incentives, or fear of sanctions. Those who are reviewing the program theory should ask whether the program theory clearly identifies the assumed mechanisms for change that underpin its selection of outcomes and activities. The mechanisms are the ways in which outcomes will occur. Statements about mechanisms are the "because" statements that underpin the "if-then" statements. An example of such a statement in relation to the extension program to change farmer practices would be that *if* farmers have a better understanding of the effects of their practices and how to change them, *then* they will change their practices, *because* their current practices are caused to a significant degree by their ignorance. A critique of the program theory should assess the clarity of description of the mechanisms and their validity. In this case, we would ask for evidence concerning the assumption about the relationship between farmers' knowledge and their practices, including whether the assumption applies to a small or large proportion of the farmer population. The very act of articulating the assumption makes it fair game for investigation as part of an empirical evaluation, a research study, or perhaps a literature review.

The adequacy of a program theory can also be assessed in terms of the sufficiency of the stated (or elicited) mechanisms for bringing about intended outcomes. Some questions that can be raised are:

- If the causal mechanisms that are relevant to the program have been identified, would successfully activating these mechanisms be expected to create movement from one level of outcome to the next in the chain of outcomes? For example, for a program theory in which ignorance is identified as a cause of behavior and reducing ignorance is the mechanism that is expected to change behavior in desired ways, then one could ask: If knowledge improves, will that be sufficient to bring about intended changes in behavior? For some members of the target audience but not others?

- If not, what other mechanisms might need to be involved, for, say, different causes of behavior for different segments of the target population or under different circumstances? For farmers, examples of other mechanisms (additional to or instead of improving knowledge and skills) might be subsidies to remove financial impediments, activities to build confidence such as engagement in joint research activities with other farmers and professionally supported piloting activities, and creating an expectation that undesirable practices will be noticed and action will be taken following inspection of farms to check compliance with land management regulations.

- How, if at all, are these alternative or additional mechanisms incorporated in the causal chain? How should they be represented in the causal chain? Does the program address these directly or depend on others to do so? Does the program try to engage with these mechanisms in any way, perhaps by influencing partners? If the program is not addressing these mechanisms at all, we would then go on to ask what else is happening beyond the program that would give some assurance that the program will not be a waste of effort—a dead-end program. We consider the latter in our discussion of the extent to which the program theory takes into consideration the context in which it operates as one criterion for external validation of the program theory.

- Are there timing and sequencing issues around which mechanisms need to be activated first and later? Does the program theory identify important feedback loops in its narrative and its logic model? For example, in the program of cash grants for victims of domestic violence, does the experience of victims with the program motivate them to use the program again or recommend it to other victims?

- Are the program activities and resources sufficient to activate the mechanisms to the degree needed to achieve the desired outcomes? This can be explored by comparing the quality and quantity of program activities and resources as described in the program theory with the desirable features of program activities and resources (according to experienced staff, experts, research, clients, and other stakeholders) that are included in that part of the program theory that describes program factors that will affect outcomes (as described in Chapter Eight).

QUESTIONING THE SUFFICIENCY OF CAUSAL MECHANISMS INCLUDED IN PROGRAM THEORY

In our discussion in Chapter Six of the program of cash grants to victims of domestic violence who report incidents to the police, we identified that the main mechanisms were reducing financial dependence on the perpetrator, increasing the immediate safety of victims, and reducing their fear. We could ask whether the operation of those mechanisms will be enough to encourage them to report incidents to the police. What about victims of domestic violence for whom money is not a major consideration? Does this program frame the issue of domestic violence as one that is confined to poor or financially dependent people? What impact might this have on reducing domestic violence overall?

For those for whom removal of financial barriers is potentially a powerful mechanism, we could ask whether the program activities will be sufficient to fully activate these mechanisms. Adequacy of cash grants for providing sufficient financial support to encourage reporting is a factor that is likely to affect the success of the program. One would expect to see reference to that factor in that part of the program theory that identifies factors that will affect success, sometimes labeled assumptions. A critique of the program theory and the program might conclude that the size of the cash payment (something the program can control) is inadequate to provide the level of support that victims need for immediate crisis purposes. As a result, the cash grants may be insufficient as an incentive to report to police given all the possible negative effects of doing so (humiliation, stigma, trauma of dealing with the police and legal system, possible retaliation by the perpetrator, and so on).

CRITERIA FOR EXTERNAL VALIDATION

External validation is about how well the theory stands up in relation to external sources of evidence, the context within which it operates, and intended uses of the program theory. Empirical evaluation studies would be one but not the only method for externally validating the theory.

Evidence Base for the Program Theory

Is the program theory consistent with available data about the program? With research evidence? Does the research evidence on which an existing program was based still hold true? Does the program theory need to be reconsidered in light of new evidence? Currency can be particularly difficult to achieve when wicked issues (described in Chapter Seven) with unstable features are the focus of a program. Information about broad trends may be more relevant than highly specific data.

Front-end judgments about the evidence base for the program theory can be about the availability and adequacy of evidence concerning the nature, extent, causes, and consequences of the problem or issue that the program is addressing. Evidence can also be about effective practices for addressing the problem. Where evidence is available, it is then possible to consider how consistent the program theory is with that evidence. For example, in the case of the example of the extension program working with farmers to change their behavior in order to reduce environmental degradation, one might ask about evidence that farmers' practices are discernibly contributing to environmental degradation.

Once a program has been empirically evaluated, the evidence from that evaluation can also be used to test the program theory as it applies in that situation. Did the program theory work in practice? If the program did not work, was this because of faults with the theory or its implementation, or both (Weiss, 1972)?

Unfortunately, demonstrated achievement of desired high-level outcomes of a program is an insufficient basis on which to conclude that the program theory is correct. A range of different types of evidence will be required to make those judgments. Looking inside the magic box of the program to see how the parts of the program theory are played out in practice can strengthen conclusions about the adequacy of the program theory. Chapters Fifteen on causal attribution and contribution and Chapter Sixteen on the use of program theory for synthesis discuss this issue further.

Consistency with Accepted Theories

This is a variation on the previous point about consistency with evidence, but here we are referring to evidence that is not specific to this program but is about broader theories of change, such as stages of change theory, diffusion theory (which we discuss in Chapter Twelve), or program archetypes (discussed in Chapter Thirteen). Use of this information can assist with ensuring that a theory of change is built on plausible solutions and includes key elements of particular types of programs.

Inability to identify a recognized wider theory of change and program archetype from which the program was derived would not necessarily condemn a program. Cross-cultural issues, for example, may render some theories quite inappropriate, and it would be a mistake to build a program around them. Many program theories are home spun and may in fact be the source of innovation and new theories. Some are worthwhile and others are not. However, substantial departure from widely accepted and relevant theories of change or program archetypes would at least be a platform from which to raise questions about the program's theory of change and the rationale for moving away from established theories. The intent of recommending the use of these theories is not to discourage innovation but simply to raise critical issues to which program managers may well be able to give good reasons for departure and for trying something different.

Whether the Program Theory Incorporates Consideration of Its Context

This is about the extent to which the program anticipates and takes account of external factors, their impacts on outcomes, and how it will work within its context. It extends the "if-then-because" statements that reflect program mechanisms to include "as long as" statements about what else needs to occur and the conditions under which the *if-then* assumptions may and may not hold.

An example of a full "if-then-because-as long as" statement in relation to the extension program to change farmer practices would be that *if* farmers change their practices, *then* environmental benefits will occur *because* farmers' current practices are contributing to environmental degradation. These benefits will occur *as long as* the climatic conditions are right and *as long as* the good that has been done by these farmers is not undone by the continuing poor practices of other farmers upstream.

Consideration of context can identify many points in an overall outcomes chain that slippage could occur because of mechanisms beyond the program that are affecting it. Conversely, some external mechanisms can boost the power of program mechanisms by, for example, removing other barriers or competing mechanisms. A program should be regularly observing and gauging the likely impacts of what is going on around it and implications for program theory and program effectiveness. These observations should be an integral part of the program theory. We argued in Chapter Eight that a good program theory identifies nonprogram factors that will affect success. These factors are an important source of "as long as" statements.

For any outcomes in the outcomes chain that require significant contribution from nonprogram activities or are substantially affected by other factors, ask: What reason is there to believe that those other nonprogram activities or required external factors will be in place? What is the program doing to address those factors or manage the risks associated with them? Is the program worth pursuing in the absence of those other enabling factors? For example, in the case of the extension program to change farmer practices that are affecting land degradation, one might ask: Given the available evidence about the causes of degradation, what else, other than changing farmer behavior, might need to happen in the operating context of the program for there to be a discernible impact? Are those other changes recognized in the program theory? Are those other things likely to occur, and, given what we know about those other things, is it worth putting significant resources into changing farmer behavior?

WHETHER EXTERNAL CONDITIONS AND MECHANISMS HAVE BEEN ADDRESSED IN A PROGRAM THEORY

For the cash grants for victims of domestic violence program discussed in Chapter Six, we identified mechanisms for change that related to developing financial independence, improving immediate safety, and reducing fear. However, victims reporting to police and then police response to their crises are only two points in the full chain of outcomes whose end point would be about safer, more fulfilling lives for victims and their families. Other mechanisms may relate to what happens (and victims' beliefs about what will happen) with respect to receiving adequate ongoing support for accommodation and achieving ongoing financial independence, being able to make suitable family arrangements, and the effective operation of the judicial system following reporting of incidents.

If these mechanisms (mostly beyond the control of the program) do not occur, then victims may suffer as a result of reporting to police, and their probability of reporting future incidents is likely to decrease. Also, they may tell others about what happened to them (for example, others in a victims' support group, key professionals who play an important role in referring victims to the police but who will do so only if it is seen to be in the interests of their clients). As a result there may be a wider negative impact that undoes much of the good that the cash grants might have initially achieved in terms of improved reporting and access to support. Indeed, a negative program theory, a concept discussed in Chapter Six, could be constructed around this possible real scenario.

The Ethical Base for the Program Theory

A program theory might be quite internally coherent and have considerable potential to produce the effects to which it aspires. However, the ethical foundation of the desired outcomes or the processes for achieving them may be questionable. A program theory may be internally consistent but ethically indefensible. For example, consider the theory for a program that requires polluting industries to purchase carbon credits. Even if this theory were internally consistent and highly effective with respect to its own objectives, some would dispute the ethical foundation of a program theory that enabled polluters to continue to pollute as long as they are prepared to pay the price. This is the age-old question of whether the ends justify the means.

Programs are underpinned by values, ideologies, and ethical assumptions, not just by empirical evidence. Shaw and Crompton (2003) draw attention to the ideological underpinnings of different program theories. They discuss shifts over the years and across countries with respect to the ideological underpinnings of different approaches to public health (in simple terms, individual versus community responsibility versus reciprocal responsibility). Other ideological differences might be around utilitarianism (greatest good for greatest number) and social justice (impacts on least advantaged in the community) ideologies.

A review of a program theory could consider the principles and ideological underpinnings of the theory. Such consideration might raise issues about which principles and ideologies are accepted by various stakeholders, whether inconsistent principles and ideologies are operating in parallel for different stakeholders, and how these might be managed. If a program has an agreed value system or set of principles, then a review might also consider the consistency of all elements of the program theory with those values or principles. For example, if it is agreed that local determination or individual control is to be a ruling principle of the program, does it flow consistently through the program processes? If gender equity is a key principle, does that follow through all aspects of the program theory, such as inclusion of gender equity considerations in the situation analysis, among attributes of outcomes, with respect to program activities and resources?

The Intended Use of the Program Theory

This book is about purposeful program theory. We have argued that a key consideration when deciding how to develop a program theory and determining what it should look like is whether it is fitted to the purpose—for example:

- If the purpose of the program theory were to clearly explicate a policy position or espoused theory, then it would need to be judged in terms of its fidelity to the policy.

- If its purpose were to represent current practice, that is, theory in action, then it would need to be checked against evidence about implementation.

- If its purpose were to communicate with external stakeholders, then it could be judged in terms of how well they understand it.

- If its purpose were to provide the framework for designing and reporting an evaluation, then it could be judged in terms of whether the evaluators find it useful for that purpose.

- If its purpose were to test a theory or parts of the theory, then it would need to be described in ways that would enable testable hypotheses to be established for exploration through evaluation or through research.

- If its purpose were team building, then it is more likely that the program theory should be judged in terms of the extent to which it generates consensus and commitment within the team.

Was It Worth the Time, Effort, and Resources Given the Purposes?

In relation to all of these purposes, one might ask whether the program theory process and product (the program theory or logic model) were a good investment of time, effort, and resources to address that purpose—for example:

- If team building were the primary purpose, would there have been more cost-effective ways of doing so?

- If designing an evaluation were the purpose, did program theory add sufficient value to the design process to justify the costs? Was it actually used? What might the evaluation have looked like without the input of program theory?

In Chapter Six we discussed some other considerations when making decisions about resourcing program theory development processes. These same considerations can be applied when critiquing a program theory. They concerned size, complexity, and budget of the program; size and significance of potential impacts of the program (including the impacts that might occur if a pilot program for which a program theory had been developed was to be rolled out on a wide scale); level of risk; and potential side effects.

To make judgments about the likely usefulness of a program theory for a future program evaluation, we may need to review the processes used to develop it. Taking short-cuts on program theory development processes can, as we show in the example that follows, result in a program theory that is not as useful as it could be for evaluation.

THE RISKS OF ACCEPTING A PROGRAM THEORY UNCRITICALLY AS A BASIS FOR DESIGNING AN EVALUATION

It is important to critically review a program theory. Program theories that draw on only one source, such as documenting stakeholders' mental models, can fail to identify important intermediate variables or plausible causal mechanisms. They therefore provide a poor guide to the selection of variables to measure or analyses to perform. For example, in an evaluation of an antismoking project, the program theory was based entirely on the project developers' mental models, which focused on engaging students through an enjoyable comic book with an antismoking message. It did not explain how familiarity with the characters and story would lead to changes in students' smoking beliefs, attitudes, and behaviors (Chen, Quane, Garland, and Marcin, 1988).

The evaluation collected data about the number of times that students read the comics and how familiar they were with the story and characters. But it did not explore issues such as identification and social norming that might have been part of a program theory informed by social science research and program archetypes in addition to stakeholders' mental models. The program was redesigned after the evaluation found that students had read the comics many times and were familiar with them but had not changed their attitudes, knowledge, or behavior. Critical review of the program theory before the evaluation could have identified its plausibility gaps and either produced an evaluation that was more useful in understanding why the program did not work or a redesign of the project before it was implemented.

ENGAGING STAKEHOLDERS IN THE REVIEW

A review or critique can be undertaken at any time in the life cycle of a program. When stakeholders have been engaged in the process of articulating the mental model, this provides especially useful opportunities for critiquing a theory of change as part of the same exercise. Articulating the theory should be followed by a process of encouraging participants to think about other factors, questioning and improving the mental model, and possibly refining program design and implementation. Sitaker (2008) reports that one of the lessons that the Washington State Heart Disease and Stroke Prevention Program learned about the process of developing and then reviewing a program theory is this:

> Evaluation and logic model development should be conducted in partnership with program stakeholders, with outcomes feeding directly back into ongoing program planning and progress monitoring. Logic models and evaluation plans are dynamic tools to guide the program in carrying out activities but should always be developed and refined in partnership with activity work plans and the key staff and partners involved in the work [p. 7].

Once an *if-then* chain of some type has been established, facilitation processes for reviewing the outcomes chain include looking for gaps in the logic, inappropriate sequencing or need for feedback loops, and conditions required for the *if-then* arguments to function effectively (see criteria relating to internal consistency). Facilitators and group members can challenge the logic and completeness of the arguments that stakeholders might give for their program theories.

Stakeholders, including staff and program managers, can also be encouraged to challenge the entire mental model: Has the right solution been selected? We have found it useful to encourage stakeholders to think outside the current solution and focus instead on client needs and intended outcomes. This shift in focus from supply-driven thinking to demand-driven thinking can throw up quite different mechanisms for change (for example, a shift from overcoming ignorance as a mechanism to motivating behavior as a mechanism) and, hence, quite different solutions (for example, incentives rather than education or a combination of different mechanisms for use with different segments of the target audience).

Staff and management may be prepared to take this alternative models approach only if they believe they have real options to do so. We have found it useful to get senior management to give explicit permission to think broadly. Whether senior management will give that permission depends on many factors, including political imperatives, where the program is in its life cycle, and whether there is an interest in having an evaluation that explores alternative models.

FACILITATING DISCUSSION OF ALTERNATIVE SOLUTIONS THROUGH CRITIQUING THE PROGRAM THEORY

The program of cash grants to victims of domestic violence who report incidents to the police is a modest and relatively simple program (financial assistance and immediate removal from danger) whose success ultimately depends on other key conditions being in place. If the other conditions across the policing, social services, and judicial system are found not to be adequately in place, are unlikely to be so in the foreseeable future, and the program is unable to influence them to any significant degree, then a facilitator assisting with the review process could encourage participants to think outside the parameters of the current program. For example, one solution in this case might be to sever the connection between cash grants and reporting to police. The cash grants could be treated simply as a form of welfare support and an incentive to come forward to receive other types of support not contingent on reporting to police. This approach would need a very different type of mental model and different program design. It shows the importance of critiquing not just the particular mechanism applied by the program but also that mechanism in the context of other mechanisms that are needed to support it.

RESPONDING TO THE RESULTS OF A REVIEW OF A PROGRAM THEORY

On occasion, a program theory will be so flawed or incomplete that we would question why anyone would ever expect the program to work. We have at various times given some form of each of the following recommendations arising from a review of a program theory that identified significant flaws:

- *Conduct additional research (literature or empirical) to enable a more informed program design.* For example, more information about needs (including size and nature of target group) and causes of needs might be collected as a basis for revisiting program objectives and strategies. Performance monitoring strategies may need to be put in place to collect baseline data.

- *Establish or negotiate more realistic objectives (intended outcomes) for the program.* For example, there may be a need to set objectives that are at lower levels in the outcomes chain. Another approach to making objectives more realistic can be to reduce expected levels of attainment, narrow or more clearly specify the scope of the objectives including the target group, and change the time frames.

- *Clarify the links between program objectives and corporate or higher-level strategic objectives.* This may require revision of the objectives to ensure that they clearly link to corporate objectives. It may also mean that corporate or higher-level strategic objectives need to be clarified or revised so that they provide a better and more realistic direction for setting program objectives and for program planning.

- *Redesign the program and adapt its approach to implementation so that it more coherently coordinates the work of a range of partners.* The need for this type of redesign is especially likely for programs with complicated aspects. We have found that even when partners have developed a joint program theory (for example, for a program requiring input from several partners around a socially complex issue), the partner programs may still be functioning as silos. They can identify the boxes in the program theory to which they are contributing, but they have not adequately thought about the arrows between the boxes showing the links and interdependencies and how they will make the links work and the arrows meaningful.

- *Redesign program strategies by adding, subtracting, or strengthening existing strategies or refining implementation of the program as designed.* Sometimes it makes more sense to refine the design of the program than to devote time and effort to undertaking an empirical evaluation of the outcomes of a program that is unlikely to succeed, given what is known about inadequacies in its program theory or disconnects between the theory and its implementation. Here are the sorts of examples of refinements to design that might be considered:

  - Add strategies to fill the gaps or redistribute effort among existing strategies.

  - Revise strategies and objectives in line with resources available.

  - Change resources to ensure that existing strategies can be implemented.

  - Investigate alternative ways of implementing strategies.

  - Change strategies altogether to respond to changes in context.

Few program theories are perfect, and so decisions need to be made about whether the program theory is good enough to be a useful starting point. Otherwise we might find ourselves trapped in a perpetual program theory development mode. If a decision is made to modify the program theory, then changes should be documented. In Chapter Six, in our discussion of the importance of documenting a program theory, we gave a stylized example of how changes might be recorded.

If the review of a program theory suggests that it is generally sound for current purposes, an appropriate response could be to proceed with using the program theory to measure and evaluate program implementation and outcomes. It may be useful to undertake a review to identify the extent to which intended strategies are being implemented and how they might be improved prior to measuring their effectiveness in achieving outcomes. In Chapter Fourteen, we discuss the use of program theory for monitoring and evaluation.

SUMMARY

It is important to critique a program theory before using it to plan an intervention, a monitoring system, or an evaluation. The aim is not to produce the ideal program theory that will hold for all time but to identify important gaps or errors and to embed processes of critical reflection and adaptation as needed. We have provided some pointers for critiquing a program theory that relate to its internal coherence and plausibility, as well as its exploration of the relationships between the program and its context, including how it relates to wider wisdom about effective programs. We have also identified some possible actions arising from a critique.

EXERCISE

Choose a program with which you are familiar and for which a description of program theory is available. Apply the guidelines for assessing internal validity and for external validation discussed in this chapter. You may want to consider doing this with colleagues.

Form a preliminary judgment about whether the program has the potential to contribute in a significant way to the outcomes it is trying to achieve. Suggest some appropriate responses to the findings of your review, for example, any changes that might improve the potential of the program to achieve outcomes.

## PART 4

Resources for Developing Program Theory

11

Some Research-Based Theories of Change

THEORIES OF CHANGE are the big theories about how change occurs for individuals, groups, organizations, and communities. We have capitalized them to distinguish them from the local theories of change that might be developed for particular programs as described in Chapter Seven. Theories of Change and program archetypes (explored in the next chapter) transcend a particular program and can help program theorists avoid the trap of presenting implausible solutions that are not well grounded in wider theories. In this chapter, we describe six well-known Theories of Change that can provide a conceptual framework for developing or critiquing a program theory.

Of the many Theories of Change that can be useful, we have selected six. Our intent in selecting these six theories is not to advocate the use of these particular theories but to use them to demonstrate how research-based theories of change can contribute to program theories. Also, different social policy areas, such as education and criminology, have their own range of theories of change, many competing with each other, but it is not within the scope of this book to address theories for each policy area.

Useful compilations, comparisons, and discussions of theories can be found in Halpern, Bates, Mulgan, and Aldridge (2004), Australian Public Service Commission (2007a), Grizzell (2003), and Web sites of some professional networks and organizations. The 12Manage Web site (http://www.12manage.com/) offers descriptions of hundreds of models and theories organized into twelve categories: Change and Organization, Communication and Skills, Decision Making and Valuation, Ethics and Responsibility, Finance and Investing, Human Resources, Knowledge and Intangibles, Leadership, Marketing, Program and Project Management, and Strategy. The Beyond Intractability Web site (http://www.beyondintractability.org/) includes "knowledge base essays by experts on many theories of change especially as they relate to addressing conflict across the globe." An article by Shapiro (2005), "Theories of Change," is one such knowledge base essay on this site that maps theories of change according to whether they are primarily about changing individuals (cognitive change; behavioral change, changing their relationships), social change, leadership, reformation versus transformation, or changing structures versus changing people.

Some theories tend to focus on individuals; some on families, groups, and interpersonal relationships; some on organizations; and some on whole communities. It is important to note that some of the theories we have chosen have been applied mainly in Western contexts and may not be appropriate for other cultures. Theories that focus on the individual as the unit for change (for example, the theory of reasoned action and stages of change theory) may be less helpful in cultures and communities where the focus is on group responsibility and group decision making. Exhibit 11.1 provides references to a wide range of other Theories of Change to draw on.
The six theories that we explore in this chapter are:

- *Theory of reasoned action and theory of planned behavior* (Fishbein and Ajzen, 1975, and Ajzen, 1985). These are closely related theories about changing behaviors of individuals.

- *Stages of change theory* (Prochaska and DiClemente, 1983). This theory is about changing behaviors of individuals.

EXHIBIT 11.1 EXAMPLES OF THEORIES OF CHANGE

INDIVIDUAL-LEVEL CHANGE

Attribution (Heider, 1958)

Classical conditioning and operant conditioning (Pavlov, 1960; Skinner, 1953)

Cognitive dissonance (Festinger, 1957)

Deterrence theory (Homel, 1988)

Fun theory (http://www.thefuntheory.com/)

Health belief model (Rosenstock, 1966; Glanz et al., 2002)

PRECEDE-PROCEED model (Green and Kreuter, 2005)

Self-efficacy (Bandura, 1997)

Self-fulfilling prophecy (Merton, 1968)

Social learning (Bandura, 1977)

Social networks and social support theories (Eng and Young, 1992)

Training (Kirkpatrick, 1959a, 1959b, 1960a, 1960b)

ORGANIZATIONAL-LEVEL CHANGE

Groupthink (Janis, 1972)

Organizational learning (Argyris and Schön, 1978)

Positive deviance (Pascale, Sternin, and Sternin, 2010)

Six change approaches (Kotter and Schlesinger, 1979)

POLICY CHANGE

Advocacy coalition framework (Sabatier and Weible, 2007)

Garbage can model (Cohen, March, and Olsen, 1972)

Incrementalism model (Lindblom, 1959)

Institutional rational choice (Ostrom, 1999)

(Continued)

EXHIBIT 11.1 *(Continued)*

Multiple streams framework (Zahariadis, 1999)

Policy diffusion (Berry and Berry, 1999)

Policy networks (Marsh, 1998)

Punctuated equilibrium theory (Baumgartner and Jones, 1993)

COMMUNITY-LEVEL CHANGE

Bandwagon effect (Bikhchandani, Hirshleifer, and Welch, 1992)

Institutional analysis and development (Ostrom, 2008)

Social loafing (Karau and Williams, 1993)

Strength of weak ties (Granovetter, 1973)

Tipping points (Gladwell, 2000)

Tragedy of the commons (Hardin, 1968; Senge, 1990; Ostrom, 1999)

Source: Adapted from discussions of different Theories of Change in Halpern, Bates, Mulgan, and Aldridge (2004), Australian Public Service Commission (2007a), Sabatier (2007), Coffman (2007), Astbury (2009), and DiClemente, Crosby, and Kegler (2009)

- *Empowerment theory* (Perkins and Zimmerman, 1995). This theory may relate to individuals, groups, or communities.

- *Diffusion theory* (E. Rogers, 1995). This theory is largely about changing community behaviors and behaviors of individuals en masse.

- *Socioecological theory* (Bronfenbrenner, 1979). This theory is about mechanisms for change for individuals, families, groups, and communities and the interplay among all those actors.

- *Network theory* (Granovetter, 1973). This theory is about how the relationships, networks, and connections among entities, and not just the characteristics of the entities themselves, affect outcomes. The entities could be individuals, organizations, special issues groups, or even whole countries.

Even theories that focus on a particular level (individuals, groups, and so on) can have implications for change at other levels. For example, stages of change theory focuses on changing the behaviors of individuals, but its therapeutic application may include recruiting the family as allies and effecting changes to the environment in which the individual lives. Also stages of change theory can be applied at a population level (by marketers, for example), not just at an individual level. Kirkpatrick's model (Kirkpatrick, 1998) is about changing the behavior of individuals to achieve organizational results.

Many programs incorporate Theories of Change that operate at several levels. In fact, socioecological theories of change draw attention to the importance of working at several different levels and, more particularly, on the interrelationships among the levels.

Each of the Theories of Change that we have selected for discussion has strengths and weaknesses, proponents and opponents, circumstances in which it is more likely to apply and others in which it might break down. It would therefore be a mistake to consider these theories as having been in some sense proven or as representing the ideal. Like theories in any other domain of inquiry, most continue to evolve. Nevertheless, given that they have been explored in many contexts and subjected to peer review, they can add some insights and avenues for exploration by practitioners.

The theories are not tightly prescribed recipes for success, and applying them as such could constrain one's receptiveness to looking at problems and options for addressing those problems. If you are going to use a particular theory, consider its applicability carefully and look at the research that has been published about it. It is also important to recognize that every Theory of Change (including homespun theories of change) carries with it particular values and ideological positions about, for example, the role and responsibility of the individual versus the state.

THEORY OF REASONED ACTION AND THEORY OF PLANNED BEHAVIOR

The theory of reasoned action and the theory of planned behavior have been applied in many contexts, such as for changing health- and safety-related behaviors and changing behaviors that affect the physical environment. The theory of reasoned action (Ajzen and Fishbein, 1980; Fishbein and Ajzen 1975, 2009) is based on the notion that humans are rational and have control over what we do. Our intentions predict our behaviors, provided that the intentions and the behaviors are defined and measured using the same concepts in relation to the nature and target of the action, and the timing and context of behavior. A person's intentions to behave in a particular way are in turn influenced by that person's:

- Beliefs about the likely consequences of behavior (the belief and knowledge component).
- Attitudes (whether they have positive or negative feelings) toward the behavior and the consequences of the behavior.
- Perceptions of norms, that is, other people's opinions about the behavior.

However, the relative influence of attitudes and norms has been found to vary from population to population.

One of the criticisms of the theory of reasoned action was that it focused on the individual and did not take into account the role of context in influencing behaviors. Accordingly, the theory of planned behavior (Ajzen and Fishbein, 1980) added, as a determinant of behavior and intention to behave, the concept of a person's perception about his or her control over opportunities, resources, and skills needed to perform a behavior. This concept is similar to the concept of self-efficacy included in social learning theory (Bandura, 1997). A simple representation of the theory of planned behavior is shown in Figure 11.1.

Some Theory Limitations

A criticism of both the theory of reasoned action and the theory of planned behavior is that in some instances, people may change their behaviors first (for example, because of laws or regulations) and then, having changed their behaviors, their hearts and minds follow suit: their attitudes and normative beliefs also change. Cognitive dissonance theory (Festinger, 1957) would be relevant to this phenomenon. On some occasions, it is more effective

Figure 11.1 **Schematic Representation of the Theory of Planned Behavior**

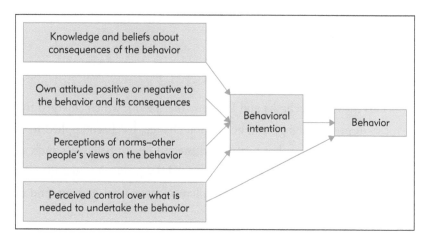

to use legislation and regulations to change behavior (especially where attitudes are very entrenched). When doing so, it would be expected that once behaviors have occurred, attitudes that are consistent with those behaviors will follow; social norms will develop and reinforce attitudes and behaviors. The introduction of affirmative action legislation was one example of this approach. Beliefs about the consequences of noncompliance (one of the factors that this theory would predict would influence behavior) were likely to have played a significant role in changing behavior.

Relevance to Building a Program Theory and Measuring Outcomes

An implication of the theory of reasoned action and theory of planned behavior is that interventions that seek to change behavior must go beyond changing beliefs. They must also change attitudes, normative beliefs, and perceptions of individuals concerning their capacity to control their behaviors (self-efficacy). Criticisms of the theory in terms of its excessive focus on the individual, with little attention to context, reinforce the importance of including those other contextual factors in the theory of change of a particular program.

Considerations about the circumstances in which attitudes are likely to follow behavior change, and vice versa, would also be relevant to choosing an appropriate program archetype (for example, education, incentives, or penalties), deciding when to measure which outcomes, and including measures of outcomes that recognize virtuous and vicious circles.

STAGES OF CHANGE THEORY

This theory (Prochaska and DiClemente, 1983), which emerged from theories in psychotherapy and behavioral change, has been applied to efforts to reduce smoking, other health and mental health behaviors such as weight control, alcohol abuse, condom use for HIV protection, use of sunscreen to prevent skin cancer, medical compliance, stress management, and organizational change.

As the name suggests, the organizing construct of this theory is that behavior change goes through stages. It has three core subconstructs:

1. *Temporal dimension* (Figure 11.2). Individual behavioral change is complex. It unfolds over time through a sequence of stages:

   Precontemplation—not intending to take action in the foreseeable future. Clients may be uninformed or underinformed about the consequences of their behavior or have tried unsuccessfully to change and are demoralized.

   Contemplation—intending to change in, say, the next six months. The person is weighing the costs and benefits of changing and may be procrastinating.

   Preparation—intending to take action in the immediate future, say in the next month. The person has a plan of action, such as joining a health education class.

   Action—has made specific overt modifications in lifestyles in the past six months or so and in the case of health problems has attained a criterion that scientists and professionals agree is sufficient to reduce risk for disease.

   Maintenance—working to prevent relapse and increasingly more confident that they can continue their change.

Figure 11.2   **Temporal Dimension as the Basis for the Stages of Change**

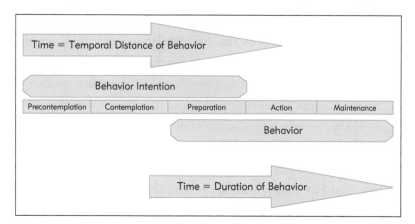

Source: From the Cancer Prevention Research Center: www.uri.edu/research/cprc

Behavior change is not necessarily linear and can be shown as a spiral. Relapse is possible at any point. More recently, the theory has been developed further and renamed the transtheoretical model. It includes a sixth stage, termination, at which point a person is sure that he or she will not return to old habits as a way of coping (Prochaska and Velicer, 1997).

2. *Incorporation of intermediate outcomes as dependent variables and measures, not just final outcomes.* This has relevance to the development of an outcomes chain for a program. Individuals engage in a series of change processes when attempting to change and as they move through each stage. These include decisional balance (individuals weigh the pros and cons of changing and self-efficacy) and individual change, which is influenced by confidence and temptation. Self-efficacy is a good predictor of relapse.

3. *Independent variables and measures are about how change occurs.* They provide important guides for intervention programs since processes are the independent variables that people need to apply or be engaged in. Processes of change can be broadly classified into those that are experiential and those that are behavioral (www.uri .edu/research/cprc/TTM/detailedoverview.htm).

Although stages of change theory is cast as a model that relates to individuals, social marketers sometimes use similar concepts applied to populations: precontemplation, contemplation, preparation and action, and maintenance. Social marketers recognize that the population will include people who are at each of these different stages and develop their campaigns to cater to all or deliberately selected groups or stages. Social marketers often target the groups with which they believe they can have the greatest impact, for example, in the case of smokers, those who are close to attempting to quit smoking and those who successfully quit but require maintenance. Similarly, a program using social marketing to change environmental behaviors targeted those who had already started to think about doing the desired behaviors and those who were doing the desired behavior some of the time (Lee and Jacobson, 2006).

Some Limitations

Some limits to the application of stages of change theory include the fact that some populations would not be expected to move through the stages of change as discussed. These include populations with severe intellectual disability, since movement through the stages requires the capacity to reflect and reason.

One of the criticisms of stages of change theory is that it gives insufficient attention to the impacts of the external environment on the behavior of individuals. Other external factors include such things as the degree of peer support, community norms, conflicting or supportive messages coming from regulations, incentives, and pricing signals.

There are other criticisms of the stages of change theory as well:

- A lack of clarity of the different stages and difficulty in differentiating among them and measuring them.

- Insufficient attention to the factors that precipitate movement from one stage to the next.

- Confusion concerning the links between particular time periods and particular stages, that is, that different time periods have been used as part of the definition of the stages but may be arbitrary.

Sutton (2001) discusses evidence concerning the validity of the theory.

Relevance to Building a Program Theory and Measuring Outcomes

Stages of change theory is likely to be particularly useful for case management programs that work with individuals and in which the individual's progression through stages can be monitored and the case management program adapted as needed. This makes it more appropriate for recruiting the entire relevant population to a program and not just those who fit a particular stage and outcome.

The attention given by stages of change theory to intermediate dependent variables, and not just final results, makes it useful for fleshing out the detail of an outcomes chain and identifying what intermediate outcomes should be measured. We used this theory to develop a program theory for a call center providing assistance to problem gamblers. The call center applied principles of case management. Stages of change were incorporated in the outcomes chain for that program, just as they were in the counseling processes for that program.

Different measures of outcomes are appropriate for clients entering at different stages. The time by which end point outcomes (for example, a person gives up smoking for good) can be expected to occur varies for clients entering at different stages. These expected variations have implications for the success criteria for each outcome in the outcomes chain (see Chapter Eight for a discussion of success criteria and their role in program theory and evaluation and their inclusion in a program theory matrix). Also the time at which measurement and evaluation occur could coincide with a time of relapse for particular individuals. So the stages of change theory would suggest that information about the pattern of development for individuals should be included among the measures of success and not just their status at a particular time.

Knowledge of the important independent variables can assist with identifying program factors that will affect the achievement of outcomes. Those variables could be incorporated in the program factors column of the matrix (see Chapter Eight for a discussion of the importance of identifying program factors that will affect success in addition to program activities and resources). So a factor that might affect the success of the program would be the extent to which it was able to flexibly provide a range of strategies to address different stages and individual needs. Table 11.1 shows some common experiential and behavioral approaches used to achieve outcomes at the various stages that are incorporated in stages of change theory.

When the program theory matrix shown in Chapter Eight is used, we would have the range of strategies in the activities column. The actual range of activities could be compared with the full range to make a judgment about the quality and likely effectiveness of the program.

Table 11.1 **Stages of Change Theories Useful Strategies for Different Processes of Change**

Experiential change processes: Most relevant during pre-contemplation, contemplation, and preparation stages	*Some possible intervention strategies*
1. *Consciousness raising* about the causes, consequences, and cures for a particular problem behavior	• Feedback • Education • Confrontation • Interpretation • Media campaigns
2. *Dramatic relief* initially produces emotional experiences followed by reduced affect if appropriate action is taken	• Psychodrama • Role playing • Grieving • Personal testimonies • Media campaigns
3. *Environmental re-evaluation* combines both affective and cognitive assessments of how the presence or absence of a personal habit affects ones social environment	• Empathy training • Documentaries • Family interventions
4. *Social liberation* requires an increase in recognition that there are social opportunities or alternatives, especially for people who are relatively deprived or oppressed	• Advocacy • Empowerment procedures • Improving opportunities and pressures for desired behaviors, e.g., smoke-free zones, salads for school lunches, easy access to condoms

Table 11.1 *(Continued)*

Behavioral Change processes *Most relevant during preparation,* *action, and maintenance stages*	*Some possible intervention strategies*
5. *Self-re-evaluation* combines both cognitive and affective assessments of self-image with and without a particular habit such as for physical fitness interventions, one's image as a couch potato, or an active person	• Value clarification • Healthy role models • Imagery
6. *Stimulus control* removes cues for unhealthy habits/undesirable behavior and adds prompts for healthier/desirable alternatives	• Avoidance • Environmental reengineering • Self-help groups can provide stimuli to support change and reduce relapse
7. *Helping relationships* combining caring, trust, openness, acceptance, and support for desired change	• Rapport building • Therapeutic alliance • Counselor calls • Buddy systems
8. *Counter conditioning:* learning healthier behaviors that are a substitute for problem behaviors	• Relaxation to counter stress • Assertiveness to counter peer pressure • Acceptable alternatives such as fat-free foods; nicotine replacement
9. *Reinforcement management* provides consequences for taking steps in a particular direction—rewards for self changers; punishments	• Contingency contracts • Overt and covert reinforcements • Positive self statements and group recognition
10. *Self-liberation*—belief that one can change and recommitment to act on that belief	• New year's resolutions • Public testimonies • Multiple choices of actions are more effective than single choices

Source: Information provided by the Cancer Prevention Research Center:
http://www.uri.edu/research/cprc/TTM/detailedoverview.htm

We have noted that one of the criticisms of stages of change theory is its lack of attention to context and external factors. If this theory is being applied to a particular program, then that program's theory of change can include reference to context and external factors to enrich the concepts provided through stages of change theory. Chapter Eight emphasized the importance of incorporating external factors and how a program will address or risk-manage those factors in a program theory.

Although stages of change theory focuses on individuals, it can also have implications for mass or population behavior change programs. These programs, while not tracing progression through the stages, can incorporate different intervention strategies in recognition of the fact that different segments of the population will be at different stages of change with respect to the social problem. For example, an antilittering strategy incorporated some changes to the physical environment through the location of trash cans. In doing so, it was catering to that segment of the population at the preparation, action, or maintenance stages with respect to changing their littering behavior and who would therefore be likely to respond to social liberation and stimulus control. The further inclusion of antilittering regulations in the strategy was a means of generating contemplation. The inclusion of community development activities in the strategy was pitched at the preparation and action stages and the activities were about cultivating social norms that would maintain reduced littering behavior.

EMPOWERMENT THEORY

Empowerment is a process by which people gain control and mastery over their lives and are able to influence others who affect their lives. It emphasizes democratic participation, improvement, and self-determination. It may be applied to individuals but is also often applied to organizations, communities, societies, and cultures. Some of the literature (Perkins and Zimmerman, 1995) distinguishes between empowering processes and empowering outcomes.

Empowerment theory (Perkins and Zimmerman, 1995) has roots in community psychology, citizen participation, and action research. Programs built using the empowerment model make the following assumptions:

- Problems are best addressed by the people experiencing them.
- People possess valuable knowledge about their own needs, values, and goals.
- People possess strengths that should be recognized and built on.
- Processes can be implemented that develop independent problem solvers and decision makers.

Empowerment theory provides the philosophical underpinnings for strengths or assets-based approaches to community development (Kretzmann, 1995; Kretzmann and McKnight, 1993). In Chapter Twelve, we discuss a program archetype, which we refer to as the community capacity-building archetype, that incorporates some empowerment theory concepts.

Some empowerment comes through a sense of being part of something bigger. This was a key factor affecting the success of a program to educate the community about environmentally sustainable behaviors. A sense of being part of the collective emerged as one of the most powerful driving forces in this program, and participants actually referred to being empowered.

Some Limitations

Empowerment concepts provide a useful general guide for developing preventive interventions in which participants feel they have an important stake (see Rappaport, 1987). However, the empowerment approach assumes a desire to participate. In community capacity-building programs that we have evaluated, we have found that this assumption is not always valid. Threats to the validity of the assumption include issues of readiness, other priorities among those who are most disempowered who may be too busy managing day-to-day crises and too weary to participate, and dangers (perceived or real) associated with participating. Selecting an appropriate type of empowerment is therefore important.

Perkins and Zimmerman (1995) identify some circumstances in which empowerment could be counterproductive:

We must be wary of restricting community psychology by concentrating too much attention on a single construct. Although empowerment does provide the field with a useful approach for working in communities and is a compelling construct clearly in need of further research it is not the only approach nor is it a panacea. Efforts to exert control in some contexts may actually create rather than solve problems in a person's life.

Consider an individual who lives in an oppressive society where organizing one's community around a social issue may result in greater authoritarian control and individual and community disempowerment. Or consider the analogous situation of an urban teenager who tries to exert some control in his neighborhood by confronting a local gang. We need to be more precise about the construct and research it as thoughtfully as other psychological constructs or it will remain a warm and fuzzy one size fits all concept with no clear or consistent meaning [pp. 571—572].

It is important to recognize that empowerment approaches are not always inclusive. By empowering some people, they have the capacity to further marginalize some members of the community who, for whatever reason, do not wish to participate or are unable to do so. There may be a need to complement the empowerment activities with safety nets for those who do not wish to participate.

Relevance to Building a Program Theory and Measuring Outcomes

Empowerment theory is particularly relevant to community-driven capacity-building programs and case management programs in which the individual is given a strong agenda-setting and control role.

Empowerment theory presents some challenges for program funders and managers in that in order to empower individuals and communities, they need to surrender some of their own power. Because the community is empowered, it is difficult to know in what direction a project or program might develop, and it could be somewhat different from what program funders and managers had in mind. Incorporating the potential for emergent outcomes and, in fact, valuing such outcomes as indicators of the success of the program is therefore important when developing the program theory for a program based on the principles of empowerment theory. Programs with complex aspects are the most likely candidates for application of this theory.

The fact that outcomes of an intervention based on empowerment theory are especially likely to be emergent rather than preordained also limits the extent to which standardized highly specific measures of outcomes will be appropriate.

DIFFUSION THEORY

Diffusion theory holds that change occurs when new ideas are invented, diffused, adopted, or rejected, thereby leading to certain consequences. Everett Rogers (1995) has been the key proponent of it, with the first edition of his book having been published in 1962.

Diffusion is a process in which an innovation is communicated through certain channels over time among members of a social system. More recent emphasis of this theory has been on information exchange among participants, networks of individuals, and groups. Diffusion is therefore a two-way process: participants create and share information.

Diffusion theory has been widely applied in agriculture, public health, nutrition, and family planning programs. It has been particularly useful for understanding take-up of technology and for diffusion of innovations (for example, see Reed, 2006).

Key Concepts and Related Theories

Diffusion theory has four key concepts and related theories.

Theory of Perceived Attributes
Innovation may be an idea, practice, or objective that is perceived as new and has perceived characteristics that will affect whether it is adopted. In particular, individuals will adopt an innovation if they perceive that it has the following attributes:

- Relative advantage over an existing innovation or the status quo.

- Compatibility with existing values and practices.

- Not too complex.

- Trialable, that is, able to be tested for a limited time without committing to adoption.

- Observable results.

Communication Channels   Communication channels refer to any of various means by which messages get from one individual to another: mass media versus interpersonal. The theory has a number of propositions about how communication channels work, including those that follow. Mass media are rapid and efficient for contacting large numbers of potential adopters and useful for increasing awareness of an innovation. Interpersonal contact is more effective in persuading people to accept a new idea. Face-to-face communication among individuals of the same socioeconomic status and educational level increases the potential for acceptance even more. Effective communication can be highly dependent on engagement of peers and opinion leaders. Different processes work better in different situations and with different social groups (Rogers, 1995). For example, heterogeneous and tolerant social systems encourage change from social norms. Opinion leaders are also more innovative, so it is possible to target them and rely on trickle-down effects. Homogeneous social systems that may be less accepting of divergence will tend toward social norms and can be more difficult to change. Ideas are less likely to trickle down from elite opinion leaders who risk being rejected because their power base comes from holding the party line. In this case, a program may need to target a wider group of opinion leaders rather than rely on trickle-down effects. The program needs to persuade opinion leaders that the innovation is consistent with social norms.

Diffusion Among Members of a Social System   Diffusion is affected by social structure, system norms, opinion leaders, and change agents. Social structure is necessary within the system to provide regularity and stability and to be able to predict others' behavior with some accuracy. Not all members communicate equally with each other. Knowledge of patterns of communication can help with predicting when and by whom an innovation will be adopted. Who talks to whom? How often? About what? Network theory, which we discuss later in this chapter, and social network analysis (Davies, 2002; Durland, 2005) can be useful for this purpose as part of program design. Also Yates (2001) points to the importance of having both macrotheory (systemic adoption, that is, organizational and structural change) and microtheory (individual change).

A Time Dimension to Diffusion   Innovation decision process theory relates to an adoption process that has five steps:

1. *Knowledge:* Potential adopters become aware of an innovation and search for information about how and why it is applicable to their situation.

2. *Persuasion:* Potential adopters search for innovation evaluation information that addresses the uncertainty about the consequences of an innovation to the personal situation of the potential adopters. They form a favorable or unfavorable attitude toward the innovation.

3. *Decision:* Potential adopters engage in activities that lead to a decision to adopt or reject an innovation.

4. *Implementation:* Adopters use and possibly reinvent the innovation.

5. *Confirmation:* Potential adopters evaluate the results of their decision and either confirm adoption or reject it.

However the process is a social one, not just an information gathering one for rational decision making.

In individual innovativeness theory, individuals differ in how innovative they are. Rogers suggested that the different categories of adopters followed a normal curve, with the percentage of each in the community being innovators (2.5 percent), early adopters (13.5 percent), early majority (34 percent), late majority (34 percent), and laggards (16 percent).

In the theory of rate of adoption, adoption of an innovation follows an S curve (Figure 11.3). It grows slowly and gradually in the beginning, then has a period of rapid growth, followed by tapering off. Each individual is a member of the social system. For most members of the social system, the innovation decision depends on the innovation decisions of other members of the social system. There is a tipping point—a point at which a trend catches on, typically marked by opinion leader adoption. Targeting opinion leaders is therefore an important strategy for a program that is based on diffusion theory. However, the types of opinion leaders that programs should target depend on the nature of the social system and in particular whether its values are conservative or innovative.

Figure 11.3   **S Curve of Diffusion Theory**

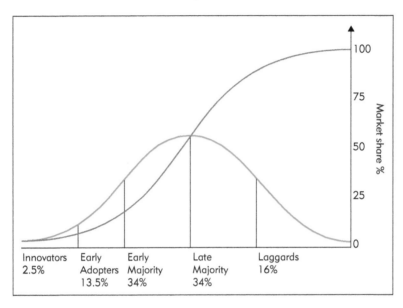

Some Limitations

Escalating use of technology and social networking since the period over which diffusion theory was developed can be expected to have significant implications for the continuing validity of its principles or the specific ways in which it plays out in practice. New forms of social networking will require continuing development of new concepts under the broad umbrella of diffusion theory. For example, lessons learned in the past about the relative importance of mass media versus interpersonal contact may need to be revisited and the implications of the burgeoning size and more inclusive composition of peer groups explored. Time spans for the diffusion process may be considerably compressed with increased speed of transmission of information and rapid exchange of experiences.

The contexts in which diffusion theory has been applied for bringing about change have largely been those in which organizations have ideas or products to sell. The theory does include the possibility that communication is two way and that adopters may adapt the product or idea during the implementation phase. However, the theory may give insufficient attention to the more extensive and active role of adopters as producers who progressively, actively, and publicly expand on and formulate ideas and products and may take them off in a different direction, like a school of fish changing direction. Urban Dictionary, in which contributors have a relatively unbridled opportunity to contribute ideas and definitions, and Wikipedia, a more controlled approach, are examples of technology that give strong agency to their audiences.

Relevance for Building a Program Theory and Measuring Outcomes

Understanding these features of dissemination and adoption has implications for developing a program theory, including undertaking the situation analysis and appreciating the time spans over which change can be expected to occur on a large scale and where to look for the changes (for example, with opinion leaders or community). Understanding where in the diffusion process a particular community stands at the time of the intervention and the nature of the community can assist with setting realistically achievable objectives about outcomes to be achieved and selecting appropriate interventions.

SOCIOECOLOGICAL THEORY

This model was developed from concepts of ecology as they apply in biological sciences. *Ecology* refers to the interrelationships between organisms and their environment. People are part of a system or systems. The model emphasizes that behaviors are influenced by the environment in which people function: social, cultural, physical, and other external environmental variables. Individuals also influence their environments. The model blends concepts from ecology with those from sociology and psychology and has been applied in a wide range of contexts, including child development, health, and community development (Bronfenbrenner, 1979, 1994).

Systems are a central concept of socioecological theory. They operate at several different levels and can be "conceived as a set of nested structures each inside the other like a set of Russian dolls" (Bronfenbrenner, 1994, p. 1644). In a microsystem, the individual, say a child, interacts with various other systems such as home and school. In a mesosystem, these various systems in which the child is involved interact with each other. In an exosystem, the person is not involved in at least one of the interacting systems (for example, a child may be involved with her school but not in the relationship between her school and her parents' workplace). The macrosystem is the overarching system of belief systems, culture, customs, and social norms that permeate systems at the other levels. The chronosystem sets the other systems within the passage of time, moving the systems through a third dimension, recognizing the changes that occur over time in the individual and the systems with which the individual interacts. We have developed Figure 11.4 to depict

Figure 11.4 **Socioecological Model for Child Development**

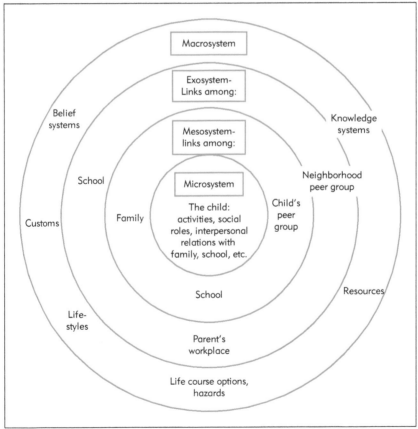

Source: The model was constructed from information included in Bronfenbrenner (1994)

these systems in simple form, drawing on Bronfenbrenner (1994). It includes elements of all systems except chronosystems.

In this example, the family, school, community, and broader society, as well as the child's own attributes, contribute to the child's development in complex, interacting ways. Child development is seen as a process in which the child's characteristics (biological, psychological, and other) interact with the environment so that the child affects its various environments (family, school, neighborhood) as well as being influenced by those environments. Parents and family remain significant influences during childhood but are gradually supplemented by other sources of influence, such as peers and the school environment. The larger social, structural, economic, political, and cultural environment has an impact on the resources available to families and children.

A program for working with at-risk children was an example of the application of socioecological theory. Its program theory incorporated known risk and protective factors for the child derived from various parts of the child's ecosystem: child characteristics, parents and their parenting style, family factors, life events, and community factors (including neighborhood, the wider community, and the social context).

Some Limitations

The general principles of socioecological theory can be applied in many different contexts. However, to be most useful, it presupposes that substantial research evidence is available about the domain in question. Its appeal in the context of interventions that relate to child development in part reflects its origins but also that much research has been done on child development, so the various factors (for example, risk factors and enabling factors) are quite well known. In less researched and understood domains or in specific aspects of various domains, the general principles of socioecological theory may be more difficult to apply directly to the development of program theory for a particular intervention. Socioecological theory could, however, help with generating various classes of research questions, such as around influential factors in the immediate environment.

Relevance to Building a Program Theory and Measuring Outcomes

For program theory development purposes, consideration of this range of environmental factors could be useful for identifying program and nonprogram factors that affect successful achievement of outcomes, as discussed in Chapter Eight. In addition, the systems approach that is inherent in socioecological theory reinforces that when a program theory is developed, many different strategies may be required to work on different parts of the system and on the interrelationships among the parts.

NETWORK THEORY

Although we cite Granovetter (1973) as a reference for network theory, it is in fact difficult to identify a single person as its creator. Granovetter is often identified as the person who began theorizing from what in the past had been a methodological approach to mapping networks rather than a theoretical focus on the impacts of networks. Network analysis methodology has consisted of a set of increasingly sophisticated practical methods for mapping relationships with transdisciplinary origins coming, for example, from sociology, anthropology, social psychology, communications, and mathematics.

Network theory posits that "actors' position in a set of relationships or networks can enhance or constrain their actions. . . . The position of actors and the type and nature of the relationships with others in the network determines the outcomes" (Keast and Brown, 2005, p. 6). Actors could be people, organizations or groups that are not clearly defined, such as interest or special issues groups (Durland, 2005), communities, or even nation-states (Knoke and Kuklinski, 1982).

Networks can function as systems that evolve over time. Davies (2002) uses the diagram in Figure 11.5 to portray that "in networks the trajectory of change is a series of consecutive states in the network as a whole. Some of these states may be recurrent, others new. National elections and economic recessions are large scale examples of the former" (p. 19). Changes could relate to various features of the networks, including the membership and the nature, direction, and strength of the relationships.

Figure 11.5　**Evolving Networks**

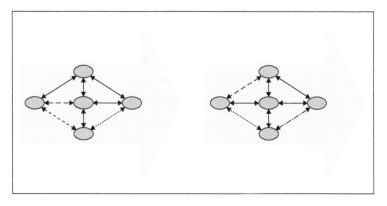

Source: Davies (2002, p. 19)

Davis goes on to say that "in CAS [complex adaptive systems] analyses the outcomes of networks of interacting agents are by definition collective constructions. Often they are described as emergent because of the difficulty of drawing out which agents, if any, had particular responsibility for which part of the final outcome."

Network theory concepts have been used as a source of both independent variables and dependent variables. An example of using networks as an independent variable is Granovetter's 1973 article on strong and weak ties that purported to demonstrate that weak ties (casual acquaintances) were more effective than strong ties in helping a person get a job. This article gave rise to a great deal of debate, theorizing, and research around various aspects of how relationships work to produce results, the circumstances under which they do and do not work, and the effects of different characteristics of relationships. That is, the practical methods of network analysis came to be used to develop and test theories rather than theories giving rise to the development of methods to test the theory.

Networks may also be a dependent variable. Increasingly organizations and programs try to achieve their objectives through achieving intermediate outcomes that involve establishing effective networks and changing their composition, functioning, and outcomes (Biggs and Matsaert, 1999). These networks are rarely an end in themselves. They usually are part of a more comprehensive theory of change that would include, for example, the types of changes that networks can effect (for example, knowledge synthesis, advocacy, and resource mobilization) to achieve longer-term outcomes (for example, in education, health, or economic productivity). Measuring the composition, structure, and functioning of networks and changes over time will therefore become an important part of evaluating the success of such programs. Diagrams such as that in Figure 11.5 could provide a useful template for the purpose of tracking desired changes when the purpose of an intervention is to transform a network and the way in which it operates.

Much research has been undertaken about the features of relationships that affect outcomes (see Keast and Brown, 2005, for a short summary of some key research studies). Network theory holds that relationships vary along many dimensions, including the direction of influence and whether it is one way or reciprocal, the value placed on the relationship, the centrality of influence by various parties in the network, and the density of networks.

The nature of the relationships may also vary with respect to purposes and context. Examples are networks whose primary purposes are economic transactions, instrumental purposes such as helping each other with tasks or integration of service delivery across organizations; purposes of exchanging or information for gaining access to other relationships, groups, or networks (sometimes referred to as bridging); social, personal, and emotional purposes; familial and caring relationships; and authority or power relationships. The articulation of the differing nature of relationships comes close to being a theory about mechanisms by which change can occur through relationships.

Some Limitations

In this discussion of limitations, we also identify some ways in which those limitations could also be seen as strengths and some emerging developments that can help to address some of these limitations.

Many hypotheses and much research have addressed how networks work, but what has been learned about networks has not been pulled together into a coherent theory of change. Indeed Everett Rogers (1987) identified theory-less research as one of the problems of network research. Since then, some advances in theory have been made in particular contexts. For example, Provan and Milward (1995) identified lessons learned from research on networks in the context of community mental health networks, and Glasbergen (2010) identified lessons learned about what makes global action networks effective.

In some regards, it is questionable as to whether network theory can really be called a theory of change rather than a methodology or a domain of research about how networks function. There has been much debate in the literature concerning this issue and whether it would be more useful and realistic for knowledge synthesis to focus on developing a series of midrange theories rather than an all-encompassing single theory (see, for example, Kilduff and Tsai, 2003, for a discussion of these issues). Nevertheless, network concepts have been usefully applied to investigations in a wide range of domains, such as business, epidemiology, community development through social capital, global action networks, and spread of ideas (and viruses) through the Internet. The breadth of their application and the extent to which network concepts have spawned theory and research in other areas such as the theory of social capital (for example Lin, Cook, and Burt, 2001) are therefore commendable. Network theory has both borrowed from a wide range of disciplines and areas of investigation and contributed to their further development.

As for many other areas of investigation, hypotheses explored under the rubric of network theory continue to produce conflicting results. Many of the findings about the influence of strong and weak ties and the circumstances under which each type is more (and less) helpful, for example, continue to be contested by empirical research and the interpretation of the research debated. The lack of clarity about what works can be an irritant to program designers and evaluators and means that a program theorist who is applying network theory needs to proceed with caution. On the other hand, it may be that the sophistication that has been achieved in the measurement of networks has enabled more definitive testing of hypotheses with more potential for clearly conflicting results. The methodology may present greater opportunities to explore more rigorously what works for whom in which circumstances than has been the case in other areas of evaluation dogged by difficulties in establishing comparable and precise measurement methods.

One of the greatest strengths of network theory is its focus on aspects of complexity. This focus can lead to complex representations using more and more sophisticated modeling and mathematical processes. These can go beyond the bounds of human comprehension and reduce the usefulness of network theory and its analytical methods for program design, monitoring, and evaluation. However, most networking software enables filters to be applied that can reduce the complexity of representation.

Debate continues around network concepts and measurement methods, and some of these issues continue to pose challenges for evaluators wishing to use network theory. However, methodological debate is not unique to this theory and is likely to be a sign of a healthy domain of investigation. An example of a controversial area of network theory and its associated measurement techniques is the theoretical discussion and empirical investigation around issues such as whether the boundaries of networks should be established from the perspective of network participants or from the perspective of researchers (see Laumann, Marsden, and Prensky, 1983). Ethical issues arising from the disclosure of relationships and how they function also need to be carefully considered when applying social network analysis techniques. Indeed the journal *Social Networks* devoted an entire issue (2005, Volume 27, Issue 2) to the topic of ethical issues in social network analysis. The issue includes an article on guidelines for network research in organizations (Borgatti and Molina, 2005).

Relevance to Building a Program Theory and Measuring Outcomes

Network theory lends itself ideally to pictorial representation as a means of generating discussion of its validity, establishing hypotheses and intuitions about relationships that can be tested. At the point of developing a program theory, social network analysis techniques can be used to test assumptions based on network theory or even just network concepts. For an intervention that seeks to influence or draw on the relationships, social network analysis techniques can be used to collect baseline data, set a direction for change, and then measure the impacts of an intervention on relationships (direction, strength, centrality, and so on), and the extent to which those relationships affect outcomes. Durland and Fredericks (2005) provide some examples of such applications.

One of the benefits of network theory is that it shifts the focus of thinking from individuals and their attributes to relationships and systems and their attributes. This is especially important in a world in which "we can no longer personally experience and acquire learning that we need to act. We derive our competence from forming connections" (Siemens, 2005, p. 4). Well-designed interventions need to capitalize on connections. Given its systems orientation, network theory is especially useful for complex programs, for considering structures, and for looking at the whole as more than the sum of its parts. Even so, Kilduff and Tsai (2003) have demonstrated that the attributes of particular actors (components of the system) have an impact on the ways in which networks function and therefore also need to be carefully considered when applying social network analysis techniques. In moving to a focus on networks, it is important not to ignore the features of and roles played by actors in networks, whether they are individuals, groups, or organizations. Evaluations need to consider components and their relationships, not just one or the other.

Network theory can be a valuable adjunct to outcome mapping and the network aspects of participatory impact pathways analysis by drawing attention to the types of considerations that should be incorporated in those approaches. It can also be used to turbocharge other theories such as diffusion theory (E. Rogers, 1987).

SELECTING AND USING THEORIES OF CHANGE

In this chapter we have discussed only a small selection of Theories of Change. A particular program theory may quite constructively borrow concepts from several Theories of Change. Moreover some concepts (for example, self-efficacy) recur in several Theories of Change. In addition, many other Theories of Change concern specific types of behavior and cognitive, affective, and decision-making processes that can add important concepts (see Exhibit 11.2).

EXHIBIT 11.2 EXAMPLES OF BLENDING VARIOUS ASPECTS OF DIFFERENT THEORIES

The stages of change theory includes a process that is called decisional balance: weighing the pros and cons. Various theories have been developed about how people go about this process; an example is Tversky and Kahneman's theory (1974) about how humans use heuristics and biases to make judgments in situations of uncertainty. These mini-theories may be a useful adjunct to the operation of larger, more comprehensive theories, such as stages of change theory.

Similarly, diffusion theory gives considerable attention to the role of opinion leaders in diffusion of an innovation or idea. Accordingly, other theories about types of authority and power (expert, legitimate, coercive, referent, and so on) that influence people's attitudes and actions could be useful for providing guidance about the circumstances under which opinion leaders may and may not be most influential and how they might be exhorted to exercise their power. As well as having considerable value in its own right, network theory is a useful adjunct to diffusion theory.

Drawing out and debating Theories of Change can be an important part of improving practice. Eyben, Kidder, Rowlands, and Bronstein (2008) reported on their practical experience of exploring and debating several different Theories of Change and concluded that "organizational decision making processes can be better informed and strategic choices made more transparent" (p. 201).

The templates that Theories of Change and program archetypes provide can improve the efficiency of program theory development processes. The templates can also raise questions about the adequacy of the ways in which the program is being conceptualized. Is it missing some ingredients that have been found to be key to the success of similar programs? Are there good reasons (contextual, for example) for omitting those ingredients, or has omission resulted from oversight? We therefore see their usefulness as a point of reference, a heuristic device rather than a recipe or straitjacket.

SUMMARY

In this chapter, we have discussed some research-based Theories of Change that, if used discriminately, can help with the development of and critical reflection on the theory of change for a particular program. These Theories of Change operate at various levels: from individuals, to groups, organizations, communities, and even nations. To demonstrate the potential usefulness of big Theories of Change, we selected six theories for discussion: theory of reasoned action and theory of planned behavior, stages of change theory, empowerment theory, diffusion theory, socioecological theory, and network theory. We summarized the features of these theories and pointed to some strengths and weaknesses of each and their potential relevance for building a program theory and measuring outcomes. We emphasized the importance of treating these theories as heuristic devices to encourage critical reflection rather than as gold standards.

EXERCISES

1. Think of a program with which you are familiar. Which of the following Theories of Change (or parts of those theories) would be relevant to that program: theory of reasoned action and theory of planned behavior, stages of change theory, empowerment theory, diffusion theory, socioecological theory, network theory?

2. In what ways would that theory or theories be relevant? What would be the implications for what might need to be included in the program's theory of change or its theory of action? How could application of the concepts associated with that Theory of Change help to refine the way you think about and measure outcomes?

12

Some Common Program Archetypes

█N THIS CHAPTER, we discuss program archetypes and how they can be used as building blocks when developing a program theory. Program archetypes are classes of interventions that are used to activate mechanisms for change. We describe five common archetypes: information (advice, information provision, and education), carrots and sticks (incentives and sanctions), case management, community capacity building, and direct service delivery programs. Each has its own generic outcomes chain.

SOME IMPORTANT PROGRAM ARCHETYPES

Many programs use a combination of mechanisms and types of intervention to cater to different aspects, causes, and consequences of the public policy issue they are addressing. The program archetypes and their outcomes chains discussed in this chapter were developed from a combination of academic, evaluator, and practitioner knowledge. They have proven useful in working with program managers on program theories over the last two decades.

Their primary use is as a heuristic device to assist managers in identifying the basic elements of the program theory, representing it in a logic model and identifying the monitoring and evaluation implications of that model for their programs. However, having chosen a particular program archetype as a guide for their program, managers need to judge the applicability of the various aspects of that particular archetype to their situation and adapt it as needed. Archetypes, like Theories of Change, provide analogies rather than straitjackets. They should not be used to discourage innovation.

We also see potential for development of other archetypes. One that strikes us as having particular potential is an archetype that draws on network theory (as we discussed in Chapter Eleven) to show how interventions can be designed to use networks and relationships to achieve outcomes. One variation of such an archetype would be the use of network theory to show how existing relationships and coalitions of interest could best be mobilized to achieve desired outcomes (networks as an independent variable). Another variation on that archetype would be for interventions that seek to influence or change networks and relationships (networks as a dependent variable) in order to achieve outcomes. Examples would be interventions addressing issues such as those relating to peace and conflict or the environment. In fact, it is likely that the two approaches and their variations will interact, particularly in complex situations where feedback loops, emergent outcomes, and opportunism are rife. Some promising sources of information for the development of such an archetype (with its own outcomes chain) would be work that is being done using outcome mapping and participatory impact pathways analysis.

ADVISORY, INFORMATION, AND EDUCATION PROGRAM ARCHETYPE

This archetype refers to programs that try to change behavior through changing attitudes, knowledge, and skills. For brevity, we will refer to it as the information program archetype. Nevertheless, these programs often go beyond simple transmission of information. They are likely to use additional processes such as demonstrations, advocacy, counseling, and facilitated learning processes to assist learners to come to their own conclusions. This archetype includes some elements from Bennett's (1975) hierarchy of objectives for agricultural extension programs to change farmer practices and deliver improved results for farms and Kirkpatrick's (1998) four levels of evaluation of training programs, which were discussed in Chapter Two.

Figure 12.1 shows the generic outcomes chain for the information program archetype and applications of it to two programs: an educational program to change the polluting practices of small businesses and a mass media campaign to boost tourism.

This program archetype has been used for such programs as the Best Practices Collection of the Joint United Nations Programme on HIV/AIDS, environmental education programs, road safety education programs in schools, applied research programs undertaken to provide policy advice to government, and various health promotion programs. Training programs and public communications programs whose intended outcomes are about reach, reaction, recall, and response are other examples of where this program archetype has been useful.

Programs that follow the information program archetype make the assumption that the behaviors that are the focus of the program are primarily the result of people knowing or not knowing, having, or not having the right attitudes and interests, and that the provision of information will affect their knowledge, their attitudes, and the way in which they behave. The guiding idea is that if people knew what to do, they would do it. There are many instances in which this assumption holds true—for example, providing technical information that people need in order to change practices where they have an interest in doing so to achieve their goals.

However, with many types of behavior, this assumption is patently untrue or applies only to a segment of the total population whose behavior is being targeted. Examples of behaviors unlikely to be changed simply by the provision of relevant information are those that involve persistent self-harm activities like smoking, overeating, and underexercising or where the perceived benefits of continuing with certain behaviors outweigh the perceived costs.

Figure 12.1 Generic Outcomes Chain for Advisory, Information, and Education Program Archetype and Applications of the Generic Outcomes Chain

Generic outcomes chain for the advisory, information, education program archetype	Applications	
	Educational program to change polluting practices of small businesses	State tourism mass media campaign
6. Desired end results occur; problem, issue, situation, or need is addressed	6. Reduction in pollution caused by target businesses	6. Increased tourism revenue to the state to be used for a variety of purposes both tourism and non-tourism related
5. Participants/target audiences make desired changes in behavior, actions, practices	5. Participating small businesses who have learned what to do from the program and how to do it, increasingly adopt non-polluting practices and serve as a role model for others	5. Increased visitation to the state by high-spending tourists who have been exposed to the campaign (response)
4. Participants/target audiences have desired changes in knowledge, skills, attitudes, and intentions to change	4. Participating small businesses have improved knowledge, skills, and access to resources and intentions to change those practices that are contributing to pollution	4. Desired number and type of potential tourists obtain more information about coming to this state and intend to visit (response)
3. Participants/target audiences have desired reactions to information/advice/learning experiences	3. Participating small businesses exhibit greater awareness of their own polluting practices and are willing to learn more	3. Desired number and type of potential tourists remember the campaign (recall) and have a positive attitude to its messages (reaction)
2. Desired numbers and types of people/ organizations, etc., are exposed to/receive advice/information/learning experiences	2. Targeted small businesses participate in various educational activities (en masse and one to one) offered by the program	2. Desired number and type of potential tourists are exposed to the campaign either directly or through friends who saw it (reach)
1. Desired numbers and types of people/organizations, etc., know about program and choose to participate	1. Targeted small businesses become aware of the program and make contact with it.	

Not all outcomes in the archetypal outcomes chain will be relevant to all programs of that general type. In Figure 12.1, the tourism mass media campaign example does not have a level 1 outcome, which concerns whether people choose to participate in the mass media program. The lowest-level outcome will be relevant to programs for which take-up by the target audience is voluntary. It will not be relevant to programs that are universally provided—for example, school education.

Moreover, for the tourism mass media campaign example, as for outcomes chains for most programs, it is clear that as one ascends the outcomes chain, many other factors come into play, and different segments of the target population will need to be reached. Social marketing approaches would also be needed to adapt the strategies and messages to reach these different segments. In the tourism example, particular segments might be targeted—such as high-spending groups that are currently underrepresented among tourists, or tourists who might be attracted to areas that are currently undervisited. Targeting concerns would help to define what is meant by "desired number and type" of potential tourists. (In Chapter Eight, we introduced a process to define these success criteria.)

Achievement of the top level of the chain of outcomes will be influenced by many factors other than the marketing campaign. Among these would be the quality and cost of the state's tourism facilities, what members of the target group have heard about the state from their friends, their own experience if relevant, and costs and competing tourism attractions. Although the impact of the marketing program may be barely discernible at these upper levels, it is important to include them in the outcomes chain so that the design of the program has a clear sense of purpose and targeting. The extent to which this program continues to be needed can be monitored in terms of the size and composition of the tourist population (visitation is a high-level outcome) and the potential population for visitation.

There will be many potential feedback effects in an information program as participants in the program and other members of the target audience see the effects of the program and the effects of changes in behavior on the social issue or problem that the program addresses. For example, visible positive effects of changes in behavior may reinforce new behaviors and encourage participants to learn further ways of producing positive effects. Previously skeptical nonparticipants who see participants or others benefiting from participating or from changing their behaviors may now choose to participate in the program or adopt those behaviors.

Key Considerations When Using Advisory, Public Information, and Education Programs to Change Behavior

Some key factors to consider when using advisory, public information, and education programs are that:

- Providing information and changing knowledge skills and understanding is not enough: behavior change must follow. This can affect program design decisions such as whether to charge fees for the target audience to participate and whether the intervention needs to be complemented by carrot or stick interventions.

- Targeting participants whose behavior most needs to be changed and engaging them in the program is important. This has implications for use of incentives to get them to participate.

- Advisory programs usually combine mass communication and one-on-one or group communication processes.

- Sources of information must be credible, and opinion leaders are often used.

- Credibility of messages can often be enhanced by presenting both sides of the story.

- These programs will be easier to implement when there is a coincidence between the outcomes desired by the target audience and those of the program.

- Effective programs use feedback processes about successes arising from behavior changes to recruit more participants for the program or audiences for its messages. Application of principles of diffusion theory can assist.

Relationship to Theories of Change

The advisory, information, and education program archetype relates most closely to the theory of reasoned action and theory of planned change. A generic outcomes chain for this archetype would more closely reflect that theory of change if it disaggregated the level 4 outcome in the outcomes chain shown in Figure 12.1 into three stages: knowledge and skills, attitudes (personal, perceptions of norms, and self-efficacy), and intention. Sometimes we have found it useful to disaggregate this level 4 outcome into three or more boxes in a logic model, especially when discretely different activities are used to develop each of the various components (knowledge, attitudes, skills, and so on). On other occasions, it may be better to keep them together for simplicity of presentation.

Diffusion theory may come into play at the point at which contagion from participants to nonparticipants starts to affect rates of change of the desired behaviors. Contagion may occur through such processes as role modeling, observing the benefits, or peer pressure. Of course, the converse could also apply if behavior change among participants is not evident or it leads to negative rather than positive results.

CARROTS AND STICKS PROGRAM ARCHETYPE

These are programs that try to influence behavior through use of incentives (carrots) to promote positive behaviors or use of penalties and threats (sticks) to deter undesirable behaviors. Some programs use a combination of carrots and sticks. Sometimes difficult choices need to be made about whether carrots or sticks will be more successful. On occasion, incentives may be appropriate for some segments of the population and sanctions for others.

Carrots and sticks are really two separate archetypes, but they share some common principles. Both are about motivating behavior, and for both types, an effective communication strategy is critical.

We present the archetype for motivational programs in two versions: one for carrots and one for sticks. Although they are presented largely as mirror images of each other, in some regards, carrots and sticks are not simply reverse sides of the same coin. For example, rewards may be given to everyone who behaves in the desired manner, whereas penalties are applied to those who are exceptions to the rule. These differences can have significant resourcing considerations. Most of the differences, however, seem to be more relevant to the detailed application of sticks and carrots than to their basic concepts.

The program archetype for carrot and stick programs has two arms in the generic outcomes chain. One relates to the direct outcomes of effective program delivery processes (application of incentives and penalties) and the other to effective communication with the target audience. The target population must know about the basis for applying rewards and sanctions and believe they will be applied before required behavior occurs in order for it to influence behavior. People cannot be expected to change their behaviors in anticipation of rewards and sanctions unless they know the rules, know about the consequences, and believe they will be applied.

For this archetype, the sequencing of outcomes relates more to the likely sequence of activities in the intervention strategy than to a necessary order in which outcomes are achieved for each member of the target audience. The target audience could in principle be simultaneously developing an understanding of the rules, a belief that compliance will be detected, and a belief that the behavior, once detected, will be appropriately rewarded or sanctioned.

The order of outcomes shown in the generic outcomes chain does, however, reflect the type of logical thinking that the carrot or stick intervention is attempting to generate in the target audience. When that logical thinking occurs, it is expected to bring about the desired changes in behavior. The change in thinking is the mechanism for change. In relation to the example we use to illustrate the stick archetype—the use of regulations to deter driving while under the influence of alcohol—the new logic that the intervention is trying to generate for potential drunk drivers goes something like this: "I know what is a legally acceptable level of alcohol to drink before driving, and I believe that if I exceed the limit, I stand a reasonable chance of being picked up for drunk driving. If I am picked up, there is a fair chance that I will be heavily fined or lose my driver's license. Therefore I had better not drink and drive." In this case, if intention does not translate into action, then one might expect that the driver who has been drinking, unless totally inebriated, will try to avoid detection and thereby avoid the string of consequences that flows from that.

In the long term, a carrot or stick program cannot function effectively unless the program delivery processes to which it relates are functioning properly and the program has credibility with the intended audience so that it affects their thinking about what they will do.

Motivational programs are typically used when it is considered unlikely that information, advisory, or educational programs alone will be sufficient to bring about behavior change for at least some segments of the population. They are also used in the short term until new social norms are established and often as a means of introducing new social norms.

On a positive note, the hope is that most people will respond to motivational programs (once they understand the rules and the reasons for them) not because of any threat of punishment or hope of reward but simply because they accept that the required behavior is desirable and thus voluntarily choose to enact it. As social norms change over time, the need for extrinsic motivators in the form of carrots and sticks is expected to decline, although occasional reinforcement may be required.

Relationship to Theories of Change

Among the Theories of Change, diffusion theory is likely to be particularly relevant to carrot and stick programs because of the importance of communication strategies and the development of norms. However, those aspects of the theory of reasoned action and theory of planned behavior that relate to the effects of beliefs about the consequences of behavior on whether that behavior occurs will also be relevant.

Carrot Programs to Motivate Behavior

Figure 12.2 shows the generic outcomes chain for incentive programs. The left-hand side of the diagram shows the outcomes chain for the carrots archetype. It has two streams: one that relates to effective actions and one that relates to effective communication processes. The arrow on the right-hand side of the generic outcomes chain shows that some people will behave in the desired manner without expectation of or desire for some form of tangible reward recognition. They have, in effect, internalized the desired behavior and may be intrinsically rather than extrinsically motivated. Figure 12.2 also shows the application of the program archetype to performance-based pay systems.

Figure 12.2   **Generic Outcomes Chain for the Archetype of Motivational Programs That Use Incentives and Applications of That Chain**

Generic outcomes chain for archetype of motivational programs using incentives: carrots

6. Desired end result is achieved

5. Target group behaves in desired ways

4. Target group is motivated to behave in desirable ways (behavioral intent)

OUTCOMES OF SUCCESSFUL COMMUNICATION STRATEGY

3. Target group believes that behavior valued by the program will be rewarded in ways that are valued by recipients and the relevant community

2. Target group believes that desirable and undesirable behaviors will be identified effectively, reliably, and fairly

1. Target group has a clear understanding of desirable behavior and accepts the standards and measures as valuable

OUTCOMES OF SUCCESSFUL ACTION STRATEGY

3. The probability of receiving appropriate rewards for behavior is high and for inappropriate behavior is low

2. Desirable and undesirable behaviors are identified effectively, reliably, and fairly

1. Desirable and undesirable behaviors, practices, and performance standards and measures are clearly identified

Outcomes chain for performance-based pay programs

6. Results for the organization are improved

5. Managers change/maintain their behavior to ensure that desired results are achieved

4. Managers desire/attempt to change their behaviors in ways that will attract rewards

OUTCOMES OF SUCCESSFUL COMMUNICATION STRATEGY

3. Managers believe that rewards are linked to and commensurate with performance and that inappropriate performance is not rewarded

2. Managers believe that processes for measuring and judging performance are effectively, reliably, and fairly administered

1. Managers accept the importance of the criteria and understand and agree with the measures, standards, and processes to be used to assess performance and decide rewards

OUTCOMES OF SUCCESSFUL ACTION STRATEGY

3. Rewards that are given are commensurate with assessed levels of performance and in line with professional standards and measures

2. Performance assessment processes are effective, reliable, and fair

1. Performance-based pay system (criteria, measures, rewards) is clearly defined

Key Considerations When Using Carrots Program Archetype

Some key conditions need to be in place for incentive programs to work:

- Social legitimacy and acceptance by the target group of the behaviors that are to be encouraged by the incentives.

- Clarity concerning which behaviors will and will not be rewarded and the conditions under which they will be rewarded.

- Effective processes for reliably identifying whether the desired behaviors are occurring.

- Effective processes for rewarding the desired behaviors, including setting rewards at an appropriate level and choosing the right types of rewards (financial, recognition, in kind, or other) to encourage behavior and consideration as to whether rewards are needed to stimulate the desired behaviors.

- Effective communication strategies about what behaviors are expected and that confirm the effectiveness of processes for identifying and rewarding those behaviors.

- Acceptance by the target group that rewards are fairly distributed according to behaviors and commensurate with variations in behaviors.

Hatry, Greiner, and Gollub (1981), in a review of local government management motivational programs, commented that motivational programs using incentives to improve productivity had run into major problems: "These appear to have been primarily problems in the program's design and implementation, not the concepts" (p. 34). Some of these problems related to the action strategy and some to the communication strategy. Problems included such difficulties as "providing an adequate number of, and size of, the monetary incentives; providing objective rather than subjective linkages between pay and performance; and overall building a system that was perceived by managers as being fair and meaningful" (p. 28) and "inadequate communication to managers about the program and about the basis for awards that were made" (p. 35).

The potential for corruption or distortion of an incentive system and the importance of selecting functional rather than dysfunctional performance measures are illustrated in the following example from Pawson (2003b).

> Pro-rata cash incentives have recently been offered to poppy grow-ers in Afghanistan to cease production in order to dry up this major source of heroin supply to the West. The result? Local officials have colluded with farmers to exaggerate their production levels. Poppy production has increased as more farmers clamour to cash in on the incentive. Rule one of carrot theory is that high levels of order and regulation are required to make incentives work. Someone should have explained this to the warlords! [p. 488]

Although most incentive or carrot programs operate by holding out the promise of a reward, a softer version exists, with members of the target audience being enticed to behave in certain ways by an up-front incentive. In other words, there is a reward in anticipation of appropriate behaviors occurring.

Pawson (2002) examined a class of programs in health, safety, corrections, transport, housing, and education that he called incentives. His definition of the generative mechanisms that operate for incentive programs is that "the incentive offers deprived subjects the wherewithal to partake in some activity beyond their normal means or outside their normal sphere of interest which then prompts continued activity and thus long term benefit to themselves or their community" (pp. 349–350). An example would be the distribution of free long-life light bulbs to low-income households to encourage their immediate and then longer-term use as a carbon-reduction measure. On occasion, incentives can be used to expedite behaviors that were going to occur anyway or to change the way in which behaviors are undertaken, by, for example, affecting their quality or the amount of effort assigned to them.

Considerable waste of resources will occur if these incentives are not followed by sustained behaviors. Waste can also occur if the target audience was going to behave in the desired way with or without the reward (as shown by the long arrow on the outcomes chain for the carrot archetypal program).

Nevertheless, even if the use of rewards in this situation is in one sense wasteful, their use can be an important means of reinforcing desired behaviors so they become norms, sending consistent messages about which behaviors are valued and showing that the system is operating fairly. As the desired behaviors move closer and closer to being the norm and are accepted by more and more of the target audience, the need for incentives and rewards should decrease. Also, inexpensive carrots such as recognition and appreciation (awards, certificates, media reports, and so on) are available and can act as an incentive or reinforce social norms from time to time, including reminding the target audience of particularly desirable attributes of behavior.

Insights that have been gained about performance-based pay systems and what contributes to and detracts from their success are relevant to many incentive programs. Flannery, Hofrichter, and Platten (1996) summarized the key features of performance-based pay systems as "ultimately about sending messages and changing behavior" (p. 115). The bottom line, they write, is this: "If performance can't be accurately measured, if employees can't understand how it is evaluated, or if they can't see the link between their efforts and the desired results, the program won't work or will be less than fully effective" (p. 249).

Stick Programs to Deter Undesirable Behaviors

The program archetype that uses sanctions and penalties to deter target groups from undesirable behavior and its accompanying generic outcomes chain (adapted from Funnell and Lenne, 1990) was developed from criminology research relating to random breath testing programs to deter driving while drunk. It is also derived from experience with other deterrence programs such as those combating fare evasion, tax evasion, and vandalism (Homel, 1990; Homel, Lenne, and Walker, 1989). As noted, the sticks archetype has many similarities to the carrots archetype.

Figure 12.3 shows the generic outcomes chain for the sticks archetype on the left-hand side and application of the deterrence outcomes chain to random breath-testing programs on the right-hand side.

Figure 12.3   Generic Outcomes Chain for the Archetype of Motivational Programs That Use Deterrence and an Application of That Chain

Generic outcomes chain for archetype of motivational programs using deterrence strategies: sticks	Outcomes chain for a random breath-testing program
6. Problem and its impacts are reduced or prevented	6. Reduced incidence of injuries, mortality, property destruction, and flow on effects in the community due to drunk driving
5. Target group behaves in desired ways	5. Reduced incidence of drunk driving
4. Target group is deterred from behaving in undesirable ways (behavioral intent)	4. Potential drunk drivers intend not to drink when driving and not to drive when drinking
OUTCOMES OF SUCCESSFUL COMMUNICATION STRATEGY	OUTCOMES OF SUCCESSFUL COMMUNICATION STRATEGY
3. Target group believes that inappropriate behavior will be sanctioned, that sanctions will hurt, and that the punishment will fit the crime	3. Drivers perceive there is a high probability that they will incur a significant penalty if detected when drunk driving
2. Target group believes that inappropriate behaviors will be detected	2. Drivers who drink believe there is a considerable likelihood they will be detected if driving while over the legal limit
1. Target group has a clear understanding of undesirable behavior and sanctions to be applied	1. Drivers and the community are aware of drunk driving legislation and of how much they can drink and still comply
OUTCOMES OF SUCCESSFUL ACTION STRATEGY	OUTCOMES OF SUCCESSFUL ACTION STRATEGY
3. The probability of receiving commensurate sanctions for inappropriate behavior is high	3. The probability of receiving a sanction such as a substantial fine or withdrawal of driving license when detected drunk driving is high
2. Undesirable behaviors are effectively detected: probability of detection is high	2. There is a high probability of being detected and stopped by police if driving with a blood alcohol level that is over the limit
1. Undesirable behaviors, practices, and commensurate deterring sanctions are clearly identified	1. Drunk driving legislation clearly specifies legal and illegal blood-alcohol levels for drivers and identifies penalties.

Key Considerations for Applying the Stick Archetype

Typically these programs use regulations and penalties as the policy tool. Successful deterrence programs have the following key components:

- Social legitimacy of the regulations. There must be community support. The deterrence aspects of implementing regulations are intended as a strategy of last resort, targeted to members of the community who are not inclined to behave in the desired manner.

- A sound understanding of community norms based on empirical evidence rather than assumptions. For example, a tax department's study of the causes of tax evasion found that members of the public saw tax evasion to the value of, say, one hundred dollars as being less serious than shoplifting goods worth the same amount. This finding about social norms had implications for strategies needed to gain acceptance that tax evasion is unacceptable.

- Effective recruitment of the community as allies in implementing the regulation. This can be done by engendering peer pressure, supporting the community to speak out against crime and demonstrating that follow-up action occurs, or getting the community to participate in problem solving (Leeuw, Gils, and Kreft, 1992).

- Effective communication processes concerning the legitimacy of the regulation and that persuade the target audience that the regulation will be effectively implemented.

- Integration of the various phases of the deterrence process: detection, prosecution, and application of sanctions.

- Effective processes for registering and licensing the target group, especially in situations where the penalty relates to withdrawal of licenses, as in the program to combat drunk driving.

- Effective screening processes to identify potential offenders.

- Effective inspection processes that detect noncompliance to a sufficient degree to act as a deterrent.

- Effective prosecution processes to ensure that offenders can be successfully prosecuted following detection.

- Penalty processes that ensure that following successful prosecution, the penalty fits the crime. Some stick programs begin with relatively small stick threats in the event of noncompliance and move progressively to larger stick threats in the event that offending behaviors continue. Sometimes the mere threat of a penalty following initial noncompliance is sufficient to bring noncompliers to heel.

Organizational capacity to apply all of the processes shown is critical. If the regulation is not enforced consistently, the message sent to the community will be that the behaviors required by the regulation are not highly valued, and social legitimacy may be eroded. Effective motivational programs using regulatory processes and penalties are expensive to operate and, like all other programs, can fail through ineffectual implementation and lack of integration of the various stages of the deterrence processes. If serious offenders escape detection or, when detected, receive only a symbolic but nondeterrent penalty, then inspection and prosecution procedures may have limited credibility and limited deterrent effects.

One of the greatest impediments to effective operation of motivational programs that use deterrence is that responsibility for the various processes that are needed to achieve the different outcomes in the chain often resides with different agencies, so integration can be difficult. For example, the justice system usually separates policing from prosecuting and penalizing. The follow-through from one step in the chain of outcomes to the next can be lost without some supervision and coordination of the deterrence system as a whole. This includes the links between the processes of detection, prosecution, and sanctions, on the one hand, and communication with the target audience, on the other hand. These programs are classic examples of programs with complicated aspects.

It is through the communication processes and changes in people's beliefs rather than through the implementation of processes leading to sanctions that the deterrence effect is achieved. On the nature of deterrence, Homel, Lenne, and Walker (1989) wrote: "Deterrence is the omission of an act as a response to the perceived risk and fear of punishment for contrary behavior. . . . Deterrence is actually a statement about what is going on in people's heads. . . . Actual levels of enforcement are important only to the extent that people are aware of them" (p. 1).

Homel, Lenne, and Walker (1989) also referred to what they called the "hole in the bucket phenomenon." We can think of the water level in a bucket as representing the pool of deterred people at any given point in time. However, there are holes in the bucket that cause leakage and result in a drop in water level. In the case of random breath testing programs, these holes could include drunk driving episodes without being caught, low visibility of police, peer pressure to drink and drive, and so on. From time to time, to fill the bucket again, there needs to be a highly visible campaign of detection of drunk drivers (a blitz of random breath testing, for example).

CASE MANAGEMENT PROGRAMS PROGRAM ARCHETYPE

This category of programs works with each case in a way that recognizes that where significant behavior change is required, there are likely to be many and varied factors, many different individualized intermediate outcomes, and many different processes (services, treatments, and the like) for getting there. The case could be an individual, a family, an organization, a business, a community, and so on. Examples given in this book relate primarily to case management programs that operate with individuals. Figure 12.4 shows the archetypal

Figure 12.4   **Generic Outcomes Chain for Case Management Archetypal Program**

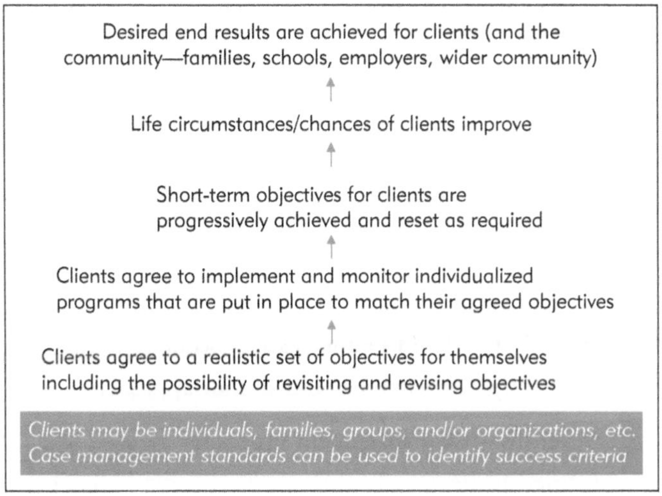

Desired end results are achieved for clients (and the community—families, schools, employers, wider community)

↑

Life circumstances/chances of clients improve

↑

Short-term objectives for clients are progressively achieved and reset as required

↑

Clients agree to implement and monitor individualized programs that are put in place to match their agreed objectives

↑

Clients agree to a realistic set of objectives for themselves including the possibility of revisiting and revising objectives

Clients may be individuals, families, groups, and/or organizations, etc. Case management standards can be used to identify success criteria outcomes chain for case management programs. Other somewhat similar program theories for case management programs are shown in Weiss (1997).

This outcomes chain was originally developed in the context of a work experience program for high school students with disabilities to develop their independent living skills and enhance their employability. Examples of other applications are a drug and alcohol program for prisoners, a call center for problem gamblers, and a rehabilitation program for injured workers. The mature workers employment program discussed in Chapters Seven and Eight is an application of the case management generic outcomes chain.

Although each case is in some sense unique, the cases are bound together in the program by the nature of the problem that the program is to address, for example, problem gambling or long-term unemployment. Cases differ with respect to both the causes of the problem and the consequences. For example, for some problem gamblers, the consequences relate to family breakdown; for others, attempted suicide; and for many, problem gambling itself results from a combination of causes and risk factors. Similarly, the "causes" differ across problem gamblers. Across many cases encountered by a case manager, there will be some recurring themes in the nature of the problem, and its causes and consequences, but the case manager needs to identify which themes apply to a particular individual and custom-tailor the service to that person.

Case management programs are labor intensive and typically require significant effort by clients (the cases) and their case managers. A decision to adopt a case management approach is therefore one that means committing substantial resources to addressing a particular issue. These decisions cannot be made lightly, especially given the doubts that have been raised by evaluation studies concerning the efficacy of such resource-intensive approaches (St. Pierre, Layzer, Goodson, and Bernstein, 1999).

Key Considerations for Applying the Archetype

Case management programs can be preventive, developmental, rehabilitative, or corrective (Funnell and Lenne, 1990). Some that address the needs of a very diverse group of clients may use all four approaches:

- Preventive case management programs are used when individuals are deemed to be at risk, through their own behavior or through the potential of their environment, to have an impact on their behavior—for example, children in deprived or hostile family environments.

- Developmental case management programs are used to develop the potential of individuals more fully than would normally occur through mainstream programs—for example, some specialized programs for gifted children and for children with disabilities.

- Rehabilitative case management programs are used when action is taken in relation to people who have experienced some separation from the mainstream community or from their usual lifestyle—for example, people leaving the hospital after serious illness.

- Corrective case management programs are used to modify behaviors of individuals that are damaging to either themselves or others—for example, phobias or antisocial behavior.

These different types of case management programs share some common implementation processes, but there are also some practical differences among them. For example, preventive and developmental case management programs are proactive and require processes to identify the most appropriate recipients, typically those at risk. Rehabilitative and corrective programs are reactive, and provided there are effective referral processes across a continuum or spectrum of services, screening may be less important. Sometimes such programs will need to have outreach activities to ensure that members of the target group know about and contact their case management services, especially when referral processes break down.

Evaluation techniques and performance measures for preventive and developmental case management programs also differ from those used for rehabilitative and corrective case management programs. Preventive and developmental approaches are trying to prevent something that has not occurred or develop something that may or may not have developed without intervention. Comparison with norms and expectations will be needed. Rehabilitative and corrective case management programs, by contrast, are eliminating or correcting responses and circumstances that have already occurred or, more positively, promoting recovery. Their outcomes therefore tend to be more readily recognizable by reference to the past history of the individuals concerned.

Some key requirements of case management programs follow:

- Following accepted standards of practice for case management programs (see Case Management Society of America, 2010).

- Typically a case management plan that is developed and negotiated with the program recipient rather than imposed.

- Having access to a range of appropriate strategies or services for each individual.

- Working with the individual in context and taking a holistic approach.

- Adaptability and flexibility.

- Case managers with expertise in the problem area.

- Feasible and productive case load sizes (often a significant problem).

- Use of measurement, evaluation, and reporting processes that are sensitive to individual differences among cases and emergent outcomes.

Relationship to Theories of Change

Stages of change theory can be useful when designing a case management program and managing and tracking the progress of a particular case. Empowerment theory is also relevant for ensuring ownership by the client. Outcomes are often emergent, though perhaps within an overall framework of some broad long-term goal or aspiration of the individual. Beyond the application of standard case management processes, the actual activities facilitated with clients may vary considerably, choosing from a wide range of options. These features place case management programs squarely in the camp of programs with complex aspects.

COMMUNITY CAPACITY-BUILDING PROGRAM ARCHETYPE

Community capacity building is a popular strategy for empowering communities, and many definitions of it have been developed (Aspen

Institute for Rural Economic Policy Program, 1996; London Regeneration Network, 1999). The archetypal community capacity-building outcomes chain in Figure 12.5 was developed to capture the most commonly recurring features in these definitions, and drew on the literature about community capacity-building programs and a review of many different community capacity-building projects.

The model is in sympathy with assets-based approaches to community capacity building (Kretzmann, 1995; Kretzmann and McKnight, 1993). Examples of projects to which this outcomes chain is relevant are community

Figure 12.5   **Generic Outcomes Chain for Community Capacity-Building Archetype Program**

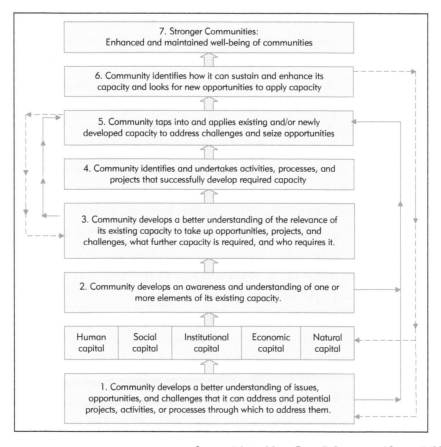

Source: Adapted from Funnell, Rogers, and Scougall (2004)

renewal programs in disadvantaged communities, community gardens projects, and various types of activity-based leadership development programs. Some community capacity building focuses on just one type of capital: human, social (including cultural), institutional, economic, or natural.

The generic outcomes chain shows community capacity building as a cyclical process with new capacity building on existing capacity. In principle, a community could enter the outcomes chain at any point in the cycle from outcome 1 to outcome 6. The feedback loop from outcome 6 to outcome 1 shows that capacity building is not a one-time exercise but rather a spiraling process. If it is working well, it will have ever-widening application or sophistication.

The feedback arrow from outcome 5 to outcome 3 in Figure 12.5 shows that communities may, in the course of undertaking a project, identify what further capacity is required or needs to be accessed. Indeed, we have found that some projects may immediately start work on a project (that is, start at outcome 5) and then become aware of a need to more systematically identify other sources of capacity or to develop further capacity (outcomes 2 and 3). They may, having started a project, put the main project on hold in order to develop the capacity that their experience in attempting the project has shown them they need (outcome 4) before they can progress.

Not all communities will achieve or need to achieve all outcomes shown in the outcomes chain. For example, some communities may be able to simply make better use of existing capacity, bypassing the need to develop additional capacity. The direct arrows from each of outcomes 2 and 3 to outcome 5 would apply to such communities. However, the generic outcomes chain can stimulate thinking about which steps are useful and which are not or which ones a particular community needs.

As for all other programs, many external factors will affect the achievement of the various outcomes in the archetypal outcomes chain and movement through it. For example, factors like whether there are opportunities to apply capacity will affect achievement of levels 5 and 6 of the outcomes chain. Competition for the time and attention of community members will also affect communities' readiness, receptiveness, and perseverance to undertake processes to achieve levels 1 to 4.

Given that many community capacity building efforts take place in at-risk or disadvantaged communities, the potential for emerging crises to derail the capacity-building process can be high. Nevertheless, these challenging situations can provide ideal opportunities for the development and exercise of community capacity, especially when communities have already achieved some degree of resilience and cohesion.

Key Considerations for Applying the Archetype

There are many different ways of approaching community capacity building: formal skills and assets audits, community consultations, and facilitated strategic planning, for example. Various manuals and resources have been prepared. A particularly useful approach is that provided by Kretzmann (1995).

The generic outcomes chain that we have shown relates primarily to approaches to capacity building that are structured around addressing particular issues, although there may also be an intent to apply that capacity to other issues as they arise.

The following sequence of questions incorporates an assets-based approach that can be used to guide community capacity-building efforts structured around addressing an issue. They start with identification of an issue requiring community capacity and move sequentially to identification of current capacity and ways of enhancing capacity:

1. What is the issue that the community would like to address or the opportunity that it would like to grasp?

2. What capacity currently exists within the community? Which elements of capacity need to be created or enhanced? Who or what needs to have capacity to act in relation to those issues or opportunities?

3. What can the community do to build on existing capacity within the community and address any gaps in capacity?

4. How can the community capitalize on new or increased capacity to address its particular issues or opportunities?

5. What processes are needed to support or diminish this new, enhanced, or existing capacity? Do they exist? What more needs to be done?

Relationship to Theories of Change

This archetype, in some ways, is much like the case management archetype except that the community is the unit of analysis and the community is typically its own case manager. For similar reasons as for case management programs, it relates most closely to empowerment theory and falls into the camp of complex programs.

PRODUCT OR DIRECT SERVICE DELIVERY PROGRAM ARCHETYPE

In some respects, the program archetype for products or services is the simplest of all archetypes. It applies when the end results are achieved simply by having members of the target audience use the product or service. These programs do not try to influence behavior except to the extent that they try to encourage members of the target group to use or not to use the program's service or to use the service in a particular way. Just using the service produces the desired result, or there is a short pathway between service use and the desired outcome. For example, increasing the proportion of a population that uses clean rather than contaminated water contributes fairly directly to improved health of the population.

Figure 12.6 shows a generic outcomes chain developed for programs that provide a product or service (adapted from Kelly and others, 1989). However, there are some variations on the product or service archetype to be considered when adapting the archetype to a particular program. These variations relate to whether the product or service is:

- Monopolistic or competitive
- Subsidized or self-sufficient
- Essential or nonessential
- A universal entitlement or based on assessed eligibility

Knowing which of these features apply to a particular service or product program is relevant to deciding which levels in the outcomes chain are most relevant to that program and for what reasons.

Figure 12.6   Generic Outcomes Chain for Service or Product Archetype Programs and Applications of the Archetype

Generic outcomes chain for programs that provide a product or service	Outcomes chain for an Art Gallery visitation program (A service program that is competitive, largely self sufficient, non-essential, universal entitlement)	Outcomes chain for a program that provides vouchers to help financially needy people to pay their electricity accounts (A service program that is fully subsidized by government to enable access to an essential service and for which eligibility is assessed)
5. End result achieved. Service/product remains viable	5. Recreational needs of the community are met and the Art Gallery is financially viable	5. Negative consequences of failure to pay electricity accounts are reduced (disconnection and its possible consequences in terms of safety, health, warmth, etc.)
4. Users are satisfied with the service/product	4. Desired numbers and types of visitors are satisfied with the exhibits and the facilities at the Art Gallery that make it a pleasurable experience	4. Recipients are satisfied with the process of using the vouchers; consider it dignified
3. Desired number and type of clients/customers use/re-use the service/product	3. Desired numbers and types of visitors come to the Art Gallery at desired times throughout the week/month/year, etc., and pay to see its special exhibits	3. Recipients use vouchers to pay electricity accounts before disconnection
2. Intended clients/customers obtain access to the service or product	2. Desired numbers and types of potential visitors consider that visiting the Art Gallery is interesting, affordable, accessible by public transport; has adequate facilities for people with disabilities, etc.	2. Members of target population receive vouchers; non-members do not 2a. Members of target population seek vouchers; non-members do not
1. Intended clients/customers know about the service or product and how to access it	1. Desired numbers and types of potential visitors know about the Art Gallery—its attractions and special exhibits	1. Members of target population know about vouchers and how to access them

Figure 12.6 shows the generic outcomes chain for service and product programs on the left-hand side and applications of this archetype to two programs that demonstrate some of the variations. One is a recreational service (an art gallery), and the other is a crisis financial assistance service for people who are unable to pay their electricity bills (an example that has been used in previous chapters).

The art gallery case example is *competitive* with other recreational activities. It is largely *self sufficient* (has to pay its own way), *nonessential*, and *universal* (all people are permitted to use it). The feedback arrows in the art gallery example illustrate principles that are relevant to services that are competitive and nonessential. The arrow from outcome 4 to 3 shows a "virtuous circle": if visitors are satisfied, they may come to the art gallery again. The art gallery wants repeat business. One of its strategies is to use special exhibits to try to encourage visitors to come more often. The downward arrow from outcome 4 to outcome 1 is used to show that if visitors are satisfied with the art gallery, they will tell others about it, who in turn might visit it. Of course, in both cases, the converse applies. If visitors are not satisfied with their experience, they might not make a return visit, and they may discourage their friends from coming.

The art gallery charges an entrance fee for special exhibitions, and this accounts for the arrow connecting the level of visitation to financial viability. This arrow would be relevant to other programs that are to some extent self-sufficient. Net profit will be affected by the types of visitors, such as full-fee paying adults, senior citizens, or school students and also those likely to make significant purchases from the art gallery shop (tourists more so than local residents). In this outcomes chain, we can see the potential for conflict between net profit and meeting the needs of the full community, a conflict that needs to be made explicit and carefully managed.

The second case example relates to the provision of emergency financial assistance to people to help them to pay their electricity bill. It is fully subsidized by government, relates to an essential service, and is given on the basis of assessed eligibility. Community welfare agencies operating on behalf of the government assess the eligibility of applicants and issue vouchers for use with the power company. Some of the factors discussed in relation to this example would be similarly relevant to other subsidized services that are based on assessed eligibility, but each program has its own particular features. For example, repeat business will be important to some programs but not others.

Because this program is a crisis financial assistance program, it does not wish to encourage repeat use by the same clients and so has another program activity that is about improving budgeting and financial management. So unlike the art gallery example, there is no feedback loop from satisfaction to repeat use. The program is, however, eager to ensure that satisfied users tell other members of the target audience about the service so that they too will use it. The downside is that the more people told about the service (some of whom will not meet the eligibility criteria), the more onerous the screening process becomes.

Key Differences Within This Archetype for Consideration

In this section we discuss some of the differences that apply to processes that need to be involved in implementing this archetype. They depend on whether the product or service is a monopoly or competitive, subsidized or self-sufficient, essential or nonessential, or universal entitlement or assessed eligibility.

Monopolistic Versus Competitive Programs
Monopolistic products or services are often less concerned than competitive services that potential users know of their existence (level 1 in the outcomes chain). Information about who knows about the service will be important for many competitive services as a means of determining what they might need to do to publicize their programs.

Competitive services will find that information about user satisfaction (level 4) is vital as a means of improving their services, products, and management processes. Because monopolistic services do not need to fight for market share, they may not be as concerned about the user satisfaction with the service. However, when they are government services (and often they are), there can be an expectation that as stewards of public funds, they will deliver a service of acceptable quality. Citizens' charters and guarantees of service introduced by governments across the globe reflect these expectations.

Governments put a high value on customer satisfaction for obvious political reasons. Also, customer satisfaction is an important outcome in its own right, even though it may not be strongly related to whether other program outcomes are achieved. For example, patient satisfaction with a doctor's services may be more about the doctor's bedside manner than the effectiveness of the treatment. Even so, the way in which the patient is treated is a valued component of the experience of the service, regardless of whether the person gets better.

Subsidized Versus Self-Sufficient Services   When services are subsidized (fully or partially) it is usually because they are recognized as performing a valuable social function while being unable to exist independent of some financial support. This may be due to inefficiencies in the way in which they operate, or it may be because most users would be unable to afford cost-reflective prices.

Subsidies are provided typically for essential services to make them affordable and in order to achieve various social or economic goals. In some cases, the subsidy is a universal entitlement. In other cases, subsidies are available to some portions of the population more than to others. This may occur through cross-subsidies or through providing the subsidy on the basis of assessed eligibility (as in the case of the financial assistance to pay electricity bills).

The social and economic goals may be quite specific and clearly articulated (for example, to provide electricity at prices that are affordable to low-income earners) or more general (such as public transport concessions to the elderly to enable them to participate more fully in community life). In either case, when subsidies are introduced to achieve social goals, there will be a concern to ensure that the subsidies are appropriately targeted. Some of the concerns that apply to carrot programs would also be relevant to subsidized services, for example, whether people would have used or been able to afford the service without the subsidy.

For products or services for which the subsidy is available to some users but not others, there will be an interest in whether, as a result of the subsidy, a significant proportion of the right people uses the service. Level 3 in the outcomes chain (service or product use) will usually be particularly important for designing and evaluating subsidies to sections of the community. Comments about assessment of eligibility will be relevant.

The level 2 outcome (obtaining access) is also important for subsidized services if access to them is not automatic or semiautomatic for the target group. Level 4 (satisfaction) is important if those who need to be reached by the service are discouraged from using it by poor-quality service or if the subsidy is insufficient.

Self-sufficient programs have a greater preoccupation with profit or breaking even as one of the end results of the program (level 5).

Essential Versus Nonessential Services   The outcomes chain for essential services such as water, electricity, and education is less concerned with levels 1, 2, and 3 (knowing about, gaining access to, and using a service). However, there may need to be a recognition of the fact that pockets of the community may miss out, especially in relation to levels 2 and 3—access and use. Examples of such pockets might be isolated rural communities, people from non-English-speaking or particular ethnic backgrounds, and other groups that may be geographically or socially isolated. An essential service may need to target these groups more actively.

For nonessential services such as museums, libraries, and other discretionary recreational and educational activities, there will be a concern with all levels of the outcomes chain. These programs need to ensure that people are aware of what the service has to offer, that they can and do access the service, and that they are satisfied with it so that their needs can be addressed (for example, self-education and recreation).

Universal Entitlement Versus Assessed Eligibility Services   Services vary in terms of whether they are available to all people in the community regardless of need or available only to those who are assessed as requiring the service. Services that are offered on the basis of assessed eligibility will be particularly concerned about issues of targeting. That is, they will be most concerned to ensure that levels 1, 2, and 3 (knowing about, gaining access to, and using the service) are achieved.

Assessment processes to achieve the level 2 outcome (gaining access) may be relatively automatic on meeting certain criteria, such as possession of social security identification. Or the assessment processes may be based on professional assessment on each occasion that the person requires assistance. The less automatic the process of assessing eligibility is, the more important the reliability and validity of processes for assessing need (level 2) will be. Issues around false positives and false negatives will arise.

Programs that assess eligibility sometimes use assessment processes that are too frequent or onerous or too costly. Costs of implementation will always need to be weighed against the relative importance of accuracy of assessment and the costs of some people incorrectly receiving a service. If a very high percentage of applicants for a service is likely to be assessed as eligible, then there may be good argument for removing the screening process and accepting that a small number of people for whom the service was not intended will receive it. In some cases, the processes that are needed to target as accurately as possible will be costly and have unintended outcomes. For example, processes that require applicants to tell their story on many different occasions, using processes that are time-consuming or costly for them, may deter potential applicants who would in fact be eligible for the service and who are the target group for that service. Program theory draws attention to the nature of the relationships among various outcomes in the outcomes chain and the potential for such unintended outcomes.

Relationship to Theories of Change

There is no direct relationship between the service archetype and any one Theory of Change discussed in Chapter Eleven. Conceptually these programs are relatively simple, although there are clearly many variations that need to be thought through. Some of the most challenging tasks for services that are not universal and have eligibility requirements may be about getting intended target audiences to use the services in the first place. For various segments of the target audience, different Theories of Change could usefully be brought into play. For example our understanding of the precontemplation stage of stages of change theory and associated useful techniques may be relevant in recruiting the target audience. Application of aspects of diffusion theory, for example, the use of opinion leaders, could be used to encourage people to take up a service. The theory of reasoned action and the theory of planned behavior may help to identify factors that would be relevant to selling the benefits of using the service to the target audience so that they develop a positive attitude about the behavior and its consequences. Empowerment theory may be useful if participants are enticed to use a service by playing an active role in molding the service themselves.

DECIDING WHICH PROGRAM ARCHETYPE APPLIES TO A PROGRAM

Generic outcomes chains associated with particular program archetypes can be a useful tool for developing an outcomes chain for a particular program or for reviewing a logic that has already been identified for a program. However, some programs do not recognize themselves as relating to any of the archetypes, and we certainly do not wish to present this as a full taxonomy of archetypes. There is room for much more development in relation to the identification of archetypes.

To determine what type of generic chain may be useful for a particular program, it is helpful to look at program objectives and activities, consider why the activities are being undertaken, and locate the assumptions underpinning the program. The title of a program is not always a reliable indicator of the nature of the program. Petrosino (2000), in his meta-analysis of programs to combat recidivism, cautioned against placing too much reliance on the banner under which a particular program operates and advised that it is important to assess the nature of the actual intervention. For example, many so-called behavior influence programs are called services, especially when promoting them to intended users. Information "services" may, for example, be delivered in order to change the behavior of the recipients rather than simply to serve them with interesting, well-presented, or useful information for personal purposes. Clearly the managers of those services would not be fully satisfied if users expressed satisfaction with the information but took no action.

Services may also be delivered as a means of making positive contact with the target audience so that the program can work with that audience to achieve other ends. For example, many programs that try to achieve their impact by changing behavior through advisory services have first to deliver another valued service to the target population in order to establish a credible relationship—that is, the foot-in-the-door phenomenon. The outcomes chain for the service component of the program is likely to differ from the outcomes chain for the behavior change component, but there will be cross-links between them. Indeed the behavior change outcomes chain may follow on from the service delivery chain.

Again, when services are used in this way to engage with a target audience in order ultimately to change behaviors, then evidence of the use of the services and satisfaction with the services would be unsatisfactory if it did not lead to other opportunities to achieve other desired outcomes—that is, if users took the service and then ceased involvement with the program. For example, a program that provided free low-carbon-emission light bulbs as a service to householders but had an agenda to encourage future use of them at users' own expense would be ineffective if recipients simply took the light bulbs, used them until they expired, and then showed no further interest in purchasing and using them in the future.

To assist program managers and staff in making the distinction between a behavior influence program and a service program, it is therefore helpful to ask whether they would consider it as fully successful as long as the clients were satisfied with the service and the benefits it delivered to them. If the answer is yes, then the program is, in all probability, a genuine service program. If the answer is no, the program is likely to be a behavior influence program rather than a simple service or product program.

Some programs are a combination of several generic outcomes chains, such as regulatory and educational ones. That is, any given program (or strategy) may consist of several interlinking outcomes chains whose collective influence is required to bring about the desired outcome.

Where combinations of archetypal outcomes hierarchies apply (say, combinations of carrots, sticks, and information), as for programs with complicated aspects, the relationships among them need to be thought out. Do they operate in parallel to address different situations, different causes, different segments of the target population, and so on? Does one archetype build on another? Are sequencing and timing of the various outcomes chains important? For example, do the outcomes for one strategy need to be effectively achieved before it makes sense to move to the next strategy?

Conversely, should the intervention mutate to another strategy following failure or only partial success of the previous strategy? For example, if education works with only some segments of the population, should regulation be introduced to change the behaviors of the intransigents? (See Bemelmans-Videc, Rist, and Vedung, 1998, for a discussion of how to package different archetypes.)

Following is an example of a litter prevention program that included several different archetypes—educational, deterrence, community capacity building, and services—and incorporated interlinked outcomes chains. In the course of doing so it also worked on developing social norms.

INTERLINKING ARCHETYPES TO ADDRESS DIFFERENT CAUSES

An antilittering campaign was established to reduce litter in public places. Reducing litter was important for many reasons, including the environmental impacts of litter and because the public regards excessive litter as indicative of a breakdown in law and order. These perceptions in turn affect feelings of safety in the public at large, and, for criminal elements, the lack of any apparent effort to control littering is thought to give general license to criminal activities.

The campaign recognized that different types of strategies and different types of outcomes chains would apply to different segments of the population. As part of designing the program, the responsible agency conducted qualitative research concerning the nature and motivation of those who litter.

The research found that some people litter because it is too much trouble to place litter in the trash can; others litter because they see nothing wrong in doing so—they do not know any better or believe it is someone else's job and source of income to pick up the litter; others litter as a means of making a statement against authority or society; nonlitterers in some circumstances will become litterers in other circumstances; conscientious nonlitterers not only rarely or never litter, but they often pick up and dispose of other people's litter.

For each of these different segments, a different strategy was required, ranging from educational strategies (for those who needed to understand impacts of littering), to the placement of trash cans (for those for whom convenience of disposal was a factor), to fines for littering (for those who could be neither educated, persuaded, nor enticed). Other strategies were to work with industry to reduce the volume of packaging materials: reducing litter at source. Even the nonlitterers became agents in the campaign because their help could be enlisted to reinforce the new social norms that the agency was endeavoring to develop around littering. They could remind litterers that the practice is undesirable and can lead to a penalty.

For this program, the outcomes chain was a series of interlocking chains showing the different intermediate outcomes that would need to be achieved with different segments of the population. It also showed the points at which the various strategies depended on and contributed to each other—for example, that the deterrence strategy using fines had to work hand in hand with the public communication strategy with respect to messages, timing, and producing publicly communicable evidence of the success of detection and prosecution strategies. The provision of more trash cans to make it easier for people to dispose of their trash was timed to coincide with the introduction of the antilitter legislation and the introduction of penalties. Over time, as social norms changed toward wider public condemnation of littering, the balance of effort across the different strategies could be expected to change.

Choosing one or more archetypes is only the first step in using them. They need to be adapted to particular programs, taking into account research that is relevant to what works for whom and under what circumstances in the context of that particular program. For example, the stick archetype gives only general advice about the types of intermediate outcomes that should be included in an outcomes chain. It does not specify such matters as the extent to which each of those outcomes needs to be achieved in order to lead to the higher-level outcomes. Among the issues that need to be resolved within the context of a particular stick program are these:

- How high does the perceived probability of being caught in the act need to be before it acts as a deterrent?

- How heavy do sanctions need to be before they will deter undesirable behavior?

- Under what circumstances will the social embarrassment of being caught in the act be sufficient to deter the behavior without the application of further sanctions?

The answers to these questions will likely vary with, among other things, the nature and prevalence of the behaviors that are to change.

As we have emphasized, an archetype should not be used to force an intervention into its structure, but as a thinking aid for identifying important outcomes that the program needs to achieve. An archetype can also be used to stimulate relevant research questions such as those posed in relation to the sticks archetype.

SUMMARY

In this chapter we have presented five program archetypes—information (advice, information provision, and education), carrots and sticks (incentives and sanctions that motivate behavior), case management, community capacity building, and direct service delivery—that have been useful when developing program theories or designing or evaluating programs that resemble the archetypes. We have described their principal features, including their generic outcomes chains, and demonstrated their application to particular programs. We have discussed considerations that should be applied when using each archetype and drawn some parallels between each archetype and one or more of the big theories of change discussed in Chapter Eleven. We have also provided some guidance about how to decide which one or more of the program archetypes apply to a particular program when seeking to use archetypes.

EXERCISE

Use a generic outcomes chain to assist with the construction of an outcomes chain for a particular program. Identify which program archetypes would be most relevant for the rail ticket inspection program described here. Using the outcomes chain for that archetype and the information supplied about the program, develop an outcomes chain for the inspection program:

Rail Ticket Inspection Program

Revenue raised by the collection of fares is essential for the operation of passenger train services in the state. Passengers are obliged to purchase tickets at the railway station where they begin their journey.

The corporation that runs the passenger train service suspects that high levels of fare evasion exist. It aims to reduce fare evasion and thereby increase revenue. One of its strategies is the use of small teams of ticket inspectors who move in an unpredictable fashion from station to station inspecting tickets as passengers leave the train. The brief of these inspectors is to identify passengers who have evaded paying the fare. Once a fare evader has been identified, the inspectors initiate the prosecution processes that will be undertaken by the corporation to deter both detected and potential fare evaders from future evasion. The inspectors also have a brief to identify those who have a legitimate reason for not having a ticket (for example, the ticket dispenser was broken) and to collect the appropriate fare from them.

The corporation has only a skeleton staff at most suburban railway stations and no staff at all at some of the smaller and less frequently used stations. The ticket inspectors concentrate on routes for which the railway stations are least likely to be continually staffed. They wear uniforms and locate themselves in positions in which they will be highly visible. In addition, media campaigns and posters at all railway stations draw attention to the penalties associated with fare evasion and the role of the inspectors. The penalties for fare evasion are substantial relative to the estimated savings made by each fare-evading passenger. However, their capacity to act as a deterrent is unknown.

13

Logic Models Resources

THIS CHAPTER PROVIDES resources to help in developing logic models. It shows variations on the pipeline and outcomes chain styles of logic models and discusses technological options. Many versions of logic models have been developed over the years, each with some unique features. It might be useful to adopt or adapt one of these for use in your organization or to add some of its features to a hybrid format.

PIPELINE LOGIC MODELS

Pipeline logic models represent a program theory as a linear process with inputs and activities at the front and long-term outcomes at the end. In this section, we set out some variations that have been developed and discuss their relative advantages and disadvantages.

Ways in Which Pipeline Logic Models Vary

Pipeline logic models can vary in terms of whether a single box is used at each level or several boxes, the number of components, the labels used, and the definitions of these labels.

Single Box or Separated Causal Strand   A very simple logic model has one box for each level, so that all the inputs are in one box and all the outputs are in another box. A useful variation is to have several boxes at each level, so that specific inputs can be linked to specific activities and then to specific outputs and outcomes (Table 13.1). It is usually better to separate them like this to avoid losing the main messages about how the intervention works in the laundry lists of items.

Number of Components   Additional components can be added to the basic four-box pipeline logic model to show context or needs before inputs or to distinguish different types of outcomes: from short term to long term, or from results for participants to broader results for the organization or the community (Table 13.2). It is usually better to separate outcomes into at least short and long term. This helps to identify other interventions and factors that are likely to influence results (that may need to be addressed in the actual intervention or at least in the evaluation) and results that might be observable during the life of the evaluation.

Labels   The labels used for the components, and the definition of these labels, vary considerably. For example, outcomes are sometimes defined as short-term results and impacts as long term, but the definitions can be the other way around. Tables 13.3 and 13.4 show the labels that are used. Table 13.3 shows

Table 13.1   **Variations on the Layout of Components in Pipeline Logic Models**

Option	Features
Single line of boxes	All inputs are in one box, all outputs in another, and so on.
Separated causal strands	The different components are in the form of columns, and there are separate boxes for different elements.

Table 13.2 **Variations on the Number of Components in Pipeline Logic Models**

Number of Components	Labels of Components
Four	Inputs, processes, outputs, outcomes
Five	Inputs, processes, outputs, outcomes, impacts
	Inputs, processes, outputs, short-term outcomes, long-term outcomes
	Needs, inputs, processes, outputs, outcomes
Six	Context, inputs, processes, outputs, outcomes, impacts
	Inputs, processes, outputs, short-term outcomes, medium-term outcomes, long-term outcomes

Table 13.3 **Variations on the Labels Used for Components in Pipeline Logic Models**

Component	Labels Sometimes Used
What is needed	Inputs, resources
What is done	Processes, activities, outputs-activities, outputs-participation
Initial results	Outputs, products
Subsequent results	Outcomes, short-term outcomes, medium-term outcomes, impacts, purpose
Final results	Outcomes, long-term outcomes, final outcome, impacts, goals

the different labels sometimes used for each component; Table 13.4 shows the different definitions that are sometimes used for each label.

Additional Elements   Some pipeline logic models show other items, such as external factors, assumptions, and risks (Table 13.5). These no longer represent the program as a closed system of inputs and outputs (like a factory production line) but show the influence of other factors and frame the program within a specific situation analysis. However, these logic models still present the program as being like a pipeline, with the activities at the front flowing through to subsequent outcomes, in contrast to outcomes chain logic models, where activities can occur at any point along the chain of outcomes.

Table 13.4   **Variations on the Definitions of Labels Used in Pipeline Logic Models**

Label	Different Definitions
Inputs	All resources needed to produce the results
	Only the resources contributed by the intervention
Outputs	Quantity of service delivered or product produced
	First change for which the program is accountable
Outcome	Intermediate result between outputs and impacts
	Final result for participants
	Final result shown on logic model
Impact	Final result for participants
	Final result shown on logic model
	Intermediate result between outputs and outcomes
	Long-term change, final result shown on logic model
	Changes in the broader community or system
	Changes that can be attributed to the intervention

Table 13.5   **Additional Components Sometimes Included in Pipeline Logic Models**

Additional Component	Definition
External factors	Factors outside the intervention (including programs implemented by other agencies and context) that influence the achievement of outputs
	Factors outside the intervention (including programs implemented by other agencies and context) that influence the achievement of outcomes and impacts
	Factors outside the intervention (including programs implemented by other agencies and context) that influence any stage of the pipeline
Assumptions	Causal links that are needed for the intervention to work but are not directly addressed
	Motivations and capacities of intended beneficiaries and implementers that are needed in order for the intervention to work
Needs	Needs of the intended beneficiaries

Variations of Pipeline Logic Models

Table 13.6 lists the variations of pipeline logic models, which we discuss in more detail in the following sections and illustrate using a community health program.

Charities Evaluation Planning Triangle   Charities Evaluation Services (U.K.), which strengthens the voluntary sector through offering free and below-cost support and services to charities and community organizations, has developed probably the simplest logic model in the form of a triangle that shows the activities, the outcomes, and the overall aim. Clearly this does not distinguish between different causal strands, but for some organizations, it provides a useful starting point for articulating the intended outcomes of activities. The example in Figure 13.1 shows an adaptation of the planning triangle by Evaluation Support Scotland to represent a community health project that has a variety of activities.

Table 13.6   **Variations of Pipeline Logic Models**

Charities Evaluation planning triangle	A three-part logic model of activities, outcomes, and overall aim.
United Way logic model	Inputs, activities, outputs (numbers of services or products delivered), and outcomes for participants (which can be separated into short, medium, and long term)
W. K. Kellogg Foundation logic model	A five-component model of resources/inputs, activities, outputs, outcomes (results for participants), and impacts (results for the broader community, organization, or system)
Bennett's hierarchy	A seven–component logic model that represents changing behavior through providing information
University of Wisconsin logic model	A six-component model of inputs, activities, participation, and short-, medium-, and long-term outcomes, with assumptions and external factors also shown
Logical framework (logframe)	A four-component model of outputs, component objectives, outcome (or purpose), and impact (or goal) widely used in international development. It is set out in a matrix showing, for each component, a description, indicators, means of verification, and assumptions.

Figure 13.1   **Charities Evaluation Planning Triangle Logic Model of a Community Health Center**

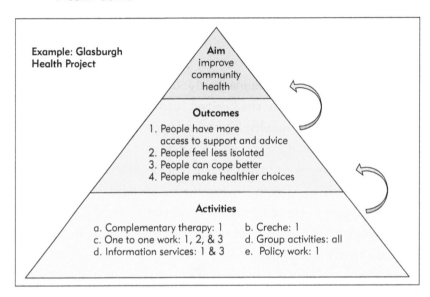

Example: Glasburgh Health Project

Aim
improve community health

Outcomes
1. People have more access to support and advice
2. People feel less isolated
3. People can cope better
4. People make healthier choices

Activities
a. Complementary therapy: 1
b. Creche: 1
c. One to one work: 1, 2, & 3
d. Group activities: all
d. Information services: 1 & 3
e. Policy work: 1

Source: Evaluation Support Scotland (2009)

United Way Logic Model   The United Way's version of the pipeline logic model shows inputs (including both resources and constraints), activities, outputs (direct services or products), and outcomes for participants (United Way of America, 1996). Outputs are shown as numbers of products or services delivered. Outcomes can be separated into initial, intermediate, and long term. Figure 13.2 represents the same community health project as in Figure 13.1 using the United Way logic model. This example shows different outcomes at each level grouped together, but they can be separated into different boxes for clarity. United Way logic models can be drawn from left to right or from bottom to top.

W. K. Kellogg Foundation Logic Model   Like the United Way logic model, this is a simple linear pipeline logic model that defines outputs as the amount of product and/or service that you intend to deliver by accomplishing the planned activities (Figure 13.3). Its main point of difference lies in the way that impacts

Figure 13.2   **United Way Logic Model of a Community Health Center**

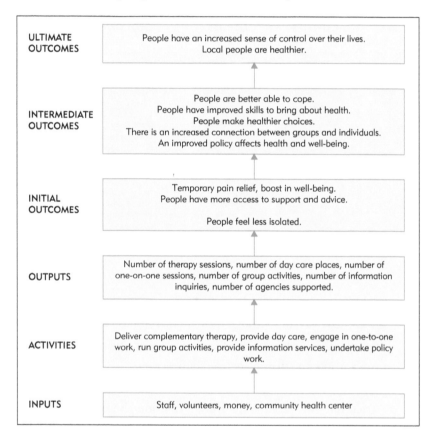

ULTIMATE OUTCOMES	People have an increased sense of control over their lives. Local people are healthier.
INTERMEDIATE OUTCOMES	People are better able to cope. People have improved skills to bring about health. People make healthier choices. There is an increased connection between groups and individuals. An improved policy affects health and well-being.
INITIAL OUTCOMES	Temporary pain relief, boost in well-being. People have more access to support and advice. People feel less isolated.
OUTPUTS	Number of therapy sessions, number of day care places, number of one-on-one sessions, number of group activities, number of information inquiries, number of agencies supported.
ACTIVITIES	Deliver complementary therapy, provide day care, engage in one-to-one work, run group activities, provide information services, undertake policy work.
INPUTS	Staff, volunteers, money, community health center

are defined—not as longer-term outcomes but as broader results for the organization, community, or system beyond those directly involved in the program.

Bennett's Hierarchy   This is an elaboration of the four Kirkpatrick levels of outcomes from training (introduced in Chapter Two). It is not a generic logic model like the United Way or Kellogg Foundation logic models but one designed for interventions that are designed to bring about behavior change by providing information (the information provision archetype we discussed in Chapter Twelve). It therefore specifies short-term outcomes in terms of changes to knowledge, attitudes, skills, and aspiration (KASA); medium-term outomes in terms of practice changes; and long-term outcomes in terms

Figure 13.3   **W. K. Kellogg Foundation Logic Model of a Community Health Center**

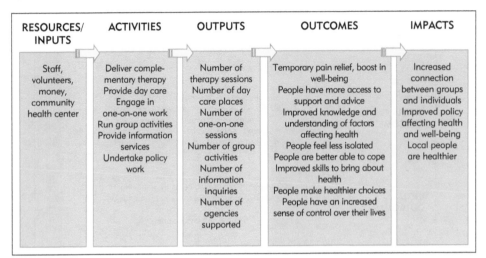

of changes to social, economic, and environmental conditions. Compared to the United Way and Kellogg logic models, it has two additional levels early on—one that relates to the characteristics of those who participate in the program and one that relates to their reactions.

Originally developed by Claude Bennett of the U.S. Department of Agriculture for use in agricultural extension (Bennett, 1975), it has been used more widely (Bennett and Rockwell, 1999; Roberts Evaluation, 2010). Figure 13.4 shows its application to the behavior change components of the community health center. If Bennett's hierarchy were to be used to represent the program theory of the whole community health center, additional elements would need to be added to cover the direct service delivery elements.

In Chapter Sixteen, we discuss an example of using Bennett's hierarchy as a reporting framework.

University of Wisconsin Logic Model   This model builds on the structure of Bennett's hierarchy and generalizes it for a wider range of interventions. It defines outputs in a particular way to distinguish between activities (what the organization does) and participation (who is engaged by the intervention). It labels subsequent results as "outcomes-impact" and distinguishes

Figure 13.4   **Bennett's Hierarchy Logic Model of a Community Health Center**

Level 7: Impacts on social, economic, and environmental conditions	Local people are healthier Reduced health expenditure
Level 6: Changes in behavior	People make healthier choices
Level 5: Changes in knowledge, attitudes, skills, and aspirations	Improved knowledge and understanding of factors affecting health, increased sense of control over their lives, improved skills to bring about health
Level 4: Reactions	Service seen as welcoming, supportive. Advice seen as credible and practical
Level 3: Participation	People with poorly managed chronic conditions
Level 2: Activities	Engage in one-to-one work, run group activities, provide information services, undertake policy work
Level 1: Resources	Staff, volunteers, money, community health center

among short-term, medium-term, and long-term results. The logic model also includes some additional features—assumptions and external factors.

Figure 13.5, a University of Wisconsin logic model of the community health program theory, developed by Evaluation Support Scotland, provides more detail about the various elements and how they are connected. This example does not show assumptions or external factors on the logic model as is sometimes done with this logic model.

Logical Framework (Logframe)   A logframe is a particular type of pipeline logic model that is used widely in international development. It has four components of results: outputs, component objectives, outcome (or purpose), and impact (or goal). A logframe is presented in the form of a matrix that sets out for each component a description, indicators, means of verification, and assumptions (Table 13.7). The difficulty of adequately representing large and complicated interventions with just four boxes (Bakewell and Garbutt, 2005) can be partly addressed by creating a logframe for each of the purposes of the program (Guijt and Woodhill, 2002).

A revised form of the logframe is now being used by the U.K. Department for International Development (2009), which, among other

Figure 13.5　**University of Wisconsin Logic Model of a Community Health Center**

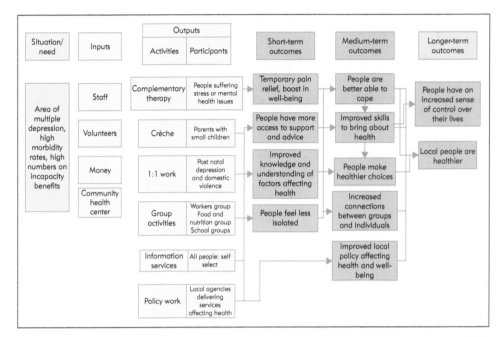

Source: Evaluation Support Scotland (2009)

things, includes more detail about the objectively verifiable indicator by separating it into indicator, baseline, and target; shows assumptions for goal and purpose only; and changes the way that resources and risks are reported. It also requires an estimate of the percentage contribution each output is likely to make to the overall purpose (U.K. Department for International Development, 2009).

VARIATIONS OF OUTCOMES CHAIN LOGIC MODELS

Outcomes chain logic models are inherently more variable than pipeline logic models. In addition to the examples shown throughout this book, two variations address particular issues (Table 13.8).

Table 13.7 Components of a Logframe

Activity Description	Indicators	Means of Verification	Assumptions
Goal or impact: The long-term development impact (policy goal) that the activity contributes at a national or sectoral level	How the achievement will be measured, including appropriate targets (quantity, quality, and time)	Sources of information on the goal indicators, including who will collect it and how often	
Purpose or outcome: The medium-term results that the activity aims to achieve in terms of benefits to target groups	How the achievement of the purpose will be measured, including appropriate targets (quantity, quality, and time)	Sources of information on the purpose indicators, including who will collect it and how often	Assumptions concerning the purpose to goal linkage
Component objectives or intermediate results: This level in the objectives or results hierarchy can be used to provide a clear link between outputs and outcomes, particularly for larger multicomponent activities	How the achievement of the component objectives will be measured, including appropriate targets (quantity, quality, and time)	Sources of information on the component objectives and indicators, including who will collect it and how often	Assumptions concerning the component objective to output linkage
Outputs: The tangible products or services that the activity will deliver	How the achievement of the outputs will be measured, including appropriate targets (quantity, quality, and time)	Sources of information on the output indicators, including who will collect it and how often	Assumptions concerning output to component objective linkage

Table 13.8 **Variations of Outcomes Chain Logic Models**

People-centered logic model	A varying model, often based on Bennett's hierarchy, that specifies who will experience each of the levels
ActKnowledge/Aspen Institute Approach to Theory of Change	Shows the precondition outcomes that lead up to the intended final result

The people-centered logic model specifies who will engage in the program and who is expected to experience changes (Dart and McGarry, 2006). It is part of people-centered evaluation (PCE), a highly participatory approach to developing program theory through a two- to three-day workshop involving diverse stakeholders that also emphasizes the use of participatory methods for evaluation. A people-centered logic model extends the notion of including participation (as in Bennett's hierarchy or the University of Wisconsin logic model) to identify the targets of each impact, that is, who will be doing things differently. Figure 13.6 comes from an evaluation of Landcare, a natural resources management program involving a partnership of farmers and other land users, industry, and government.

This ActKnowledge/Aspen Institute Community Builders Theory of Change logic model is an outcomes chain, but it is somewhat constrained in the number of levels and interactions it represents. Through the use of annotations, it distinguishes between causal links that are expected to happen through the domino effect and those where an additional intervention is needed. It maps these additional interventions and important assumptions onto the logic model and also lists them separately. Figure 13.7 represents the program theory for a collaborative project where a social service provider, a nonprofit employment training center, and a domestic violence shelter worked together to assist survivors of domestic violence to secure long-term employment at a livable wage.

Figure 13.6   **Example of a People-Centered Logic Model**

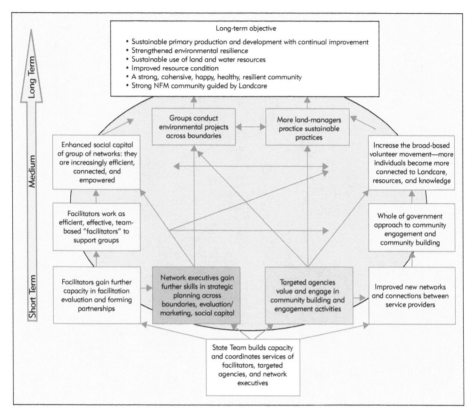

Source: From Dart and McGarry (2006, p. 17)

TECHNOLOGY FOR REPRESENTING
PROGRAM THEORY

Technology does not necessarily mean using computers. Logic models can be drawn on paper, boards, or sticky notes. The right technology can make it easier to develop logic models that are clear and readily accessible and to revise them as necessary. The wrong technology can make the process of drawing logic models cumbersome, and therefore discourage revision of them

Figure 13.7   **Example of ActKnowledge/Aspen Institute Theory of Change Logic
Model with Assumptions and Interventions**

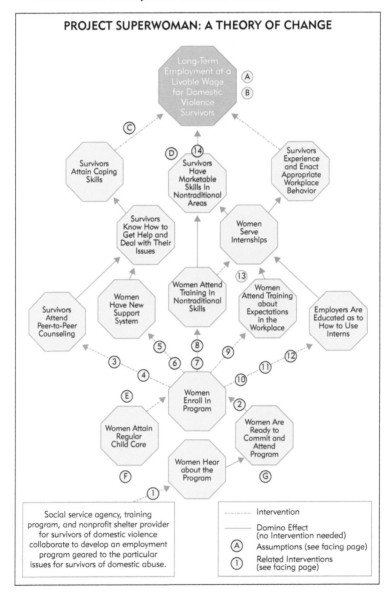

Source: ActKnowledge (2003); Aspen Institute (2005)

Figure 13.7 *(Continued)*

Assumptions	Interventions
(A) There are jobs available in nontraditional fields for women.	(1) Implement outreach campaign
	(2) Screen participants
(B) Jobs in nontraditional areas of work for women, such as electrical, plumbing, carpentry, and building management, are more likely to pay livable wages and are more likely to be unionized and provide job security. Some of these jobs also provide a ladder for upward mobility, from apprenticeship to master, giving entry-level employees a career future.	(3) Set up counseling sessions
	(4) Lead group sessions
	(5) Provide help for short-term crises, such as housing evictions or court appearances
	(6) Provide one-on-one counseling
	(7) Develop curricula in electrical, plumbing, carpentry, and building maintenance
	(8) Conduct classes
	(9) Develop curricula and experiential learning situations
(C) Women who have been in abusive relationships need more than just skills; they need to be emotionally ready for work as well.	(10) Conduct classes
	(11) Identify potential employers
	(12) Create employer database
(D) Women can learn nontraditional skills and compete in the marketplace.	(13) Match women to internships
	(14) Help women secure permanent jobs

Sample Indicator

OUTCOME:
Long-term employment at a livable wage for domestic violence survivors

INDICATORS:
Employment rate

TARGET POPULATION:
Program graduates

BASELINE:
47% of program attendees are unemployed
53% are earning minimum wage

THRESHOLD:
90% of the graduates remain in job at least six months and earn at least $12 per hour

(E) The program cannot help all women, and so entry into the program must include screening so that women who have sufficient literacy and math skills to take the training and have lives stable enough to attend classes are admitted. The program does not have the resources to handle providing basic skills or major social services.

(F) Women who have left abusive situations are often single mothers and therefore cannot work unless they have child care.

(G) Women must be out of the abusive situation. The program assumes that women still in abusive situations will not be able to attend regularly, may pose a danger to others, and will not be emotionally ready to commit.

when it is needed, and produce logic models whose form and style is not helpful. Sometimes one form of technology might be appropriate to work with a group of people to develop a logic model, and another type might be appropriate to develop a version that can go into a formal report. In all cases, the form of the logic model must be driven by the specific needs of the evaluation and the people who are developing the program theory. The technology should be one that engages the relevant participants. When workshops are used to articulate stakeholder mental models, the use of sticky notes (give each person a handful of them) can provide more stakeholders the opportunity to contribute to the logic model than does computer-aided logic model development, which can too easily be monopolized by one person. Table 13.9 sets out the technology options that we examine in the following sections.

Drawn Logic Models

Drawing logic models doesn't always require computers. Sometimes low-tech approaches can be more appropriate for easily engaging stakeholders in the process without being distracted by technical problems or barriers.

Large Piece of Paper   It is possible to draw four or five columns representing the components of a pipeline logic model and fill in items in each column or draw a realist matrix and fill in each cell. It is harder to develop a results chain logic model using a large piece of paper because it is likely to need frequent revision as it is drawn.

Table 13.9   **Technology for Drawing Logic Models**

Drawing logic models without computers
    Large piece of paper/whiteboard
    Sticky notes or index cards
    A "magic wall": Plastic shower curtain and spray adhesive

Standard computer packages
    Word processing packages
    Spreadsheet packages
    Presentation packages
    Drawing packages
    Project management software

Specialist visual representation packages
    Visio

Specialist logic model packages
    DoView
    InnoNet's Logic Model Builder

Advantages: This needs little equipment or preparation and can be used anywhere.

Disadvantages: It is likely to get very messy when changes need to be made, which means a risk that it will discourage trial and error and iterative review and revision. The number of people who can participate in the development of the logic model will be limited by how many can gather around the sheet of paper.

Likely to be useful: This technology is best used as an adjunct to other methods. It can be useful, for example, when discussing program theory with someone to sketch out some components of it on paper and review it with them, particularly if there are important causal arrows to discuss. But more work will be needed to develop a clear diagram for later use.

Whiteboard   Draw four or five columns representing the components of a pipeline logic model, or draw boxes and arrows representing the components of a pipeline logic model or a results chain logic model.

Advantages: Revising a logic model on a whiteboard is more easily done than on a piece of paper, and it can be easily seen, and added to, by all participants in a group process. The diagram can be made permanent by photographing it or printing from whiteboard (if available).

Disadvantages: The diagram usually needs to be subsequently typed up on computer for use in a report. The result may be a diagram that does not well represent the thinking of the group, or engage them.

Likely to be useful: This lends itself to use in small groups where the program theory is being developed largely through articulating the mental models of key informants. In most cases, the logic model will need to be subsequently drawn up using a computer to produce a neater version.

Sticky Notes or Index cards   Write each item on a separate sticky note or index card (with temporary adhesive on the back) and then move them to a suitable place on a large piece of paper, for example, from a flip chart. A single color can be used or multiple colors. When the position of the components is finalized, add arrows to show links among the components. Adding a number to each sticky note or photographing the flip chart paper can be useful in case the sticky notes become detached during transportation.

Advantages: Each participant can be encouraged to add sticky notes when they suggest outcomes to be included. This facilitates participation by all group members, and moving the components around is easy. It can be used for results chain logic models, as well as pipeline logic models.

Disadvantages: An occasional disadvantage is that in the effort to be concise with the information included on each sticky note, there may be insufficient detail to finalize the logic model after the workshop or remember the full rationale and prepare a narrative statement. We suggest developing working titles for groups of sticky notes that are about similar outcomes. Retaining the sticky notes that contributed to the working title can help with filling in the detail later. Another disadvantage can be the risk of notes falling off when the material is moved. It is a good idea to photograph the sheet as a backup copy or to replace the hard copy before moving it. A digital photograph can also be readily shared with participants.

Likely to be useful: Like a whiteboard, this lends itself to use in small groups where the program theory is being developed largely through articulating the mental models of key informants. In most cases, the logic model will need to be subsequently drawn up using a computer to produce a neater version.

A Magic Wall   More flexibility can be achieved through using a magic wall: a cheap plastic shower curtain sprayed with art adhesive (Dart, 2008). Then any type of paper can be used for the components of the logic model. Single or multiple colors can be used as above.

Advantages: This approach brings the advantages of a whiteboard to any room.

Disadvantages: Care must be taken to use sufficient adhesive so that items do not fall off and information about content or relationships is lost. Photographing the logic model before attempting to move it is wise. Conversely, if the adhesive is too strong, there is less potential to move items around when in a workshop situation. The disadvantages with respect to level of detail that can be captured are similar to those discussed for sticky notes.

Likely to be useful: Like a whiteboard, this lends itself to use in small groups where the program theory is being developed largely through articulating the mental models of key informants. In most cases, the logic model will need to be subsequently drawn up using a computer to produce a neater version. It is likely to be more appropriate when working with community groups where a low-tech approach will make the process seem accessible.

Standard Computer Packages

Computer applications can be used after a logic model has been drawn manually at a group event, or they can be used in real time with a group and displayed using a data projector.

Word Processing Packages Packages such as Word and the open source equivalents are often used to develop logic models. There are two main ways of drawing logic models in Word: as a table (Table 13.10) or as a diagram of boxes and arrows (Figure 13.8). To draw the logic model as a table in Word, choose the Insert Table option in the Insert menu and create a table with the required number of columns and rows.

To draw the logic model as a diagram of boxes and arrows in Word, use the Picture AutoShape option in the Insert menu to access boxes and connectors of different types. Text can be added to the boxes, and the format of boxes and connectors can be modified as required and varied in terms of style, color, and weight.

Advantages: This is likely to be a readily accessible option, Most office and home computers already have a version of proprietary software, such as Microsoft Office, that can be used. Alternatively, open source software can be used to perform similar functions. This means that logic models can be readily shared in an editable form.

Disadvantages: Not everyone has the skills to use the packages effectively to produce logic models. Tabular versions are easier to produce than a diagram but lack the arrows that show causal links, and they have the usual disadvantages of pipeline logic models. If those developing the logic model are not able to draw a diagram and enlist the help of staff, there will be less opportunity to develop the logic model iteratively, checking that it is focusing on key messages and meeting good design guidelines.

Likely to be useful: This option will suit situations where there is existing skill in using word processing packages and little opportunity to acquire more specialist packages that would be easier to use.

Table 13.10   Tabular Pipeline Logic Model in Word for Meals on Wheels

INPUTS	PROCESSES	OUTPUTS	SHORT-TERM OUTCOMES	LONG-TERM OUTCOMES
Fresh, frozen, canned, and dried food Trained dietician Cook Food deliverers Motor vehicles Food preparation and storage equipment Knowledge of suitable potential clients Government funding and co-contribution from clients	Assess clients' eligibility and suitability for service Assess clients' nutritional and cultural needs Develop menus Buy, store, and prepare meals Recruit and train deliverers (volunteer or paid) Develop meal delivery schedules Deliver meals Receive co-payments and deposit money Assess clients' well-being	Nutritious and culturally appropriate meals prepared Meals delivered Meals eaten Clients have regular social interaction Hazardous conditions identified and addressed	Maintained physical well-being Maintained social and emotional well-being	Reduced incidence of premature or inappropriate hospitalization of elderly people or people with a disability Reduced cost to government of funding nursing homes and other institutional care Improved quality of life for elderly and people with a disability

Figure 13.8 **Picture Pipeline Logic Model in Word**

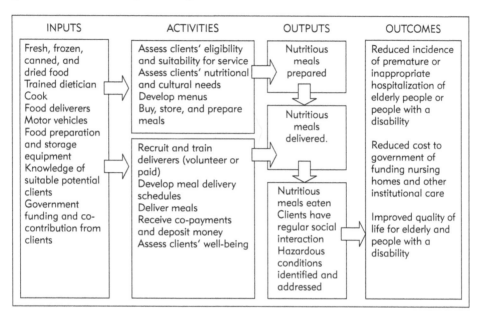

Spreadsheet Packages   Spreadsheets such as Excel are not designed for drawing logic models, but some people find them useful, usually because their expertise in Excel is much greater than their knowledge of Word or their willingness to learn a specialist package.

For a logic model in Excel, create a cell for each component (for a simple pipeline logic model), for each strand of the component (for a pipeline logic model with different causal strands), or each result (in a results chain logic model). Adjust the cell height and width as needed, and leave a column between the cells for the arrows. Use alt-enter to insert line breaks in the cell. Use Insert Picture Autoshapes to insert an arrow. Use shading and borders to outline the cell (see Figure 13.9).

Advantages: Like word processing packages, this is likely to be a readily accessible option. Most office and home computers already have a version of proprietary software, such as Microsoft Office, which can be used. Alternatively, open source software can be used to perform similar functions. This means it is easy to share logic models in an editable form.

Figure 13.9   **Picture Pipeline Logic Model in Excel**

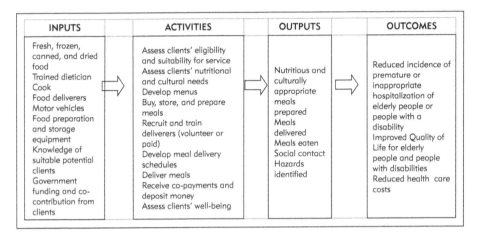

Disadvantages: There is limited capacity to develop complicated diagrams or revise the layout of them to communicate more effectively.

Likely to be useful. This is likely to be useful only when those using computers have very limited skills and a very simple logic model will be sufficient.

Presentation Packages   Packages such as PowerPoint are often used to develop logic models. As with word processing packages, there are two main ways of drawing logic models in PowerPoint: as a table or as a diagram of boxes and arrows.

To draw the logic model as a table in PowerPoint, choose the Table option in the Slide Layout menu, and create a table with the required number of columns and rows.

To draw the logic model as a diagram of boxes and arrows in PowerPoint, use the Picture AutoShape option in the Insert menu to access boxes and connectors of different types. Text can be added to the boxes, and the format of boxes and connectors can be varied in terms of style, color, and weight. The resulting logic model can then be pasted into a Word document or used in a presentation. The Centre for Community Research (2008) has produced a detailed guide to using PowerPoint to draw logic models.

Advantages: Like word processing packages, this is likely to be a readily accessible option. Most office and home computers already have a version of proprietary software, such as Microsoft Office, that can be used. Alternatively, open source software can be used to perform similar functions. This means that logic models can be readily shared in an editable form.

It is perhaps easier to develop a diagram in these packages than in word processing packages.

One of the advantages of using PowerPoint is that you can use custom animation to build up the program logic for presentation purposes so that people do not have to take in the whole picture at once. When using PowerPoint in a presentation, start by showing top-level outcomes and then work up from the bottom and add other features, such as showing which particular activities relate to which particular outcomes.

Disadvantages: As with word processors, not everyone has the skills to use the packages effectively to produce logic models. Tabular versions are easier to produce than a diagram but lack the arrows that show causal links, and they have the usual disadvantages of pipeline logic models. If those developing the logic model are not able to draw a diagram and enlist the help of staff, there will be less opportunity to develop the logic model iteratively, checking that it is focusing on key messages and meeting good design guidelines.

Likely to be useful: This option will suit situations when there is existing skill in using presentation packages and little opportunity to acquire more specialist packages that would be easier to use.

Specialist Visual Representation Packages

There are many drawing packages available, such as CorelDraw or Smart Draw, and diagramming packages such as Inspiration, which is designed to draw mind maps and can be adapted to draw logic models. These are likely to be suitable only where the additional investment in a visual representation package, in terms of purchase price and learning, can be justified.

We have focused on a commonly used add-on to Word called Visio. Visio is a Word add-on that provides templates and standardized components for drawing diagrams (Figure 13.10). The diagrams can be saved in

Figure 13.10   **Picture Pipeline Logic Model in Visio**

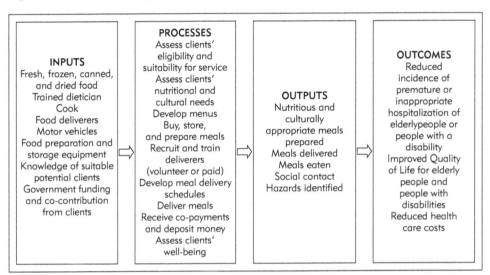

various graphic formats (including GIF, JPEG, and an enhanced meta-file) and then inserted into a document.

Advantages: The package has more options to support customized diagrams, and in skilled hands diagrams are quicker to develop and can look more visually appealing than a diagram developed in a word processing or presentation package.

Disadvantages: The additional visual elements in the wrong hands can lead to cluttered diagrams. With a specialist package, there is more risk of having one staff member allocated to produce all the logic model, reducing the opportunity to develop the logic model iteratively, checking that it is focusing on key messages and meeting good design guidelines. The biggest disadvantage is that although it can produce diagrams in a pdf format that can be shared easily, editable versions of the diagrams can be opened on computers only with the software. This can be a problem when working collaboratively with other groups or organizations that do not have the software.

Likely to be useful: This option will be suitable where this is sufficient investment in making the package and adequate support available to all who will need to use it.

Specialist Logic Model Packages

There are two types of specialist packages for developing logic models. Some, such as DoView, discussed below, and Theory of Change Online, have considerable flexibility in the components and the layout. Others, such as the InnoNET Logic Model Builder and the U.S. Department of Health and Human Services' Logic Model Builder for Child Abuse and Neglect Prevention/Family Support Program (http://toolkit.childwelfare.gov/toolkit /home.do), are more prescriptive and take users through a process of filling in information in response to prompts. The latter also includes links to research evidence about effective strategies and relevant measures to use to collect performance information.

DoView DoView is a specialized package for drawing logic models. It is designed primarily to draw outcomes chain logic models but it can be used for pipeline logic models. It includes the potential to have layered documents that can provide more detail of the program theory or links to supporting evidence. In Chapter Sixteen, we show how this can be used for reporting findings using the program theory as a reporting framework. Figure 13.11 represents the program theory of an advocacy program.

Advantages: The package has a shorter learning curve than a more comprehensive drawing package and produces diagrams more quickly than a word processing or presentation package can. This does open up the option of using it live during group sessions. The program theory can be drawn at a number of levels, with a high-level overview on one page and links to more detail of various sections.

Disadvantages: Like other specialist packages, the biggest disadvantage is that although it can produce diagrams in a pdf format that can be shared easily, editable versions of the diagrams can be opened only on computers with the software. This can be a problem when working collaboratively with other groups or organizations that do not have the software, although the purchase and learning costs are considerably lower than visual representation packages.

Likely to be useful: This option will be suitable where there is enough use being made of logic models to justify a small investment in specialist software or when interactive use of the computer is particularly important.

Figure 13.11   Outcomes Chain Logic Model in DoView

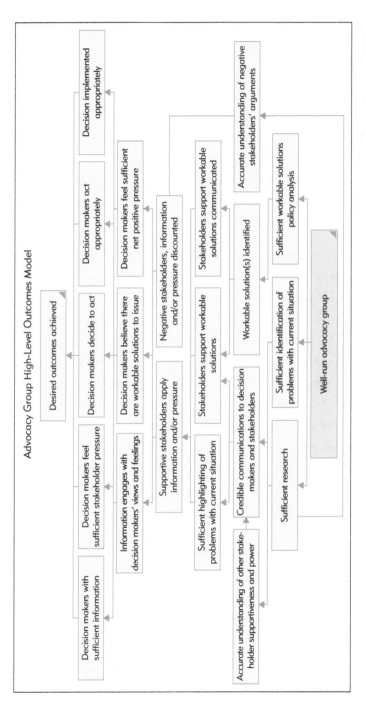

Advocacy Group High-Level Outcomes Model

Source: Duignan (2009a)

Figure 13.12 InnoNET Logic Model Builder Example

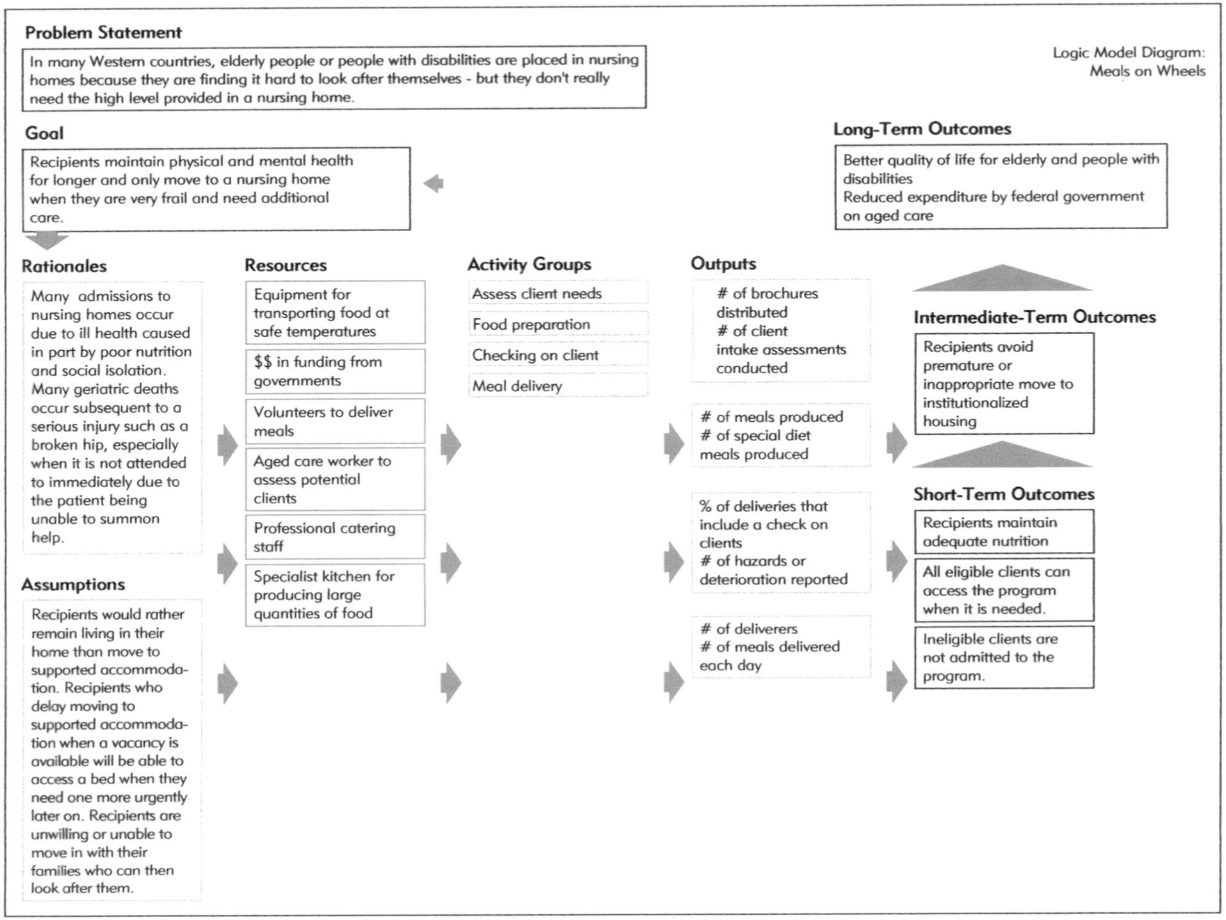

Problem Statement

In many Western countries, elderly people or people with disabilities are placed in nursing homes because they are finding it hard to look after themselves - but they don't really need the high level provided in a nursing home.

Goal

Recipients maintain physical and mental health for longer and only move to a nursing home when they are very frail and need additional care.

Rationales

Many admissions to nursing homes occur due to ill health caused in part by poor nutrition and social isolation. Many geriatric deaths occur subsequent to a serious injury such as a broken hip, especially when it is not attended to immediately due to the patient being unable to summon help.

Assumptions

Recipients would rather remain living in their home than move to supported accommodation. Recipients who delay moving to supported accommodation when a vacancy is available will be able to access a bed when they need one more urgently later on. Recipients are unwilling or unable to move in with their families who can then look after them.

Resources

Equipment for transporting food at safe temperatures

$$ in funding from governments

Volunteers to deliver meals

Aged care worker to assess potential clients

Professional catering staff

Specialist kitchen for producing large quantities of food

Activity Groups

Assess client needs

Food preparation

Checking on client

Meal delivery

Outputs

# of brochures distributed
# of client intake assessments conducted

# of meals produced
# of special diet meals produced

% of deliveries that include a check on clients
# of hazards or deterioration reported

# of deliverers
# of meals delivered each day

Long-Term Outcomes

Better quality of life for elderly and people with disabilities
Reduced expenditure by federal government on aged care

Intermediate-Term Outcomes

Recipients avoid premature or inappropriate move to institutionalized housing

Short-Term Outcomes

Recipients maintain adequate nutrition

All eligible clients can access the program when it is needed.

Ineligible clients are not admitted to the program.

Logic Model Diagram:
Meals on Wheels

InnoNET Logic Model Builder   The Logic Model Builder guides users through an interactive process of thinking about program goals, activities, and resources and how those elements work together to produce the program's intended outcomes. It does not have flexibility about how the program theory is represented. The example in Figure 13.12 represents the program theory of a Meals on Wheels program.

Advantages: The package is easy to use as it simply involves filling in the answers to a series of questions.

Disadvantages: There is no flexibility to change the format of the diagram to suit particular needs. It requires an Internet connection.

Likely to be useful: This option will be suitable where there is insufficient technical expertise to use a more flexible package and the style of logic model it produces is appropriate.

SUMMARY

There are many different ways of representing program theory. Both pipeline and outcomes chain logic models can be structured in different ways and using different technologies. Think carefully about which will work best for you.

EXERCISE

1. Find a logic model that represents an intervention you know fairly well. Redraw it using one of the options in this chapter. In what ways is this an improvement on the original version? In what ways is it less effective?

## PART 5

Using Program Theory for Monitoring and Evaluation

14

Developing a Monitoring and Evaluation Plan

AMONG THE PROGRAM THEORY traps that we have identified is the possibility that a program theory that was developed for conducting an evaluation is not then used for such purposes. Sometimes a critique of the program theory suggests that the program be redesigned before monitoring and evaluating its implementation and outcomes. However, sometimes the failure to follow through from program theory identification to monitoring and evaluation occurs because of a lack of understanding about how to use program theory for monitoring and evaluation. This chapter identifies some ways to do so.

Program theory can provide a focal point for making decisions about which aspects of program performance are to be measured or evaluated through performance monitoring (routine, regular, frequent) and which through occasional in-depth evaluation studies to answer cause-and-effect questions, assist with program improvement, and assist with accountability.

We look first at the use of program theory for performance monitoring. In this discussion, we also look at the role of comparisons as a critical part of useful performance information. Our discussion of considerations for choosing performance information and comparisons for monitoring is also pertinent to planning and undertaking an evaluation.

We look as well at other aspects of planning an evaluation to which program theory can contribute. These include establishing useful evaluation purposes, identifying key questions that an evaluation will address, and identifying relevant information to address those questions. We do not propose that the program theory be the sole basis on which a program evaluation is designed, and there are some circumstances, which we discuss, in which it may be less appropriate to use a program theory for such purposes.

Table 14.1 gives an overview of various monitoring and evaluation purposes at different stages in the life cycle of a program, how program theory can contribute, and where it is situated.

USING PROGRAM THEORY FOR PERFORMANCE MONITORING

One of the purposes of describing a program theory is to encourage the identification of measures that can be used for monitoring. Monitoring, which is usually less expensive and easier to do than evaluation, typically shows what is happening but not necessarily why. For example, it can:

- Look for trends or signals that things may be changing or not changing in the problem the program is addressing, the program itself, or the context within which it operates.

- Provide an early warning system, that is, leading indicators.

- Use lagging indicators to show what has already happened with respect to implementation and intended outcomes.

- Check that progress is on track against plans, milestones, targets, and expectations.

- In the light of any of the above information, identify a need for an evaluation study.

Table 14.1   Some Types and Purposes of Evaluation at Different Stages of a Program and How Program Theory Can Be Situated

Type of Evaluation	Timing of Evaluation		
	Before Implementation (Planning)	*Early Implementation*	*Mature or Completed Program*
Evaluation of program theory	*What's done:* Check proposed design for adequacy of cause-and-effect reasoning as shown in program theory and likely appropriateness, effectiveness, and efficiency.	*What's done:* Compare early implementation with planned implementation as in program theory, and develop an understanding of why any discrepancies occur: poor theory or implementation difficulties? Look for early warnings of unintended outcomes.	*What's done:* Periodically assess whether there has been a change in the situation to be addressed, in factors that may affect outcomes, or in the evidence base available to the program. Check whether the outcomes that occurred were as predicted by the theory.
	Purpose: Decide whether to implement proposed design, alternative designs, or modifications needed before implementation.	*Purpose:* Make any necessary revisions to program theory and identify implications for implementation and outcomes including, as appropriate, unintended outcomes.	*Purpose:* Decide whether the program theory still stands and identify any implications for whether to continue with the program. A new program theory may be a product of the evaluation.

(Continued)

Table 14.1   (Continued)

Type of Evaluation	Timing of Evaluation		
	Before Implementation (Planning)	Early Implementation	Mature or Completed Program
Routine monitoring of the problem, program implementation, and outcomes	What's done: Monitor preprogram conditions such as social indicators, and identify trends and projections that enable judgments to be made about whether the problem is improving or declining. Collect information that can provide baseline data.	What's done: Monitor early intended outcomes, especially lower-level enabling outcomes and key program features identified in the program theory, compare with expectations and identify fine tuning required.	What's done: Continue to monitor the problem, intended outcomes, key aspects of implementation, and critical program and nonprogram factors, identified in the program theory.
	Purpose: Decide what type of intervention is required. Develop program theory and design the program, including its monitoring and evaluation system. Establish baseline data.	Purpose: Make any necessary revisions to program implementation or program theory.	Purpose: Make judgments about the continuing appropriateness and likely future effectiveness of program. Identify more in-depth evaluation studies or other types of investigation that may be needed to better understand monitoring data and test the program theory.

| Evaluation studies | *What's done:* Collect or review evidence concerning effectiveness of programs similar to the proposed one. Study relevant program theory (for example, Theories of Change). Conduct needs assessments and feasibility studies. | *What's done:* Pilot studies of implementation and impact. Trial periods to assess strengths and weaknesses and what works under different conditions. Fast track evaluations. | *What's done:* Occasional in-depth evaluation studies of appropriateness, effectiveness, and efficiency with attention to causal inference. Program theory helps to generate hypotheses and identify what information is needed for evaluation studies. |
| | *Purpose:* Contribute to the development of program theory: determine objectives and program design. Decide whether to implement proposed program. | *Purpose:* Decide whether to expand program to full scale and whether to continue with full implementation. Revisit and revise program theory as needed. | *Purpose:* Formative: make improvements to the program. Summative: make decisions about whether to terminate, continue, or expand the program. Knowledge building: contribute to the evidence base. |

Monitoring all important components of the program as displayed in the program theory is useful. It can help to overcome some of the difficulties that have arisen from magic box approaches that measure outcomes but fail to develop any understanding of the role of the intervention and other factors in contributing to outcomes. Accordingly the routine measures should include a careful selection of measures representing each of the following aspects of the program theory:

- *The situation*—the nature of the problem, its causes, and its consequences and whether these are changing

- *What is happening in relation to the intended outcomes as shown in the outcomes chain*—note that until the causal relationship of the program to these measures of intended outcomes can be demonstrated (for example, through an evaluation study), they cannot be claimed as outcomes of the program but simply interesting occurrences or trends.

- *Program resources, activities, and outputs*—how the program is being implemented, including substantial variations that may affect outcomes.

- *The most important relationships among the various components of the program*—for example, correlations between levels or duration of program implementation and measures of intended outcomes if feasible to do so on a routine basis.

In Chapter Eight we introduced the program theory matrix as a tool for including relevant details about each of the various important features of a program theory. This same matrix can be converted into a performance monitoring (or performance indicator) matrix that mirrors the program theory matrix or selects from among the items in the program theory matrix. Table 14.2 shows a list of possible performance indicators that could be useful for monitoring the employment program for mature workers. For illustrative purposes, we show performance indicators for the desired outcome that program clients retain the jobs they have gained as a result of the program.

In this instance, we have included performance indicators across just one row of the matrix for the particular outcome that we used in Chapter Eight from the full chain of outcomes. In practice, it is more likely that one will

Table 14.2   Performance Information Matrix for One Outcome in the Outcomes Chain

Outcome: Clients Retain the Jobs They Have Obtained Through the Mature Workers Program

Success Criteria	Program Factors	Nonprogram Factors	Outputs and Throughputs	Activities, Processes, and Principles	Human and Financial Resources
Percentage of those in different age groups and with different demographic characteristics who retain a job	Percentage of those who do and do not retain their jobs for whom the adequacy of each of the following is judged as relevant to their retention:	Percentage of those who do and do not retain their jobs for whom each of the following factors is judged as relevant to their retention:	Number of support visits made to employees in their workplace	Description of what was done to:	Percentage of total program expenditure on follow-up support
Compared with ages of all participants in the program, those who obtained jobs, and with the general population	• matching of person to the job	• change in employee circumstances	Number and percentage of clients receiving a support visit within one month of placement	• match persons to jobs	Compared with target of 10 percent of program delivery funds
Percentage of retained jobs that are paid or unpaid	• management of client expectations	• employer expectations of employee		• manage employee and employer expectations	En
Compared with similar population and other employment programs	• follow-up support after placement	• business circumstances (for example, that require lay-offs)		• provide follow-up support	sr
				• set and apply performance targets	

Table 14.2 (Continued)

Outcome: Clients Retain the Jobs They Have Obtained Through the Mature Workers Employment Program

Success Criteria	Program Factors	Nonprogram Factors	Outputs and Throughputs	Activities, Processes, and Principles	Human and Financial Resources
Percentage of those with various levels of job satisfaction at six months	• performance targets that encourage sustainable placements	• receptiveness of other staff to mature workers	Compared with expectation that each person will receive at least one visit or contact within one month of placement	Compared with expected processes and accepted best practices (relates to column 2)	
Compared with target of 80 percent reporting moderate to high job satisfaction	Percentage of clients for whom other unanticipated program factors apply and description of those factors	Percentage of clients for whom other unanticipated factors apply and description of those factors		Relationship (for example, correlation) between nature and extent of implementation with particular clients and client outcomes (dosage considerations)	
Percentage of those employed for zero to six, seven to twelve, or more than twelve months, and in same versus different jobs	Comparison of the combined impact of program factors with combined impact of nonprogram factors	Comparison of combined impact of nonprogram factors with combined impact of program factors			
Compared with similar population, targets, and similar programs					

make a selection within each row in the matrix. One would also include performance indicators for the situation that gave rise to the program as a whole. Giving consideration to the full matrix even without having completed entries for all cells can help to ensure that a balanced range of performance measures is selected to represent different levels in the outcomes chain and the other components of a program theory.

MAKING CHOICES ABOUT WHAT TO MEASURE WITHIN THE PROGRAM THEORY

Given evaluation purposes, limited resources, and other practical constraints, choices still need to be made from among the full set of items concerning the situation and aspects of context to be monitored, intended outcomes, outputs, activities, and resources. We cannot be measuring everything all of the time. Selection will be influenced by considerations such as the following:

- The need for a restricted and manageable number of routine performance monitoring measures

- Whether the measures are to focus on effectiveness, continuing relevance of the program, or its efficiency

- Whether there is interest in developing a measurement and evaluation system that includes both program evaluation studies and performance monitoring data

Some performance monitoring systems look more like a laundry list of items of data than a strategic and coordinated selection of items of performance information. They may be a collection of available measures rather than of relevant and important measures. Program theory places a strong emphasis on the relationships among all of the components of the program and not just on the components themselves. To make the best use of program theory and for sense-making purposes, we want to choose a cohesive package of performance information rather than a random or uncoordinated collection. If we choose particular outcomes to measure, it would make sense to accompany those measures with some that relate to factors that are likely to affect those outcomes. Exhibit 14.1 identifies some of the choices that

EXHIBIT 14.1 CHOOSING WHICH PARTS OF THE PROGRAM THEORY TO MEASURE ROUTINELY

Ask these questions when deciding which parts of the program to measure:

- Which are the most important aspects of the initiating situation to monitor?

- Which are the most important outcomes in the outcomes chain to monitor (rows in the program theory matrix)?

- For the important outcomes, which will be the most important (not just the easiest) success criteria to monitor?

- For the important outcomes, which will be the most important program and nonprogram factors and aspects of program implementation to monitor?

need to be made from within a program theory. Making these choices in a considered way will foster the development of a cohesive package of indicators. We then discuss considerations for making these choices.

Which Aspects of the Situation to Measure

Typically there will be some key indicators that provide contextual and baseline data for monitoring and evaluating program outcomes. For the mature workers employment program, we would want to monitor the proportional representation of mature-age people among the ranks of the unemployed, including information about their length of unemployment and whether the situation for mature workers in the population as a whole (not just those served by this program) was tracking closely with the situation for the general population of workers or converging or diverging further.

Which Outcomes in the Chain of Outcomes to Measure

When an outcomes-focused approach is being taken to performance measurement and evaluation, a key decision is which levels in the chain of outcomes should be measured. A number of considerations bear on this decision. Patton (1997) recommends that "the decision about where to enter the means-end hierarchy for a particular evaluation is made on the basis of what information would be most useful to the primary intended evaluation users" (p. 218). Other factors include the relative influence of program versus nonprogram factors. Evaluation users may be most interested in high-level outcomes in the chain. However, through spelling out the program theory, it may become apparent that these outcomes could not be achieved by the intervention alone or that it is too difficult to disentangle the impacts of the intervention from other more influential factors. Those intended outcomes can still be monitored for the purpose of tracking the problem, but care should be taken in reporting higher-level outcomes without any information about other contributing factors because this might imply direct and complete causal attribution.

For different evaluations, and sometimes for the same evaluation, users will vary with respect to their interests in making judgments about program effectiveness, the continuing relevance of the program relative to need, or efficiency. Some will be more interested in using evaluation to assist with program improvement. Others will have an interest in accountability. In Exhibit 14.2,

---

EXHIBIT 14.2 SOME SUGGESTIONS FOR CHOOSING WHICH OUTCOMES IN THE OUTCOMES CHAIN TO MEASURE

- Information about outcomes at the top of the outcomes chain tells us about the situation that gave rise to the program and helps us to make judgments about the (continuing) need for or the appropriateness of the program as a whole.

- Information about outcomes at all levels of the chain, and especially the levels that the program can influence most, can help us make judgments about program effectiveness.

- Information about outcomes at low and middle levels of the chain can help us to interpret and understand the program's impacts (or lack thereof) on higher-level outcomes and for program improvement purposes. These are typically lower and middle levels over which program management has greatest control.

we make some general observations about how information at different levels in the outcomes chain might align with these different interests.

Which outcomes are to be included in a routine monitoring system will also depend on feasibility and cost of measurement. This includes the costs to those who will be required to routinely provide the data and whether there are opportunities to complement performance monitoring with occasional in-depth evaluation studies.

Selection of outcomes can also be affected by whether accountability for outcomes is an issue to be addressed through monitoring and/or evaluation. How far up (and down) the outcomes chain should measures of outcomes be used for accountability purposes, given the extent of program control or potential program control? What else needs to be measured (for example, external factors) that might provide an alternative or additional explanation for results at all levels of the chain? If programs have been established with unrealistic expectations and objectives, the accountability requirements will exceed what can reasonably be delivered. In some cases, the evaluator can do little more than point out the anomalies. The following example and Figure 14.1 look at the use of program theory principles for discussions about accountability:

USE OF PROGRAM THEORY PRINCIPLES FOR DISCUSSIONS ABOUT ACCOUNTABILITY

Through the use of program theory principles, this intervention was held accountable for the middle levels of an outcomes chain but neither the lower nor the highest levels of the outcomes chain. The intervention was a crisis call center for problem gamblers.

EXCLUSION OF LOWER LEVELS OF OUTCOMES CHAIN

The contractual obligations of a particular call center for problem gamblers initially required the center to demonstrate performance in relation to a measure of level of use of the service (in this case, new users) by various groups in the target population (gender, ethnicity, age, and so forth). This was a lower level of outcome in the outcomes chain, preceded by an outcome concerning whether potential participants knew of the service and how to contact it, appreciated why they should contact it, and so on. However, the inclusion of the measure of numbers of new users in the contract for the call center was questionable since the call center service had specifically been relieved of responsibility for marketing and promotion activities that might have affected level of use.

For accountability purposes, the measures of performance should have started at the first point at which the target audience made contact with the service, not whether they made contact (the result of marketing efforts). The measure of level of use (a lower-level outcome in the outcomes chain) was an important measure of the performance of the program as a whole; the service provider was in the best position to collect and provide these data to program funders and designers. However, this information was not useful as an accountability measure of the performance of the service provider except to the extent that take-up of the service occurred as satisfied clients encouraged others to use it. Given the nature of the clientele, such referrals were not expected to be a major source of new users.

EXCLUSION OF HIGHER-LEVEL OUTCOMES IN THE OUTCOMES CHAIN

These higher levels related to long-term recovery of problem gamblers. As a crisis service, one of the main roles was to refer contacts to other counseling services with which they could establish a longer-term relationship with a view to longer-term solutions. Therefore, for accountability purposes, the accountability of the call center ended at a point in the outcomes chain that was below outcomes relating to long-term improvements. The call center nevertheless needed to conduct its activities in a way that was mindful of these longer-term objectives by, for example, ensuring that referrals were appropriate and undertaking research about whether contacts were following through with referrals. The call center was accountable for take-up of its referrals to counseling services to the extent that take-up was related to the appropriateness of those referrals from the perspective of clients. Again program theory was used to establish more realistic accountability requirements.

Figure 14.1 is a simplified outcomes chain that shows the range of outcomes for which the call center was held contractually accountable following the application of program theory.

Choices Concerning Which Success Criteria to Monitor and Which to Evaluate

In Chapter Eight, we emphasized the importance of specifying success criteria for each outcome as part of the process of comprehensively thinking

Figure 14.1   **Simplified Outcomes Chain Showing Levels for Which the Crisis Call Center for Problem Gamblers Was Held Accountable**
Note: Outcomes in bold are those for which the call center was held contractually accountable.

Reduced impacts of problem gambling among callers to call center.

↑

Reduced levels of problem gambling among those assisted by the crisis call center.

↑

Problem gamblers referred by the crisis call center receive appropriate assistance from gambling counseling services to which they are referred.

↑

Problem gamblers take up referrals to counseling services given to them by the crisis call center.

↑

Problem gamblers receive referrals to counseling services from the call center.

↑

Immediate crises of problem gamblers are averted through assistance received from the crisis call center.

↑

Problem gamblers who call the crisis call center receive immediate response following agreed principles and standards.

↑

Problem gamblers in a state of crisis call the gambling crisis call center.

↑

Problem gamblers acknowledge that they have a problem and are willing to contact the crisis call center.

↑

Problem gamblers know of the existence of the gambling crisis call center and how to contact it.

about and representing a program theory. Success criteria are the bridge between broad statements of outcome and relevant performance measures.

When the outcomes to measure are being selected, performance information should reflect the most important success criteria, not just those that are easily measurable. This helps to guard against goal displacement, where targets are met at the cost of actually achieving the stated goal. Performance in relation to some success criteria might be measured through routine monitoring. For other outcomes that may be too difficult to monitor routinely or that can be expected to show little change from one monitoring period to the next, it may be more appropriate to conduct a more in-depth and resource-intensive program evaluation.

It is important to flag those success criteria that will require additional evaluation work so that they are not overlooked. For example, for the mature workers employment program, it was relatively easy to use monitoring data in relation to success criteria about whether participants obtained jobs. It was not so easy to routinely monitor levels of satisfaction with those jobs and whether people felt they were appropriately employed given their skills and experience and how valued they felt in their jobs. These more qualitative aspects are more likely to be the subject of an evaluation or an occasional survey than routine monitoring. Including them among the success criteria in a program theory is a reminder to consider them as part of an evaluation.

Choices Concerning Factors That Affect Success

Monitoring important factors and aspects of program implementation that might affect whether outcomes will be achieved can provide important early warnings about troubles ahead that should be managed. Knowing which factors are the most important ones to measure can be difficult, the more so if a program is working in relatively uncharted territory. Program staff and clients will often be able to assist with identifying the most important factors. Programs that are grounded in research, have a history of collecting information to monitor performance, have conducted past evaluations, and are relatively stable will be better positioned to know what these factors are.

Programs that do not have that history but have some similarities to archetypal programs (discussed in Chapter Twelve) can take some cues from those programs. For example, standards have been developed for case management programs (an archetypal program) that offer some insights into some key types of factors such as case load that should be considered as part of an evaluation of any program that uses case management.

Programs with many complex aspects will find it more difficult to identify these factors. For those programs in particular, it is important that monitoring and evaluation processes be sufficiently open-ended to enable new factors to emerge.

Choices Concerning Aspects of Program Implementation to Measure

Using information for purposes of program improvement will have implications for choices about which activities to measure. Are program activities in place in the manner intended? Are they sufficient to produce results? Are activities being conducted in an efficient manner? Measures should be taken of the level and quality of implementation of those activities that are considered to be critical to the success of the program in achieving outcomes before attributing outcomes to the program.

Known variations among sites will provide pointers for other relationships that might be explored to tease out the critical factors. Measurement or review of program implementation usually shows variation in implementation from site to site, occasion to occasion, client to client, and, of course, one member of staff or group of staff to another. It is important to be aware of these differences when interpreting results and understanding cause-and-effect relationships with outcomes and factors that affect outcomes. For example, in relation to the mature workers program, an example of site-to-site variations was that some funded projects were much more active than others in working with the community by, for example, working with mass media and engaging in Rotary Club activities. Even though this meant that in some cases they spent less time with individual clients, they had better track records with client outcomes than sites that focused almost entirely on the clients themselves and individual employers, giving no attention to employers en masse.

Choices Concerning Links in the Program Theory to Measure

Links among outcomes and among the various components of the theory of change and the intervention theory are critical to program theory. A theory is not a theory without an articulation of the links. These links need to be measured routinely and/or when conducting occasional evaluation studies. Weiss (2000) identifies the following considerations for choosing which links to evaluate: how critical to the success of the program the links are, the plausibility of the linkages, and the degree of uncertainty about the linkages.

Both relevance and effectiveness can be affected by the strength of relationships between lower-, middle-, and higher-level outcomes in the outcomes chain. In some situations, the relationships among higher-level outcomes will be well supported by the research literature, and there is no need to demonstrate the relationship time and time again through each evaluation of each relevant program. Measures of lower levels of outcomes can be used as proxy measures for higher-level outcomes. The following example addresses the use of proxy measures:

USING MEASURES OF LOWER-LEVEL OUTCOMES AS PROXIES FOR HIGHER-LEVEL OUTCOMES WHEN LINKS ARE SUPPORTED BY RESEARCH

There is substantial evidence from the research literature that unemployed people over the age of forty suffer in relation to various aspects of quality of life: health, financial, and others. There is also evidence that while such issues as poor health can adversely affect employment prospects, at least some of this degradation of quality of life is itself caused by being unemployed. That is, unemployment is both cause and effect.

In the evaluation of the mature workers employment program, some anecdotal evidence emerged concerning the impacts of the program on quality of life in two ways: through the employment that was secured and simply through participating in the program that, among other things, widened their social networks and developed their self-confidence. However, given scarce evaluation resources, the evaluation stopped short of rigorous assessment of the impact of becoming employed through the program on quality of life. It took the research evidence about the inverse relationship between involuntary unemployment and quality of life as a given, supported by brief testimonials from some of those who had participated in the program.

Looking at links can also assist with program improvement. They can identify where the links in the outcomes chain may have broken. In some cases, a link may have broken, but a bypass of that breakage may also have occurred, and the higher-level outcomes continue to be achieved. It is desirable but not always feasible to monitor relationships among the various levels of outcomes in an outcomes chain. Correlations and conversion rates are examples of measures of relationships. For example, in the case of the mature workers program, conversion rates from acquiring skills through the program to getting a job could be measured: Of those participants who develop improved job skills through the program (and also have appropriate levels of motivation and job-seeking skills), what percentage are successful in getting a job within, say, three months of acquiring those skills? This approach obtains some preliminary but inconclusive data about the cause-and-effect (or at least sequential or contiguous) relationship between developing job skills and getting a job. Like correlations, conversion rates do not demonstrate causality, but they can signal what further investigation might be needed. (We explore the issue of causality in Chapter Fifteen.)

INCLUDING COMPARISONS AS PART OF THE PERFORMANCE INFORMATION SYSTEM

Too often the data collected as part of performance monitoring never become useful information because no one has ever thought very seriously about how the data might be used to draw conclusions. The use of comparisons can help.

Hatry, Winnie, and Fisk (1981) reminded us that when it comes to development of outcome monitoring systems, "comparison is the name of the game." Common types of comparisons used for monitoring and for evaluation include the following:

- Comparisons of actual with planned implementation.

- Comparisons of performance with standards—quantitative and qualitative.

- Time-related comparisons such as before and after the project; time series, especially interrupted time series; and rates of change—for example, for different programs or client groups within a program.

- Comparisons with objectives and targets. These could be quantitative, such as reaching a specified number of clients, or qualitative, such as reaching a particular milestone or standard.

- Program with nonprogram comparisons, for example, experimental and quasi-experimental, norms, benchmarks, or population statistics.

As noted in Chapter Eight, success criteria in a program theory should, where possible, include standards or targets provided they are nonarbitrary and can be justified in relation to the intended outcomes of the program. These standards provide one basis for making comparisons that can assist with judging and interpreting the findings. However, even without that specification as part of program design, the evaluator can apply a range of comparative tools.

Evaluators are often enjoined to use multiple methods for triangulation purposes. Similarly, use of multiple types of comparisons can shed different perspectives on results. For example, comparisons of repeated measures of the performance of a given subgroup of a target population over several years, such as scores of a minority group on literacy tests, can be useful. This information could be made even more useful if it included comparisons with the rate of change over the same time period of other subgroups or the population as a whole. Comparison with desired standards of literacy may yield a different perspective altogether.

When choosing the types of comparisons to be used, it is helpful to identify whether performance information is to be used primarily to describe, judge, or interpret performance. The distinctions among comparisons for the purpose of describing, judging, or interpreting performance are important but are often overlooked in program evaluation and performance measurement more generally. Some types of comparisons can be used for all three purposes.

Following are some examples of the different types of comparisons (for description, judgment, or explanation or interpretation) that might be considered for inclusion in a program theory matrix and accompanying performance information matrix or more generally as part of program evaluation and performance measurement. Performance information used to *describe* programs (their outcomes, activities, and so on) typically employs comparisons over time, say, several months or years, among different parts of the program—different sites of program delivery, with the operation of other programs, and among different client groups, where no judgments are made about performance or interpretations drawn. However, these comparisons can also be used to judge performance when accompanied by standards or some indication as to what constitutes desirable performance. For example, a standard in relation to comparison with other programs could be that the program in question should perform at least to the level of the comparison programs. Or it might specify that performance should exceed that of other similar programs or expected outcomes for the relevant population.

Performance information used to *judge* a program (its outcomes or activities, for example) typically employs the following types of comparisons:

- Program outcomes with program objectives

- Performance with targets and milestones

- Actual implementation with intended implementation

- Current performance against past performance and whether better or worse

- Performance relative to national or international standards

- Performance relative to norms or benchmarks

- Program activities compared with accepted professional standards or codes of conduct

- Performance compared with client service standards and client expectations

All of these comparisons typically involve some type of gap analysis between actual and desired, desirable, or expected performance. These comparisons should not be made without serious thought. For example, if an unemployment program had a particular target, not reaching it is a problem, but the program might have been doing as well as could be reasonably expected given an economic downturn. Or the target might have been met not through a successful program but due to other local developments, such as the opening of a new factory. In such a situation, we might mistakenly judge the program to be effective.

Performance information used to *interpret and explain* program outcomes and, in particular, to attribute outcomes to a program, may involve the following types of comparisons:

- Before-and-after comparisons and all the variations of before-and-after designs (such as simple pre-post with or without controls or interrupted time series) where there is some intention to attribute differences to the intervention. Does any evidence suggest that differences are attributable to the program? Are there alternative explanations? Is counterfactual evidence available?

- Between people or sites receiving the program and those not receiving the program. Does any information suggest that any differences can be attributed to the program?

- Between different levels and quality of program implementation with a view to answering such questions as whether dosage affects outcomes.

Issues of attribution and contribution are discussed further in Chapter Fifteen. Examples of a range of different types of comparisons used for the evaluation of the employment program for older people follow.

EXAMPLES OF COMPARISONS USED IN EVALUATING THE MATURE WORKERS EMPLOYMENT PROGRAM

- Comparisons of outcomes for different subgroups to determine whether results were equitable

- Comparisons with the employment rates of similar groups in the wider population

- Qualitative comparisons of practices of service providers with best practice in case management

- Qualitative comparisons of overall program management practices with international best practice for this type of program

- Comparisons among the various service providers in terms of the outcomes achieved with various types of clientele

- Client-provided comparative perspectives concerning their experience with this program relative to other employment programs in which they had participated

- Client-provided judgments about whether participation in the program had made a major, minor, or no contribution to their outcomes

Some comparisons were to describe, some to judge, and some to interpret. The evaluation compared outcomes for program participants with those of the relevant population in order to make judgments about whether apparent outcomes were better or worse than might be expected. For similar reasons, it made comparisons with past studies of employment programs, taking account of the economic climate at the time (an external nonprogram factor).

For descriptive and judgmental purposes, it made comparisons of outcomes that were achieved with the various subgroups, such as age, gender, length of unemployment, ethnicity, and industry classification (for descriptive purposes) and relative to what might be expected for these subgroups in the general population (for judgmental purposes). From these comparisons, an evaluation was also able to infer that it was unlikely that the program was achieving its success through selecting as participants only those who were most likely to succeed given known risk factors. Use of these comparisons assisted with interpreting the findings.

We have located our discussion of comparisons alongside our discussion of performance monitoring systems because we so often see performance monitoring systems that give inadequate attention to comparisons and produce a lot of relatively useless data. However, the same general principles apply to evaluation studies.

The discussion has focused on using program theory as a framework for choosing what to measure. However, the principles of program theory can also be used to assess performance information choices that have already been made and the quality of existing performance monitoring and reporting.

USING PROGRAM THEORY TO PLAN AN EVALUATION

In Chapter Ten, we described the process of critiquing a program theory. Such a critique can lead to conclusions about the likely validity and feasibility of the theory, at least on paper, and in itself constitutes a type of evaluation of the program. Sometimes these conclusions contribute to decisions as to whether to proceed with an evaluation of outcomes or some other type of evaluation.

In this chapter, we are discussing the use of program theory to plan an empirical evaluation. One of the main advantages of using program theory is that it can make the process of selecting what hypotheses to explore and what to measure more systematic. It can also provide a more coherent conceptual framework for interpreting findings and reporting results. Among the ways that program theory identification processes and program theories themselves can be used to assist with planning an evaluation are the following:

- *Assisting with decisions about the type of evaluation that would be appropriate given the stage of program development.* If program theories are developed early enough, they can play an important role in ensuring that relevant monitoring data will be available. One of the greatest laments of evaluators arises from the failure to build monitoring and evaluation into the design of the program from the outset. Fortunately, we find that we are increasingly being brought in earlier in the life of a program to assist with monitoring and evaluation. Still, because programs are often initiated in great haste, this often occurs at the early implementation stage rather than the early stages of design. This situation is not ideal because it makes collecting baseline data more difficult, though it is still better than in the past.

- *Identifying whether assumptions about all aspects of the program as shown in the program theory are met.* In particular, an evaluation can explore whether key factors that affect success have been successfully addressed and if not, whether and how that has affected performance. The program theory can also assist with identifying where disconnects occur within the chain of outcomes so that these disconnects can be investigated. Program theory used in conjunction with realist approaches to evaluation can be particularly helpful in this regard.

- Following on from the previous point, *identifying where there may be alternative and possibly competing theories that could be explored in an evaluation.* These could be for the program as a whole or for subgroups within the program. Making program theory assumptions explicit alerts evaluators to circumstances in which the theory might break down. In Chapter Seven, we gave some examples of alternative scenarios that could be explored for a program that used radio in Vietnam to improve democracy. We gave another example in Chapter Ten about the beginnings of a negative program theory for providing cash grants to victims of domestic violence.

• As for developing a monitoring system, *identifying the important aspects of the program (problem, outcomes, implementation) for which performance information should be collected.* This would include identifying which aspects of program implementation are critical to the achievement of outcomes and why (the links among the components in the program theory). Those aspects can be explored during an evaluation, and the evaluation effort can be distributed in a way that is commensurate with the relative importance of different features rather than assessing the fidelity of implementation in all its particulars for its own sake as a measure of compliance.

• *Highlighting issues that require investigation.* Among the many uses of program theory, one of the most common is its use as a source of hypotheses or questions to be addressed in an evaluation. These could relate to aspects of the program and the theory behind the program about which there are uncertainties.

• *Ensuring that stakeholder perspectives concerning single issues are kept in perspective in an evaluation* by seeing how, if at all, they relate to the overall program as shown in its program theory or whether they are peripheral to program effectiveness or relate to unintended outcomes.

When planning evaluations, we find the four-step process (as shown in Figure 14.2) useful. Although the steps are shown as being sequential, they are to some extent iterative. For example, feasibility considerations that ultimately affect choice of methods in step 4 may require changes to steps 2 and 3. The feedback arrows show iterative effects that can occur during the initial planning. Also, as an evaluation progresses, we learn more about the program and its context, including additional issues, what types of data can be collected, and what else might be useful. Indeed during the evaluation, our ideas about the program theory may also change. Evaluation plans evolve over time, but having a carefully thought-out plan from the start is a good starting point and a useful touchstone to enable more deliberate choices to be made if changes to the plan are contemplated.

Program theory can help at each step, but it is not a self-contained stand-alone methodology. It can be used in conjunction with many different

Figure 14.2 **The Logic of Planning an Evaluation**

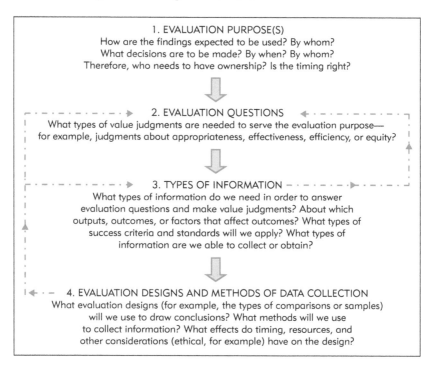

models of or approaches to evaluation, for example, decision-making models, utilization models, accountability-based models, and case study models.

Step 1: Timing and Purposes of an Evaluation

Program theory can be particularly helpful in identifying whether the overall timing of the evaluation is right, too soon, or too late, and the general type of evaluation that is appropriate at this time.

Evaluation can be undertaken too early, especially if it measures later outcomes in the outcomes chain before anyone could have expected them to have been achieved. Many programs have been prematurely condemned as a failure because their sequencing and timing had not been well thought through when the evaluation was being planned. Program theory helps us to see the sequence of the outcomes and estimate when it might be appropriate to measure outcomes at different points in the chain of outcomes.

It may caution against measuring later outcomes in the chain until earlier enabling outcomes have been achieved to a satisfactory level. An evaluator may be able to use program theory to persuade others to postpone the evaluation of some outcomes. This is also part of establishing realistic expectations with evaluation audiences concerning what they can reasonably expect from an evaluation of the program at this time and, in some cases, ever.

Evaluation can also be undertaken too late to be useful for program improvement purposes. If the absence of various preconditions for achieving outcomes (for example, aspects of implementation, factors that affect outcomes, and early enabling outcomes such as getting people to participate in the program) has not been identified early, then opportunities to rectify those preconditions before measuring outcomes may be lost. Program theory can help to identify what those critical conditions are and ensure that they have been monitored and corrected in a timely fashion. An evaluator who is engaged early in the program's life can identify some of this preemptive monitoring that needs to occur, as well as establishing what baseline and monitoring data need to be in place.

Understanding the current status of a program—whether it is in the design phase, early implementation, mature, or completed—can also assist with identifying the types of evaluation that can be done and how program theory can help. Table 14.1 summarized some roles that various types of evaluation (evaluation of program logic, routine monitoring and evaluation studies) can play at different times.

Evaluators of programs that are in the design phase can use program theory to ensure that preconditions for its success are in place. They can help to:

- Ensure that key stakeholders have a clear and reasonably cohesive idea about what the program is all about and that there is sufficient buy-in to the program for it to be successfully implemented

- Evaluate the likely future effectiveness of the program by ensuring that the situation and its causes have been adequately identified and incorporated in the program theory

- Make sure the objectives are achievable and important and the design is appropriate to the situation and feasible to implement

- Develop a monitoring, evaluation, and reporting framework for use during program implementation

Evaluators of programs that are currently being implemented can use program theory in these ways:

- Identify the critical aspects to monitor with respect to inputs, processes, outputs, immediate outcomes, and any other key factors that might affect outcomes to ensure that the program is on track and performing as well as possible, especially if this has not been done during the design phase

- Identify key evaluation questions

- Conduct an assessment of a program's readiness for measuring and evaluating outcomes at various points in the outcomes chain

- Conduct a formative or process evaluation concerning program implementation

Evaluators of programs that are mature and for which continuation or adaptation is proposed can use program theory to describe the program and decide what information needs to be collected for these purposes:

- Demonstrate effectiveness in achieving desired outcomes and addressing needs (additional to any that might already have been included in a routine performance monitoring system)

- Demonstrate that outcomes are attributable to the program

- Ensure that the design continues to be appropriate

- Determine what changes, if any, are needed to improve effectiveness and efficiency

Evaluators of programs that have finished or are nearing completion can use program theory for accountability purposes to describe their programs and identify what information needs to be collected or what should have been collected to establish these ends:

- What outcomes were achieved and how well they addressed program objectives and needs

- The extent to which outcomes were attributable to the program

- Lessons learned for the future about, for example, factors that affect success, unintended outcomes, or new program theories

A new program theory may be the product of an evaluation of a mature or completed program. In some cases, the production of a new program theory will be specified as a purpose of an evaluation. In other cases, it may emerge as a by-product.

Step 2: Identifying Key Evaluation Questions

By "key evaluation questions," we mean the relatively small number of important evaluative questions that lead to value judgments being made rather than a lengthy list of items of requested information.

Terms of reference for evaluations are often cast in general terms such as, "Evaluate the effectiveness of the program in achieving its objectives." It often falls to the evaluator to identify the specific evaluation questions that need to be addressed, the information required to address those questions (step 3), and the methods and designs that will be used to collect the information (step 4). A program theory can help put the flesh on the bones of the skeleton provided as part of the terms of reference. It can help convert very broad evaluation questions into the specific questions needed to form the bridge between the terms of reference and the evaluation methodology. Program theory can, for example, help to translate general questions about program effectiveness into more specific questions that relate to particular outcomes. These include immediate, intermediate, and ultimate outcomes shown in the outcomes chain and questions about the factors that are most likely to affect those outcomes.

Sometimes program theory can be used to recast evaluation questions altogether. For example, suppose that an initial evaluation question was, "How effective has the program been in achieving its objectives?" Suppose further that an analysis of the program theory has shown that those objectives are quite unrealistic. In this situation, rather than conduct an empirical evaluation of performance against objectives, it might be more appropriate to evaluate the program theory as having been a poor one and suggest what outcomes could be evaluated so that the actual achievements of the program could be assessed. The evaluator could also identify how relevant those actual achievements are likely to be to the formal objectives of the program, regardless of how realistic those objectives are themselves.

In this case, the evaluator may need to take an active role in persuading potential users to revise what they should ask of an evaluation. In some cases, this is more about shifting the emphasis of the evaluation so that the program's success does not stand or fall completely on whether it can be shown to achieve those higher-level outcomes. This also increases the likelihood that its success in relation to more achievable outcomes will be assessed by the evaluation and the program given credit for what it has achieved rather than just condemned for what it has not achieved.

Of course, one possible consequence of trying to set realistic expectations of an evaluation is that the evaluation will be called off altogether because what the program theory shows is likely to be achievable by the program is not worth spending valuable time and resources on. Or the shoot-the-messenger phenomenon may apply: the person or organization requesting the evaluation may seek an alternative evaluator who will be prepared to measure and invalidly attribute outcomes as desired.

Program theory should not be the only source of evaluation questions. Other stakeholder interests will need to be addressed through an evaluation, but they may not derive directly from program theory or may not stand out as priorities from the program theory. These might include unintended consequences, effects of the program on staff morale and working conditions, and so on. Throughout, we have suggested ways in which various perspectives can be incorporated in the program theory. Inevitably, however, some important ones will have been overlooked or new issues will arise, as for all other aspects of an evolving program theory.

Sometimes further probing about concerns that are absorbing stakeholder attention will show that these concerns are actually important aspects of program performance that have been overlooked in the program theory. For example, staff morale is important in its own right, but the absence of good staff morale can also be a major impediment to program performance and perhaps should have been included in factors that affect achievement of outcomes. Often these factors are thought about only once they become a problem, and they may not have been incorporated in the program theory.

Sometimes an evaluation brief may include detailed terms of reference that specify exactly what outcomes and processes are to be measured. If these questions reflect sound thinking about the program theory, then such detail can be helpful to an evaluator, and no further development of program theory may be required. On occasion, however, it will become apparent that the questions are more like the product of a brainstorming session, with no real pattern to the questions and no implied conceptual framework for organizing the answers.

The evaluator can assist with the way in which such evaluation questions are organized and prioritized. Program theory can be used to formulate sensible hierarchical sets of stakeholder questions, with each set having its own logic (rather than shopping lists of curiosities) while still being responsive to stakeholder interests. Program theory can give greater clarity about whether certain aspects of the program are to be assessed as important outcomes in their own right or whether they are being assessed as a means to an end. For example, is staff morale to be investigated because it is important in and of itself or because there are suspicions that it is detracting from program delivery and outcomes? If so, are there particular aspects of the action theory or outcomes in the chain that are being especially affected? Depending on the answer to this question, we might conduct the evaluation in different ways.

For example, the terms of reference for an adult education program included descriptive questions about the practices used by the educators and evaluative questions concerning the extent to which those practices were consistent with guidelines about how the educators should operate. Program theory was used to put the questions into perspective so that the practices of the educators were more clearly linked to the intended outcomes. Opportunities therefore arose to identify a range of different educator practices that worked with varying degrees of success in achieving outcomes in different contexts. The findings led to recommendations that a less doctrinaire one-size-fits-all approach to the role of the educator be adopted to ensure that the processes were team- and outcome-centric rather than model driven.

We now identify some common general evaluation questions. For each question, we identify whether the program's theory of change or its theory of action, or both, is more likely to be relevant when applying the general question to a specific program.

Are we doing the right things (appropriateness and relevance)?

- Do we have the right sort of program given the situation or problem we are trying to address and its causes? Is the logic right? [theory of change]

- Is the design right? Does it have the crucial elements? Has it given consideration to all key factors? This relates mainly to theory of action and the links between program resources, activities, and outputs, on the one hand, and factors that will affect outcomes (such as staff quality) on the other hand. A theory of action helps us to be clear about how the program is expected to be implemented in order to achieve outcomes.

Are we doing things right (fidelity and efficiency)? Program theory is more useful for identifying evaluation questions that relate to fidelity of program implementation relative to planned implementation than to efficiency. It can assist with such questions as these:

- Is it being implemented according to plan, or is adaptive management occurring when needed? [theory of action]

- What are the variations in implementation, and how do they affect our assumptions about the program? [theory of action]

Is it working (effectiveness)?

- To what extent is it contributing to the desired immediate and intermediate outcomes? This relates mainly to the theory of change for outcomes achieved and to the relationship between the theory of action and the theory of change for assessing the contribution of the program to those outcomes.

- Are those outcomes contributing to addressing the situation or problem that gave rise to the program? [theory of change]

- What aspects of the program have worked well and not so well, and under what circumstances? [theory of action but also the relationship between the theory of action and the theory of change]

- What else has affected the outcomes achieved by the program? This relates to nonprogram factors that affect outcomes. [theory of action]

Examples of other important evaluation questions that might not come directly from the program theory are:

- What other effects is it having (value of other effects)?

- What spin-offs have there been (for example, benefits to partner organizations)?

- Have the unintended outcomes been positive or negative?

Using the mature workers employment program as an example and referring to the various descriptions of the program throughout the book, we show how particular aspects of a program theory can be used to generate some specific questions about program effectiveness in relation to intended and unintended outcomes:

SOME GENERAL EVALUATION QUESTIONS WITH EXAMPLES FROM THE MATURE WORKERS EMPLOYMENT PROGRAM THEORY

- *General question:* To what extent have the intended outcomes of the program been achieved? Refer to the outcomes chain to identify outcomes.

  Mature workers program (MWP) questions: How effectively has the program reached its target audience of unemployed people over the age of forty-five across all demographics of the community? How effective has the program been in helping unemployed people to obtain jobs?

- *General question:* To what extent has achievement of the lower-level outcomes contributed to achievement of higher-level outcomes? This is a question that is not always asked, but it is an important one implied by taking a program theory approach. It is about the degree of association among outcomes in the chain.

  MWP questions: To what extent has the effectiveness of the program in obtaining jobs for mature workers depended on its effectiveness in developing their job skills? Job-seeking skills? Job motivation?

- *General question:* To what extent have the outcomes that have been achieved made a noticeable contribution to overcoming the causes of the problem, reducing the problem itself, mitigating the consequences of the problem, and thereby addressing the ultimate outcomes? Look at the high-level outcomes in the theory of change and the causal relationships between immediate, intermediate, and ultimate program outcomes.

> *MWP question:* What benefits in terms of improved quality of life have there been for people who acquired a job through the program?

- *General question:* What features of the program in practice have helped, hindered, or made no difference to achievement of outcomes? Refer to program factors, activities, and resources in the theory of action to identify aspects that should be assessed.

    > *MWP question:* How important and cost-effective have follow-up support visits been in helping people to retain jobs that they obtained through the program?

- *General question:* What other factors have affected the outcomes? Look to program and nonprogram factors in the theory of action for insights about what to investigate.

    > *MWP question:* What factors other than the program (health, emerging family responsibilities, location and travel convenience of job, or workplace restructures, for example) have affected whether people have been able to retain a job they have obtained through the program?

- *General questions:* Has the program been more effective under some conditions or in some circumstances more than others, and if so, why? With some target groups more than with others? At some sites more than others? Refer to assumptions about program and nonprogram factors and variations in activities and resources identified in the theory of action and outcomes in the theory of change for insights about what to investigate.

    > *MWP questions:* Has the program been equally effective with older versus younger mature-age workers relative to what would be expected given population statistics for these different age groups? Have there been variations in effectiveness according to the different approaches or models adopted at different sites?

- *General questions:* What have been the unintended outcomes, and have they been positive or negative? How important are they to the program? Some unintended outcomes that can be anticipated may be included in assumptions about program and nonprogram factors.

MWP questions: To what extent are employers benefiting from participation in the MWP through, for example, access to a skilled mature workforce? To what extent is the success and sustainability of the program dependent on employers' benefiting from it? What are the most important benefits to employers on which the program can capitalize (for example, bottom line, corporate citizenship, mentoring of younger staff) to promote the cause of older workers?

Note that the main program theory for the mature workers program was built around the outcomes for unemployed people. The direct intention of the program was not to improve the lot of employers, but they could not be recruited to the program unless there were some benefits for them or at least that the costs to them were minimized. Identifying these important enabling outcomes that are not the direct focus of the program draws attention to the fact that they are also an important part of the factors that will affect program effectiveness and should be drawn into the theory.

Another way to organize questions is by outcome in the outcomes chain of a program theory. For each outcome selected for investigation by the evaluation, ask:

1. To what extent is this outcome being achieved, and what variations are there?

2. What is being done to achieve this outcome, and how well is it being done?

3. What else is affecting this outcome?

4. What are the demonstrable links between this intended outcome and other intended outcomes?

5. Are there any unintended consequences associated with achieving this outcome?

The far left-hand column of Table 14.3 shows examples of these questions for the mature worker program.

Step 3: Identifying Information Needed to Answer Evaluation Questions

Considerations concerning selection of performance information that we have addressed in relation to performance monitoring also apply to evaluation (for example, which outcomes in an outcomes chain will be more useful for judging effectiveness, assisting with program improvement, or accountability?).

Additional aspects of program theory that are more likely to be addressed through an evaluation than through performance monitoring are those relating to relationships and causal inferences. These include program factors and their effects on outcomes; nonprogram factors and their effects on outcomes; the effects of program implementation on the program and nonprogram factors that are thought to affect outcomes; the relationships among the various outcomes in the outcomes chain; and conversion rates from one level of outcome to another, including potential for multiple paths, and what lies behind those conversion rates and paths in terms of causal mechanisms.

Step 4: Evaluation Design and Methods of Data Collection

A decision to use a program theory approach to conducting an evaluation is neutral with respect to choice of evaluation design (experimental, quasi-experimental, nonexperimental, sampling considerations, and so on) and choice of methods of data collection. However, inclusion of particular types of comparisons in the success criteria as part of the theory of action can provide some direction as to the type of design that might be appropriate—perhaps one that demonstrates improvement over extended time periods or demonstrates superior performance to other programs or no intervention. Also, once particular data collection methods have been chosen, program theory can be useful for designing those methods. For example, a structured interview or questionnaire could be constructed around an outcomes chain or relevant parts from it.

Evaluation questions derived from the program theory can be placed in a column with adjoining columns that identify relevant performance information and methods of data collection (Table 14.3).

Table 14.3 **Examples of Evaluation Questions, Performance Information, and Methods of Data Collection for the Outcome: Job Retention for Mature Workers**

Evaluation Questions About:	Some Performance Information	Examples of Methods and Sources of Data			
		Program Records	Population Norms	Client Surveys	Employer Interviews
The outcome and its variations To what extent have participants placed in paid jobs by the program retained their jobs? Moved into other jobs? Has retention been higher for mature-age workers coming through this employment program than through other means?	Percentage of placements that are retained in employment for zero to six, seven to eleven, and twelve or more months in the same or different jobs compared with expectations from population statistics	X	X		
Has the success rate been higher for some types of participants than for others?	Percentage of placements for various demographic characteristics compared with participant profile and population statistics	X	X		
Have their jobs proved to be suitable and satisfying?	Percentage of placements with low to high levels of job satisfaction compared with target			X	

Evaluation question / indicator				
What the program is doing to achieve the outcome and how well				
What activities has the program undertaken to assist people to retain their jobs? At what cost to the program? How successful have these various activities been?				
Nature, extent across participants, outputs, and cost of processes to improve retention compared with expected processes	X			
Extent to which employers and employees cite these processes as having affected retention			X	X
What else is affecting this outcome				
What other factors have affected whether people have retained their jobs? How influential are these relative to program factors?				
Percentage of those who do and do not retain their jobs and percentage of their employers who cite various nonprogram factors as having affected success or failure of retention	X		X	X
The demonstrable links between this outcome and other intended outcomes				
Is likelihood of retention associated with the extent to which the program played a direct role in getting the person a job?				
Of those retained and not retained, percentage who had low, moderate, or high levels of program assistance to obtain the job	X		X	
Is likelihood of retention related to the extent to which, through the program, individuals have improved motivation and job skills?				
Of those retained and not retained, percentage who reported considerably enhanced job motivation or job skills through the program	X		X	
Percentages of workers and of employers who attribute retention to job skills versus other factors				X

(Continued)

Table 14.3 (Continued)

Evaluation Questions	Some Performance Information	Examples of Methods and Sources of Data			
		Program Records	Population Norms	Client Surveys	Employer Interviews
Any unintended consequences associated with achieving this outcome	Percentage of employers who report positive effects of increased retention such as reduced hiring costs or contribution of mature workers to service delivery				X
To what extent have any changes in retention rates for mature-age workers and/or the processes used to increase retention rates affected employers either positively or negatively?	Percentage of employers who report negative effects and nature of effects such as costs in terms of hours required of employers to improve retention				X
Do the positives outweigh the negatives or vice versa?	Percentage of employers who say benefits outweigh costs				X
Are there implications for continuing employer support?	Percentage willing to continue with the program	X			X

Once such a matrix has been developed, it becomes easier to design the particular methods that will be used as part of the evaluation. The X's in each column identify what is to be addressed through that method. So, for example, we can see from the employer survey column that when we are designing that survey, we would address issues relating to employer perspectives on these topics:

- Program processes such as matching processes and postplacement support and how they affect retention.

- Other factors (for example, personal circumstances of employees, business factors that require layoffs) that have affected retention and the relative importance of these program- and nonprogram-related factors.

- Extent to which employee skills have affected retention. Employers would not necessarily know which skills had been developed by the program, so the evaluation would need to match those skills cited by employers to what was actually achieved by the program with respect to developing those skills.

- Positive and negative effects for employers of increased retention and processes to improve retention, whether costs of processes outweigh benefits or vice versa, and employer willingness to continue involvement with the program given the costs in terms of their time to improve retention.

CONSIDERATIONS WHEN USING PROGRAM THEORY TO DESIGN EVALUATIONS OF COMPLICATED AND COMPLEX PROGRAMS

Before identifying these considerations, we present a simple overview of some characteristics of simple, complicated, and complex programs in Table 14.4 as a reminder of the discussion in Chapter Five. We then show what challenges these differences in characteristics pose for evaluation and some general approaches to address these challenges.

Table 14.5 explores possible implications for evaluating complicated and complex aspects and some strategies for addressing them. Some of these

Table 14.4 Overview of Characteristics of Simple, Complicated, and Complex Programs

	Simple	Complicated	Complex
What it looks like			
1. Focus	Single set of objectives	Different objectives valued by different stakeholders Multiple, competing imperatives Objectives at multiple levels of a system	Emergent objectives
2. Governance	Single organization	Specific organizations with formalized requirements	Emergent organizations working together in flexible ways
3. Consistency	Standardized	Adapted	Adaptive
How it works			
4. Necessariness	Only way to achieve the intended impacts	One of several ways to achieve the intended impacts	
5. Sufficiency	Sufficient to produce the intended impacts: works the same for everyone	Works only in conjunction with other interventions (previously, concurrently, or subsequently) Works only for specific people Works only in favorable implementation environments	
6. Change trajectory	Simple relationship that is readily understood	Complicated relationship; needs expertise to understand and predict	Complex relationship (including tipping points); cannot be predicted; can be understood only in retrospect
7. Unintended outcomes	Readily anticipated and addressed	Likely only in particular situations; need expertise to predict and address	Cannot be anticipated

Table 14.5 Some Implications of Various Characteristics for Monitoring and Evaluation

Issue		Possible Implications for Evaluation	Possible Strategies
1. Focus	Complicated: Different objectives valued by different stakeholders	Need to negotiate agreement about evaluation parameters and processes	Negotiate either consensus on evaluation focus or agreed divergence
	Complicated: Multiple, competing imperatives	Need to ensure adequate coverage of the different imperatives	Ensure they are included in evaluation plan and sufficiently prioritized
		Synthesis of evidence to form an overall evaluative judgment will not be simply arithmetic	Could use qualitative weight and sum
	Complicated: More than one causal strand involved	Risk of undermining one of them if it is not sufficiently emphasized in the evaluation	Ensure both are included in evaluation plan and sufficiently prioritized
	Complicated: Objectives at multiple levels of a system	Need to ensure adequate coverage of the different levels	Ensure they are included in evaluation plan and sufficiently prioritized
	Complex: Emerging objectives	Specific measures may not be able to be developed in advance, making pre- and postcomparisons difficult; need to be flexible enough to address emerging objectives if they are important	Partially emergent evaluation design with some resources set aside to address emerging issues

(Continued)

Table 14.5 *(Continued)*

	Issue	Possible Implications for Evaluation	Possible Strategies
2. Governance	Complicated: Specific multiple organizations with formalized requi rements	Need to ensure the evaluation matches their reporting requirements	Harmonize reporting requirements between organizations; set up data systems to produce reports in different formats
	Complex: Emergent set of organizations working in flexible ways	May have new organizations becoming involved during the evaluation and new responsibilities and accountabilities for existing partners	Need to have adaptive evaluation design that can accommodate new intended users and evaluation questions
3. Necessariness	Complicated: One of several ways to achieve the intended impacts	With or without comparisons might erroneously suggest the intervention does not work	Need to investigate the experience of nonparticipants and compare costs and benefits of different ways of achieving intended outcomes
4. Sufficiency	Complicated: Works only in conjunction with other interventions (previously, concurrently, or subsequently)	The average effect might hide the effectiveness of the intervention when the entire causal package is there; to inform replication of an effective program, evaluation may need to understand the context that supports it	Need to disaggregate data to compare the results in situations where the entire causal package is present and document the contexts in which it is effective
	Complicated: Works only for specific people		
	Complicated: Works only in specific situations		

5. Consistency	Complicated: Intervention needs to be adapted in specific ways for different situations Complex: Intervention needs to be permanently adaptive and responsive to new situations	Can be difficult to classify when implementation has been adequate	Need to identify the essential elements (either components or theory of change that is involved)
6. Change trajectory	Complicated: Not a simple linear relationship Complex (including tipping points): Cannot be predicted, understood only in retrospect	Need expertise to understand and predict A small initial effect may lead to a large ultimate effect through a reinforcing loop or critical tipping point	May need to add explanation of nonlinear patterns Evaluation needs measurement over time, not at one point
7. Unintended outcomes	Complicated: Likely in particular situations Complex: Cannot be anticipated	Will not be addressed by focusing on only the intended outcomes Cannot be predicted	Use expertise to identify possible important unintended outcomes and include in evaluation plan Add data collection processes to collect unexpected outcomes

implications relate to developing an agreed-on evaluation plan, some to actual data collection and analysis, some to issues of causal inference (which we discuss in more detail in Chapter Fifteen), and some to issues of synthesis and reporting (which we discuss in more detail in Chapter Sixteen).

Applying These Ideas to Designing and Implementing an Evaluation

Many of these issues arose when we worked together on an evaluation of a federal government funding initiative in Australia that supported 635 diverse community-based projects. The Stronger Families and Communities Strategy 2000–2004 had many important complicated and complex aspects (Table 14.6).

We used a range of strategies to address these complicated and complex aspects of the intervention:

• *Overarching program theory.* This was developed before the evaluation began based on a synthesis of the research evidence about strengthening families and communities and a meeting of central stakeholders in the early stages of planning the strategy (SuccessWorks, 2001). The seven outcomes in an outcomes chain are shown in Figure 14.3.

• *Nonlinear program theory.* During the evaluation, the overarching program theory was adapted to show and document the iterative development of projects over time and to elaborate the focus of different projects (identifying existing capacity, building new capacity, developing opportunities to apply capacity). Figure 14.4 shows the final version of the overarching program logic.

• *Emergent evaluation design.* Although the evaluation was undertaken as a large external evaluation under contractual arrangements, the evaluation plan (SuccessWorks, 2001) included scope to address emergent outcomes and explore complicated causal paths through identifying four levels of the evaluation as shown in Table 14.7. Resources were allocated for each of these levels, but the specific focus of the level 2 papers and the level 3 studies were

Table 14.6   **Complicated and Complex Aspects of the Stronger Families and Communities Strategy Evaluation**

	Issue	As Evident in the Stronger Families and Communities Strategy
1. Focus	Complicated: Multiple, competing imperatives	There were sometimes competing imperatives, such as the need to demonstrate some tangible progress to the community and local organizations and the need to take time to adequately consult.
	Complicated: More than one causal strand involved	Many different factors were needed for success; there was no silver bullet.
	Complicated: Objectives at multiple levels of a system	
	Complex: Emerging objectives	Specific objectives of projects were evolving and responsive to emerging needs and opportunities. Many projects involved building community capacity in response to community-identified issues. Specific objectives were locally determined.
2. Governance	Complicated: Specific multiple organizations with formalized requirements	Networks and partnerships were formally encouraged, and all successful funding applications involved multiple organizations.
	Complex: Emergent set of organizations working in flexible ways	In some cases, additional partner organizations were needed as the focus of a project was clarified.
3. Necessariness	Complicated: One of several ways to achieve the intended impacts	Other initiatives aimed at strengthening families and communities (by state and local governments and nongovernmental organizations) were under way in communities that did not have Stronger Families and Communities Strategy projects.

(Continued)

Table 14.6   *(Continued)*

	Issue	As Evident in the Stronger Families and Communities Strategy
4. Sufficiency	Complicated: Works only in conjunction with other interventions (previously, concurrently, or subsequently)	Many projects had built on the foundations of previous projects, or their success was (positively or negatively) influenced by previous projects.
	Complicated: Works only for specific people	Cultural and personal characteristics of families and community members affected their participation in projects and their outcomes.
	Complicated: Works only in specific situations	Organizational and geographical factors affected the success of projects.
5. Consistency	Complicated: Intervention needs to be adapted in specific ways for different situations	The strategy provided funding for 635 projects that were both diverse and changing in response to community needs.
	Complex: Intervention needs to be permanently adaptive and responsive to new situations	
6. Change trajectory	Complicated: Not a simple linear relationship	Recursive causal relationships: Initial success or failure had a significant effect on the level of investment by all parties, including community members.
	Complex (including tipping points); cannot be predicted but understood only in retrospect	
7. Unintended outcomes	Complicated: Likely in particular situations	Some potential unintended outcomes could be anticipated and managed—for example, the risks of using short-term funding projects to achieve long-term changes for families and communities.
	Complex: Cannot be anticipated	Some potential unintended outcomes could not be anticipated.

Figure 14.3 **Overarching Program Theory**

7. Stronger families and communities

This is about both improved and maintained well-being, and how families and communities apply the strengths from levels 1 to 6 to improve their well-being. Outcomes at this level include the various domains of stronger families and communities.

6. An environment where communities participate in and drive their own solutions to strengthen their families and communities

Participation at level 6 transcends the participation that occurs in relation to a particular project—level 1. It is about being opportunity hungry, identifying issues that need a solution and taking initative. It goes to the issue of sustainability of community participation and self-determination.

5. Family and community trust/resilience/adaptability

This is about trust that would transcend the particular project whereas level 1 might be about trust developed on a smaller scale through a particular strategy project. It goes to the issue of sustainable levels of trust, improved family relationships, willingness to cooperate in the future, optimism, and adaptability as ways of addressing issues as they arise.

4. Demonstration/application of greater understanding, skills, and capacity

Application includes not just the application of skills during the life of the project but also the transfer of skills to other family and community issues and problems during and after participation in the strategy project. It implies some sustainability of understanding, skills, and capacity.

3. Greater choice, understanding, skills, and capacity for initiative

This includes not just the particular skills, confidence, etc., that might have been the direct target of a project but also the understanding, skills, confidence, and capacity acquired by the participants in the course of planning and managing the projects. Greater choice could include access to a wider range of services or more appropriate services through greater availability of services arising from the project, including any resources that are produced by the project, e.g., manuals.

2. Greater awareness

Awareness includes awareness of strategy, its principles and values as well as subject-specific awareness to be developed by projects. It also includes awareness of and improved access to services through awareness of services, links to services, and service directories.

1. Participation and enhanced trust

This includes direct participation in the strategy and/or the processes of the strategy, including the application process, even if the application itself is unsuccessful. It refers to the extent, range, nature, and quality of participation and consultation at the level of communities and individuals in communities. It also includes participation engendered by the strategy (e.g., of volunteers).

Source: CIRCLE (2008)

Figure 14.4   **Nonlinear Program Theory**

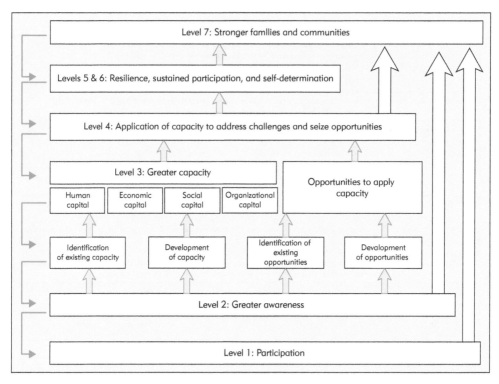

Source: CIRCLE (2006)

progressively identified during the three years of the evaluation as issues arose from the level 1 data, earlier papers and studies, and the changing policy context.

• *Nonarithmetic synthesis.* Diverse evidence from the projects was synthesized without simple aggregation. We discuss this in more detail in Chapter Sixteen.

• *Causal inference.* We explored factors that may have been necessary but not sufficient, or sufficient but not necessary, to produce the intended outcomes, with particular attention to the contexts within which they worked. We discuss this in more detail in Chapter Fifteen.

Table 14.7   **Levels of the Evaluation That Provided Opportunities to Address Emergent Issues**

Level 1 data	Data collected from all projects—progress and final reporting in terms of performance indicators and separate reports—and through initial and final questionnaires for the evaluation
Level 2 papers	Issue-focused papers that linked research evidence, policy frameworks, and data from a cluster of projects, largely involving analysis of available information and illustrations from Strategy projects
Level 3 studies	Case studies of specific projects, communities, initiatives, or issues involving collection of additional data as well as analysis of available data
Level 4 synthesis	The strategy overall, including synthesis of other levels

When It Might Be Inappropriate to Use Program Theory to Design an Evaluation

For programs that are known to have been initiated and implemented with little logical development, the process of trying to develop a program theory as a precursor to conducting an empirical evaluation can be frustrating, to say the least. (See our discussion of challenges in Chapter Six.) It can also be counterproductive if the exercise is seen as futile or impossible: trying to impose a logic that simply does not exist. In such cases, a program theory for the further development and future evaluation of the program may emerge as a product of the evaluation rather than being used as a framework to guide the evaluation.

Sometimes evaluators want to use program theory, and in particular logic modeling, simply as a tool for organizing their own thinking. How much time they spend on this process will depend on the budget and time available to do so. Evaluators have to be careful to avoid the trap of building elaborate and elegant models that may be of intellectual interest to themselves but of minimal use to others and not seen to add value to the evaluation process.

Evaluators also have to be careful lest they become so absorbed by the program theory exercise that they forget what initially prompted the evaluation. It is one thing to use program theory to clarify and improve the thinking around particular issues. It is quite another to steer the evaluation off in a totally different direction from what was intended.

When an evaluation has been motivated solely by a crisis situation that requires a rapid response, there will not be time for developing a detailed program theory. In cases where the focus of the evaluation is on processes (such as suggestions of impropriety or wasteful use of resources), the program theory might not be relevant, since the focus of the evaluation will be on compliance with required processes. In cases where the focus of the crisis evaluation is on outcomes, it can be very useful to quickly develop a program theory (by reviewing an existing program theory for the intervention and revising it as needed or by quickly articulating the tacit knowledge of key informants). The program theory can help to keep a crisis in perspective by showing how it detracts from or contributes to the achievements of the program as a whole.

Of course, many requests for evaluation prove to be requests for pseudo-evaluations, including those whose intent is to postpone action, whitewash or kill a program, or diffuse a crisis in an organization. But here the question is not so much whether program theory should be used as whether an evaluation is the right response at all.

SUMMARY

This chapter has provided some examples of different ways of applying program theory at the various stages of the life cycle of a program to identify and select a package of performance indicators, use comparisons, and plan a program evaluation. We have given particular attention to how program theory can be used to generate evaluation questions. We have emphasized that typically program theory is used in conjunction with other approaches, such as stakeholders' issues identification. Moreover, using program theory holds no particular implications for such decisions as the right combination of qualitative and quantitative methods or whether to use quasi-experimental or nonexperimental designs. We have also discussed some considerations that may apply when using program theory to design evaluations of programs that have complicated or complex aspects.

EXERCISES

1. *Timing of an evaluation:* What stage is your program at? Design? Early implementation? Bedded down, mature? Completed or nearing completion? Considering the stage, what will be the most relevant evaluation purposes and types of decisions to which evaluation can contribute? How would you use program theory? Remember to consider time lags to achieve higher-level outcomes and feedback effects.

2. *Using program theory to identify evaluation issues and questions:* What advantages and disadvantages would you see in using program theory to identify issues that should be addressed through an evaluation?

## 15

Causal Inference

EVALUATIONS OF INTERVENTIONS need to explain, to the extent possible, the causal links between the intervention and the outcomes. In this chapter, we set out a framework for causal inference that is both scientific and pragmatic.

Causal inference is not something that should be left until the final stages of an evaluation but should be considered and planned from the beginning wherever possible. It is more difficult, and sometimes impossible, if it is not planned from the beginning. Program theory that incorporates testable causal hypotheses can also be an important part of the planning process.

THE NEED TO BE SCIENTIFIC AND PRAGMATIC

An important and often difficult task in evaluation is to understand what difference an intervention makes. Causal inference in program theory evaluation involves drawing appropriately from a range of techniques and designs, taking into account the needs and constraints of the situation. That is, it requires being both scientific and pragmatic. In this section, we clarify some terminology, provide a brief history of thinking about causal inference, and explain what we mean by being both scientific and pragmatic.

Myths About Causal Inference

There are two common myths about causal inference for program theory evaluation. One is that evaluations do not need to address the issue, only report that outcomes have been achieved in a way that is consistent with the program theory. Sometimes the label *outcomes* is used rather than *impacts* to indicate that there has been no investigation of whether the observed conditions have been in any way caused by the intervention. As Jane Davidson (2010a) has pointed out, this is likely to be misunderstood by evaluation users, as inevitably the report will be read as implying that the intervention has produced the observed results.

At the other extreme is the second myth: that program theory evaluations can address causal inference only if they incorporate a randomized controlled trial (RCT), something that is not possible for many evaluations.

We believe that good practice avoids both of these myths and concur with Davidson, who advocates for all evaluations to address the issue of causal inference: "There are various fairly low-cost ways of getting an approximate answer to the causal question, and I think any genuine evaluation— yes, including those operating under serious budgetary constraints—should deliver on that rather than opt out" (Davidson, 2010a).

Clarifying Terminology

As is common in evaluation, varying terminology is used to discuss this issue, including *causal attribution, causal contribution, causal analysis,* and *causal inference.* Sometimes the term *causal attribution* is used to refer to causal inference. But this can imply that the outcomes have a single cause or that it is always possible to identify the effect of an intervention above and beyond the effect of other interventions and other factors. Sometimes the term *causal contribution* is used instead to acknowledge that outcomes usually have multiple contributing causes, but for some people this implies a less rigorous analysis approach than causal attribution.

To be clear about what we mean, because many interventions have important complicated and complex aspects, and therefore interventions cannot always be solely attributed to interventions, we use the term *causal inference.*

Being Scientific

A scientific approach to causal analysis does not imply the use of any particular research design or type of data. Rather, it refers to an approach that is systematic, draws on a range of evidence, and critically reviews and synthesizes this evidence.

In recent years, some organizations have used the terms *scientific evaluation* or *rigorous evidence* to apply to evidence from a narrow range of designs. For example, the What Works Clearinghouse (2008) of the Institute of Educational Sciences of the U.S. Department of Education, which has a stated mission of being "a central and trusted source of scientific evidence for what works in education," has until recently classified only well-designed and well-implemented RCTs as "strong evidence."

It is worth remembering that the physical, biological, social, health, and political sciences use a variety of research designs to build credible evidence about the causal contribution of interventions. We believe that evaluation should do the same.

David Freedman (2008), a professor of statistics at the University of California, Berkeley, reviewed how the scientific evidence was gathered, reviewed, and synthesized for nine success stories in medical research, including Jenner and vaccination and Fleming and penicillin. In each case, he found that knowledge was built incrementally from diverse sources of evidence, not from a single study. Freedman concluded, "The examples show that an impressive degree of rigor can be obtained by combining qualitative reasoning, quantitative analysis, and experiments when those are feasible. The examples also show that great work can be done by spotting anomalies, and trying to understand them" (p. 301).

Similarly, a recent review of causality in econometrics by Asad Zaman (2008), professor at the International Institute of Islamic Economics in

Islamabad, while acknowledging the value that RCTs can provide, emphasized the process of combining different types of evidence and careful reasoning rather than relying on a particular research design: "Establishing causality requires piecing together evidence from different sources, and out of the box reasoning. Experimental and qualitative evidence provides strong supporting evidence, but no single piece of evidence may be conclusive" (p. 18).

This chapter therefore sets out a wide range of methods and designs for causal analysis. Strategies for causal inference need to be applied with careful thought about how cause and effect works in simple, complicated, and complex relationships. Classic approaches to causal inference focus on constant conjunction (a particular outcome is always preceded by a particular intervention) or a constant dose-response relationship. Not all of these work in complicated situations. In the absence of sufficiency, A will not always be followed by B, even when it is a critical part of the causal package, unless other factors are also in place. In the absence of necessariness, B may not always occur with A, even when A is a direct cause. In a complicated causal relationship, more A may not lead to more B.

We draw particularly on John Mayne's work on contribution analysis (2001, 2008) with its emphasis on systematically identifying and investigating alternative explanations for results. We draw as well on Davidson's (2006, 2010b) list of strategies for causal inference that should be used in combination rather than relying solely on one strategy.

Being Pragmatic

In addition to being scientific, we also strongly advocate taking a practical approach to causal analysis. Although some level of causal analysis is required for evaluation, there should be a considered investment of resources. Practical evaluation involves careful risk management. What would be the consequences of a Type I error (wrongly concluding there is an effect)? What would be the consequences of a Type II error (wrongly concluding there is no effect)? How can these risks be mitigated and appropriately balanced? In some cases, it might be reasonable to look for congruence with the program theory and do some minimal checking of alternative explanations. In other cases, a carefully constructed counterfactual will be needed, along with systematic investigation of alternative explanations.

Jane Davidson (2006, 2007) has discussed the need for evaluators to be appropriately pragmatic rather than taking the traditional, tentative academic approach of prioritizing the reduction of Type I errors through setting statistical significance levels routinely at .05 or .01, and disregarding results where definitive proof is missing. For evaluation, there should be a careful assessment of the level of evidence required. Should it be on the balance of the evidence? Or beyond reasonable doubt? Or above a high level of certainty?

This approach should not be misunderstood as advocacy for sloppy evaluation or premature claims that an intervention has caused certain outcomes. In high-risk cases, for example, where the costs of an intervention are high in terms of funds expended or potential negative effects, the standards of proof should be appropriately high. However, we might be more realistic about the degree of certainty that any single evaluation can provide, and therefore aim for more modest questions in any one evaluation, as Mark Lipsey (1993, 1997) has suggested, and accumulate more comprehensive knowledge about causes over many evaluations.

A FRAMEWORK FOR CAUSAL ANALYSIS USING PROGRAM THEORY

In this chapter, we set out a systematic approach to causal analysis for program theory evaluation that consists of three components: congruence, comparisons, and critical review:

- *Congruence with the program theory: Do the results match the program theory?* This refers not only to whether the final results were achieved, but also whether the pattern of the results matched the theory in terms of intermediate results. When there are important complicated aspects, results often need to be disaggregated to identify if they have been achieved in favorable circumstances, even if not overall.

- *Counterfactual comparisons: What would have happened without the intervention?* These compare what happened with one or more estimates of what might have happened in the absence of the program. Different designs and techniques can be used to produce these kinds of comparisons, including informant assessment, experimental design, quasi-experimental designs, and qualitative comparative analysis. A good program theory identifies both program and non-program factors that might affect successful achievement of outcomes (see Chapter Eight). These factors can be a rich source of hypotheses for counterfactual comparisons. Being aware of complicated and complex aspects of programs (see Chapter Five) and understanding associated concepts such as sufficiency and necessariness can also assist with refining counterfactual comparisons.

- *Critical review: Are there other plausible explanations of the results?* This component involves systematically identifying and ruling out alternative explanations, following up, and seeking to explain exceptions. As for counterfactual comparisons, factors identified in a program theory and complicated and complex aspects of a program can assist with undertaking a critical review of results.

Even in small evaluations, some attention to each of these is possible and will improve the rigor and credibility of conclusions.

The scale of each level of causal analysis should match the situation in terms of the size of the investment and the risks of wrong conclusions. In some cases, adequate credibility can be achieved relatively easily and quickly, but other times it will involve several additional cycles of data collection and analysis.

Table 15.1 lists the methods and techniques that can be used for each of these components. They are described in detail in the following sections.

CONGRUENCE

The starting point for causal inference is whether or not results are congruent with the program theory. In this section we outline several ways to check for congruence, and how to interpret results.

Table 15.1   **Methods and Techniques for the Three Components of Causal Analysis**

Congruence	Counterfactual Comparison	Critical Review
Comparing achievement of intermediate outcomes and final outcomes	Control group or comparison group	Identifying alternative explanations and seeing if they can be ruled out
Disaggregating results for complicated interventions	Comparing the trajectory before and after the intervention	Identifying and explaining exceptions
Statistically controlling for extraneous variables	Thought experiments to develop plausible alternative scenarios	Comparing expert predictions with actual results
Modus operandi		Asking participants
Comparing timing of outcomes with program theory	Asking participants	Asking other key informants
Comparing dose-response patterns with program theory	Asking other key informants	Making comparisons across cases
Comparing statistical model with actual results	Making comparisons across cases	
Comparing expert predictions with actual results		
Asking participants		
Asking other key informants		
Making comparisons across cases		

Congruence with the program theory means more than that the intended outcomes were achieved. There are a number of ways to check more carefully for congruence. We discuss a number of these that are particularly relevant for program theory in more detail.

• *Comparing achievement of intermediate and final outcomes.* The evidence is congruent with the program theory if program participants who achieved the intended outcomes have also achieved intermediate outcomes—for example, knew about the program, engaged with it, learned new skills, and applied them—while those who have not achieved the intended outcomes have also not achieved intermediate outcomes. We discuss an example of this below.

• *Disaggregating results for complicated interventions.* The caveat with the previous analysis is that it is predicated on the intervention's sufficiency to produce the outcomes. In cases where the intervention will be successful only in favorable contexts, this analysis needs to be disaggregated to do so. We provide some examples below.

• *Statistically controlling for extraneous variables.* Where an external factor is likely to affect the final outcome, it needs to be taken into account when looking for congruence. For example, the rate of motor vehicle fatalities per thousand vehicles is affected by the number of miles (or kilometers) driven, which in turn is affected by economic conditions. An evaluation of the impact of road safety measures would need to take this into account when looking at the congruence in the timing of expected changes.

• *Modus operandi.* Some interventions have a distinct pattern of effects that can be used as evidence for causal inference (Scriven, 1974). The classic crime example is the smell of almonds, indicating cyanide poisoning. In educational interventions, the terminology or conceptual frameworks participants used in their work might provide a similar trail.

• *Comparing timing of outcomes with program theory.* Program theory might predict not only how long before final outcomes are evident, but whether these are likely to be maintained, increase, or decay over time. An effective monitoring and evaluation system that considers temporal issues (see the discussion in Chapter Fourteen) is needed to provide data that can allow this sort of checking.

• *Comparing dose-response patterns to match program theory.* Program theory might predict whether increased exposure to an intervention is expected to have a positive, negative, or curvilinear relationship to the intended outcomes.

• *Comparing the statistical model and actual results.* For very complicated situations, simple inspection of results might not be possible, and comparison with a statistical model will be needed. We discuss an example of this below.

- *Comparing expert prediction and actual results.* For evaluations conducted over a period of time, it is possible to make predictions based on program theory or an emerging theory of wider contributors to outcomes, and then to follow up these predictions over time (Miles and Huberman, 1994).

- *Asking participants.* While participants might sometimes have their own reasons for attributing or not attributing changes to an intervention, detailed accounts of their change trajectory can be credible. We discuss an example of this below.

- *Asking other key informants.* Similarly, other key informants can sometimes provide evidence that links participation plausibly with observed changes.

- *Making comparisons across cases.* The method of qualitative comparative analysis (Ragin, 1987) compares the configurations of different cases to identify the components that produce specific outcomes. Program theory can help to identify the variables that should be included in this analysis.

Comparing Achievement of Intermediate and Final Outcomes

If the final results have been achieved and results have also been achieved for each of the intermediate outcomes identified in the program theory, this lends weight to the conclusion that the intervention has produced the results intended. The technique of process tracing "attempts to uncover what stimuli the actors attend to; the decision process that makes use of these stimuli to arrive at decisions; the actual behavior that then occurs; the effect of various institutional arrangements on attention, processing, and behavior; and the effect of other variables of interest on attention, processing, and behavior" (George and McKeown, 1985, p. 35).

In addition to looking at final results, it is important to check for success all along the causal chain to identify where it has broken down (if intended impacts have not been achieved) and analyze the extent to which achievement at one level of the results chain has then converted into achieving the next level. Table 15.2 shows how to interpret findings in terms of different outcomes. Figure 15.1 returns to our Apple a Day example from Chapter One to show what these different options might look like.

Table 15.2   **Interpreting Findings in Terms of Their Congruence
with Program Theory**

Was the Intervention Adequately Implemented?	*Was There Sufficient Engagement, Uptake, and Adherence?*	*Were Intermediate Outcomes Achieved?*	*Were Final Outcomes Achieved?*	*Interpretation*
✗	✗	✗	✗	Implementation failure
✓	✗	✗	✗	Engagement or adherence failure (first causal link)
✓	✓	✗	✗	Theory failure (early causal link)
✓	✓	✓	✗	Theory failure (later causal link)
✓	✓	✓	✓	Consistent with theory
✓	✓	✗	✓	Theory failure (different causal path)

Following this program theory, in addition to measuring implementation and impact, the evaluation would also collect data about whether people ate the apples and whether their nutritional status improved. If the final intended outcome were not achieved, the evaluation would be able to identify where the causal path in the theory broke down.

The starting point is to check for the quantity, quality, and timing of implementation. This cannot be taken for granted. Michael Patton (2008)

Figure 15.1   **Results Chain Logic Model for An Apple a Day**

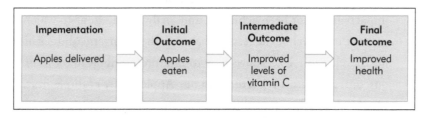

discusses the importance of checking the quality of implementation—and whether implementation has occurred at all, pointing to examples where even a total delay in implementing a program was not evident to those who had commissioned the evaluation. Table 15.3 shows a pattern of results where the final impact of improved health was not achieved and the problem was in the initial stage of implementation: for most people, the apples were not even delivered.

Although the intervention has not worked in terms of achieving the intended impact, the program theory has not been disproved. In fact, the results suggest that where implementation is effective, the program theory works well, as the conversion rate across the different levels is high. If implementation can be improved, it might be worth trying again. This time the researchers would build in proper monitoring of implementation so that problems could be detected and corrected early on to avoid high implementation failure.

Table 15.4 shows a situation where the program was implemented (the apples were delivered), but most people did not eat them.

This sort of pattern comes when intended participants either do not engage with the program (for example, they do not attend meetings or participate in activities) or do not change their behavior in the expected ways (for example, they do not adhere to recommendations for diet and exercise). Results like this do not disprove the program theory. Instead, there needs to be an investigation of whether it is likely that engagement and adherence can be improved, which might require a rethinking of the action theory. For example, rather than delivering apples directly, it might be more effective to work through a trusted local agency, or rather than providing apples it might be better to provide apple juice.

Table 15.3 **Pattern of Results Suggesting Implementation Failure**

Apples Delivered	Apples Eaten	Improved Levels of Vitamin C	Improved Health
35/1,000 people	30/1,000 people	25/1,000 people	27/1,000 people

Table 15.4 **Pattern of Results Suggesting Engagement or Adherence Failure**

Apples Delivered	Apples Eaten	Improved Levels of Vitamin C	Improved Health
1,000/1,000 people	30/1,000 people	25/1,000 people	27/1,000 people

Table 15.5 **Pattern of Results Suggesting Theory Failure: Intermediate Outcome**

Apples Delivered	Apples Eaten	Improved Levels of Vitamin C	Improved Health
1,000/1,000 people	960/1,000 people	25/1,000 people	27/1,000 people

Table 15.5 shows the causal chain breaking down at an intermediate outcome: improved nutritional status.

If the data on adherence (eating the apples) are credible, and not due to misreporting by participants keen to appear to be compliant or misunderstanding how much, or which part, of the apple should be eaten, the indication is theory failure at the point of the intermediate outcome. This might reflect a poor situation analysis—for example, the participants did not have a vitamin C deficiency. Or it might reflect insufficient dosage—an apple a day does not provide sufficient vitamin C to achieve the recommended daily allowance. Or it might be that apples do not improve vitamin C levels—either because they do not contain vitamin C or because vitamin C it is not available when digested.

Given the knowledge available from other studies, the last of these is less likely, and so we would look to the other possible explanations. For many human service programs, however, the link between adherence and an intermediate outcome has not been established, and a real possibility is that the program does not work the way predicted by the theory.

Table 15.6 shows the causal chain breaking down at the point of final impacts: the apples were delivered and eaten and people's nutritional status improved, but their health did not improve.

For some human service programs, where there is little information about this final causal link, it would suggest that the theory does not work in this way. For this example, given the existing knowledge about the importance of vitamin C in health, there may be a problem with the situation analysis: the health problems are not related to vitamin C deficiency and are not likely to be alleviated by them (maybe the problem is injuries from traffic accidents, for example).

Table 15.7, the final one in this series, shows success in terms of the final outcome but a lack of congruency with the theory. The apples were delivered and health did improve—but not because of adherence or improved vitamin C.

Two explanations for this are possible. It might be that the improved health has nothing to do with the intervention but has been caused by something else, such as improvements in water supply. Further investigation would be needed to rule this out, such as comparing participants to nonparticipants or looking in detail at the types of health improvements that have occurred (for example, whether it stemmed largely from a reduction in traffic injuries). It is also possible that the intervention has worked but not in the expected way . Perhaps it was the delivery process itself that has started a causal chain leading to health improvement, in the same way that Meals on

Table 15.6 **Pattern of Results Suggesting Theory Failure: Final Outcome**

Apples Delivered	Apples Eaten	Improved Levels of Vitamin C	Improved Health
1,000/1,000 people	960/1,000 people	890/1,000 people	27/1,000 people

Table 15.7 **Pattern of Results Suggesting Theory Failure: Alternative Causal Path**

Apples Delivered	Apples Eaten	Improved Levels of Vitamin C	Improved Health
1,000/1,000 people	38/1,000 people	25/1,000 people	890/1,000 people

Wheels services can be beneficial for some participants even if they do not eat the meals, because the benefit comes from the social contact.

Table 15.2 presents a scenario in which the findings of an evaluation support both the theory and its implementation. An example of such a scenario follows.

CHECKING CONGRUENCY OF RESULTS WITH PROGRAM THEORY IN A RURAL HEALTH PROMOTION PROJECT

The Sustainable Farming Families program was a health promotion project that operated through three workshops over two years. Farmers (usually couples) had health checks and attended information sessions about diet, exercise, and safety. The evaluation found improvements over the first and second years in a number of intended final results, such as cholesterol levels and waist circumference, consistent with the program theory. There was also evidence that intermediate outcomes had been achieved, such as reading food labels and using them to select low-fat food, adding daily exercise, and buying equipment such as kitchen scales and running shoes. Participants had identified in personal action plans specific changes to health behaviors that they intended to make, and many of these plans had been fully implemented. In addition participants stated that the program had helped them to change their health behaviors and could specify how it had done this (Boymal and others, 2007). This information about the intermediate outcomes helped to develop a more credible case that it was the intervention that had produced the impacts.

Methods for Investigating Implementation and Theory Failure

To investigate implementation failure, techniques such as beneficiary assessment (Salmen, 2002), participatory rural appraisal (Chambers, 1997), conversational interviews, focus group interviews, observation, and participatory observation can be used to identify the reasons intended beneficiaries identify for implementation failure. For example, a beneficiary assessment of a population, health, and nutrition project aimed at improving maternal health care and family planning services for the poor in Madras and Bombay found serious barriers to implementation. It turned out that the location and hours of the services were not convenient for families, and family planning was not a priority for them. Few of them had access to television, which meant a planned social marketing campaign on television was unlikely to succeed (World Bank, 2005).

INVESTIGATING IMPLEMENTATION FAILURE OF A NUTRITIONAL PROGRAM IN BANGLADESH

An evaluation of a nutrition project in Bangladesh that provided nutritional counseling and supplementary feeding for mothers and children found that it had not produced improvements in overall nutritional status, although the most malnourished children had shown improvements (White, 2009). Further analysis was undertaken to understand these results, beginning by developing a program theory and identifying where the program theory had broken down.

The first stage (participation by target group, identified as mothers of young children) was largely successful, although a significant minority (10 percent) did not participate.

There were problems at the other stages. The identification of the target group had been flawed because mothers were often not the family decision makers and not the household members who went to the market to buy food. For these reasons and because of resource constraints, there had been few behavior changes.

There had also been problems in terms of correctly identifying children to enroll in the program. Program staff could not correctly identify which children were faltering in growth or malnourished. They had enrolled children who did not need the program and failed to enroll children who did. Finally, there was leakage of the program (the food had been given to someone else) and also substitution (it had replaced a meal that would otherwise have been given), leading to no net increase in the amount of food the women and children received. This analysis enabled the evaluators to identify areas where the program could be improved.

Another evaluation investigated a nutritional program in Ghana:

INVESTIGATING THEORY FAILURE OF A NUTRITIONAL PROGRAM IN GHANA

This nutritional program in Ghana had provided vitamin A supplements to over 200,000 women in Ghana (Kirkwood and others, 2010) in order to reduce maternal mortality. It had drawn on an earlier program in Nepal where maternal mortality had been reduced by 44 percent (West and others, 1999). But in Ghana, the program failed to produce improvements in maternal mortality, and the evaluation examined where the program theory had broken down.

Had there been a problem with the capsules? The capsule content was checked by testing fifty-six randomly selected batches of capsules, covering all yearly consignments, and including fourteen batches of unused capsules returned from the field. This showed that most of the capsules had all of the required content of retinol, a form of vitamin A, and all had more than 95 percent of the content, apart from some that had been stored for several years.

Had there been a problem with adherence? This was checked during home visits undertaken every four weeks, where women were asked about adherence and also asked to show how many capsules they had left, and through visits to a random sample of forty women each week. There was some lack of adherence, where capsules had not been distributed during home visits or women had not take all four of the weekly capsules provided, but not enough to explain the shortfall.

The real problem lay in the next step in the results chain: the serum retinol level. In the Nepal study, the serum retinol levels of pregnant women receiving the treatment had been significantly higher than those receiving a placebo, with very few having serious deficiency. In the Ghana study, tests from a sample showed that many pregnant women continued to have low levels of serum retinol despite adhering to the treatment. Further investigation of possible explanations would require further research. Despite the large scale of this study, little follow-up has been undertaken of these surprising findings in an effort to explain the results.

Disaggregating Results for Complicated Interventions

The process of checking that results match the program theory should not assume the pattern will be a simple one. If the intervention works only in favorable contexts, that is, the intervention is not sufficient by itself to produce the results, disaggregating the data to show the results for cases where the entire causal package was completed is important. It is possible to take advantage of natural variation within an intervention to make comparisons that strengthen the causal analysis. This involves developing testable hypotheses about specific types of cases where particular outcomes are likely to be evident. The specificity of these hypotheses can add considerably to the causal analysis.

The British Road Safety Act implemented in 1967 introduced a number of measures to reduce traffic accidents, including stopping vehicles to test for drunk drivers. Time series data of car fatalities before and after the introduction of the act, however, showed no change over time (Glass, 1997). So did the act have no effect? The types of road fatalities it was likely to affect were those associated with driving when intoxicated, which is concentrated on Friday and Saturday nights and early mornings. An analysis of the data just for these times showed a large decline in the number of fatalities around the time of implementation, followed by a partial increase, perhaps as offenders realized the detection rates were lower than anticipated. The evidence was consistent with the program theory that the legislation would reduce the incidence of drunk driving and associated fatalities.

At the same time, the comparison between the aggregate and disaggregated data suggested that although there had been reductions in the incidence of drunk driving, these had not been not enough to produce a large reduction in the overall number of fatalities. Although the first part of the program theory (reducing drunk driving) worked, the final part (reducing road deaths) had not. Either fatalities related to drunk driving were a relatively small component of total road fatalities or reductions in this type were balanced out by increases in other types.

This classic example of evaluation detective work should be remembered whenever simple trend data have the potential to hide effects behind the average effect.

DISAGGREGATING RESULTS FOR A COMPLICATED JUSTICE INTERVENTION

An evaluation was undertaken in Australia to assess the impact of a significant investment in correctional services that was aimed at reducing reoffending (Elsworth, 2006). One of the intermediate outcomes, an important part of the program theory, was the increased use of community correctional sentences by judges instead of custodial sentences. An initial time series analysis of the proportion of offenders who were sent to prison did not show any apparent impact from the intervention.

Like the British Road Safety Act, however, the intervention was likely to have an effect only in particular circumstances. It was unlikely that the changes would have any effect on sentencing patterns for first-time offenders for minor crimes, as they were already likely to receive a community corrections order. Nor was it likely to have an effect on those convicted of a serious crime, for whom prison would be still considered appropriate.

When the data were disaggregated by offender type and offense type, striking differences became apparent. Imprisonment rates were unchanged for first-time offenders and for those convicted of serious crimes. However, for repeat offenders convicted of drug-related crimes, such as possession and use of drugs and theft, the proportion sent to jail, which had been increasing steadily before the intervention, fell after the intervention was implemented. The specificity of the change made it less likely that it was due to other factors such as historical changes in overall sentencing patterns.

Comparing the Statistical Model and Actual Results

In some cases, it is possible to construct statistical models to explain changes in outcomes over time and to use these to identify important contributing factors. Multiple regression and econometric modeling are some of the analytical methods that can be used.

This approach was used by the Intergovernmental Panel on Climate Change (IPCC), which recognized that "controlled experimentation with the climate system in which the hypothesized agents of change are systematically varied in order to determine the climate's sensitivity to these agents . . . [is] clearly not possible." Different models of climate were developed with different assumptions about possible contributing factors:

> The observed patterns of warming, including greater warming over land than over the ocean, and their changes over time, are simulated only by models that include anthropogenic forcing. No coupled global climate model that has used natural forcing only has reproduced the continental mean warming trends in individual continents (except Antarctica) over the second half of the 20th century" [IPCC, 2001, p. 39].

The IPCC has used critical review of alternative explanations and exceptions:

> Attribution of observed climate change to a given combination of human activity and natural influences requires . . . statistical analysis

and the careful assessment of multiple lines of evidence to demonstrate, within a pre-specified margin of error, that the observed
changes are:

- Unlikely to be due entirely to internal variability;

- Consistent with the estimated responses to the given combination
  of anthropogenic and natural forcing; and

- Not consistent with alternative, physically plausible explanations
  of recent climate change that exclude important elements of the
  given combination of forcings [IPCC, 2001].

The continuing debate about the existence and causes of climate change
indicates that this approach to causal inference can be unsuccessful in contentious, high-stakes policy issues.

Asking Participants

Detailed narrative accounts from participants can provide compelling evidence that the intervention has contributed to the observed outcomes. In
addition to articulating their mental models (a useful process we discussed
in Chapter Six), these discussions can provide evidence that can be compared
for congruence with the program theory.

In an evaluation of a criminal justice project that we were involved in,
semistructured interviews with offenders provided rich data that demonstrated they understood how the project was supposed to work and that it
had contributed to changes in their behavior:

> In the eighteen years that I have been a heroin addict I have never
> had this much support. I didn't think there was that much support
> or services available in the community.
>
> Someone like me needs a lot of assistance in working through
> problems, and unless that assistance is there then I'm just doomed to
> the circle of jail–drugs, jail–drugs, jail–drugs. Do you know what I
> mean? Now, I've come off that circle and I'm starting to create my own
> new one, my own new road. And I'm always going to get to intersec-
> tions where I can go one way or the other way. And that is where this
> program comes in. They help me choose which path to take.

It has given me the chance to actually live normally for just long enough to realize that there's other ways to live and it's better [CIR-CLE, 2003].

COUNTERFACTUAL COMPARISONS

The second key element of causal analysis is making counterfactual comparisons to develop an estimate of what would have happened in the absence of the intervention. Where it is possible to make appropriate counterfactual comparisons, it can add to the credibility of the evaluation (Cummings, 2006).

Several methods can be used to develop a credible estimate of what would have happened in the absence of the intervention. We discuss those that are particularly relevant to program theory in more detail:

• *Creating or locating a control group or comparison group.* Sometimes it is appropriate and possible to create or find a group of people who did not receive an intervention. A control group is created through random assignment. A comparison group might be created through matching to participants or statistically created through propensity scores. We discuss these options below.

• *Comparing the trajectory before and after the intervention.* While a simple before-and-after measure does not provide a compelling estimate of the counterfactual, time series data can provide a credible estimate of the counterfactual in fairly stable situations.

• *Thought experiments to develop plausible scenarios.* Evidence about policies and procedures and other opportunities in some cases can be used to develop a realistic scenario of the chain of events in the absence of an intervention.

• *Asking participants.* Participants might be able to describe a plausible scenario of what would have been likely to happen in the absence of an intervention. We provide an example below.

• *Asking other key informants.* Experts, including experienced practitioners, may be able to provide a plausible scenario based on their observations of patterns in previous cases.

- *Making comparisons across cases.* Depending on the configuration of variables across cases, it might be possible to use these to develop a credible scenario.

Creating a Control Group or Comparison Group

Control groups and comparison groups are both intended to provide a counterfactual—an estimate of what would have happened in the absence of the intervention. Both options can be improved by using program theory to identify mediators and moderators to incorporate in sampling design and data analysis (Cook, 2000).

Control Groups   A control group is formed by randomly assigning potential participants in an intervention to either a treatment or a control group, which receives either no treatment or the current standard treatment. This is the model on which drug trials operate.

CONTROL GROUP COUNTERFACTUAL FOR EVALUATION OF WORM MEDICINE PROGRAM

A large RCT was undertaken to evaluate the impact of providing worm medicine to school children in Kenya. Given the likelihood of reinfection if only some students in a class received the treatment, randomization was done at the school level. The study found that the treatment group not only had lower infection rates but also significantly reduced school absences (Miguel and Kremer, 2004). A subsequent study compared take-up of the medicine when a subset of schools was randomly assigned to institute fees for the service rather than providing them free and found that take-up fell from 75 to 19 percent (Holla and Kremer, 2009).

Comparison Groups   Quasi-experimental designs involve creating a comparison group in other ways—for example, matching cases on relevant variables for which data are available, using propensity scores, using cohort studies to track forward a group that has been involved in a program and another that has not, or using a case control design to track back from an outcome to identify patterns among those who have achieved the result and those who have not.

COMPARISON GROUP COUNTERFACTUAL FOR
EVALUATION OF AN EDUTAINMENT PROGRAM

Tsha Tsha is a television drama series developed in South Africa to communicate messages about HIV prevention. The impact evaluation of the program could not simply compare the knowledge and attitudes of those who watched it with those who did not because the two groups differed considerably. The technique of propensity scores was used to statistically create comparable groups based on an analysis of the factors that influenced people's propensity to watch the show. The evaluation found that viewers were more likely to practice HIV-preventive behaviors, such as abstaining from sex, being faithful to one partner, having sex less often, and using a condom (Kelly and others, 2004).

Debates About the Use of Control Groups   Some organizations favor evaluations that incorporate a control group. For example, the review process for the What Works Clearinghouse of the U.S. Department of Education stated in the 2008 version of its procedures and standards handbook: "Currently, only well-designed and well-implemented randomized controlled trials (RCTs) are considered strong evidence, while quasi-experimental designs (QEDs) with equating may only meet standards with reservations; evidence standards for regression discontinuity and single-case designs are under development."

RCTs can be a feasible and credible option for some interventions. However, in many situations, they are not appropriate in these situations:

- When interventions are not focused at individual change but at system-level change (for example, national anticorruption initiatives)

- When interventions cannot be rolled out gradually but have universal application immediately (for example, national policy)

- When access to the intervention cannot be controlled or randomized (for example, information campaigns with a lot of leakage or secondary transmission, or selective programs such as scholarships or advanced training)

- When population numbers are too small for adequate sample size

- When the intervention is responsive and adaptable so the intervention is not constant

- When an experimental design has not been established at the beginning and a credible comparison group cannot be constructed
- When there are no resources to undertake an experimental impact evaluation

In situations such as these, alternative comparisons are needed, in conjunction with analysis of coherence and critical review.

In some cases, it is possible to construct a double-blind RCT, where neither participants nor staff know whether they are in the treatment or control group. For example, the nutritional supplementation program in Ghana we discussed earlier (Kirkwood and others, 2010) provided capsules that contained either vitamin A or a placebo, and ensured that neither the staff providing the capsules nor participants knew which group they were in.

In most cases, however, it is evident which is the control group, even when it is receiving the current best practice or the existing intervention rather than no treatment. Care should be taken to consider possible contamination (people in the control group also receive the treatment) and the placebo effect (people's responses are partly caused by their belief that they are receiving special treatment).

In addition, the choice of research design does not necessarily demonstrate the quality of the evaluation. In a keynote address at the annual conference of the American Evaluation Association, Ernest House outlined the ways in which drug trials, which have been used as an exemplar for impact evaluations using RCTs, have been found to have been manipulated to produce favorable results:

- Choice of placebo as comparator
- Selection of subjects (Bodenheimer, 2000)
- Manipulation of doses (Angell, 2004)
- Method of drug administration (Bodenheimer, 2000)
- Manipulation of timescales (Pollack and Abelson, 2006)
- Suspect statistical analysis
- Deceptive publication (publishing the same study in different journals so that it is counted multiple times in meta-analysis)

- Suppression of negative results (Mathews, 2005)

- Selective publishing (Mathews, 2005, Armstrong, 2006; Harris, 2006; Mathews, 2005; Zimmerman and Tomsho, 2005)

- Opportunistic data analysis (Bodenheimer, 2000)

Comparing the Trajectory Before and After the Intervention

In many cases, it is not possible to create or identify a comparable group of people who have not participated in an intervention. Examples are national policy, universal programs, or where those who have not participated are systematically different from those who have (for example, competitive scholarship programs). In such cases, comparing changes over time might provide a credible estimate of the counterfactual.

A single before-and-after measurement by itself does not provide compelling evidence of impact even if it shows a change. This evidence can be strengthened by including the third component of causal analysis by exploring alternative explanations. It can also be strengthened by adding more data points. If an outcome measure had been stable for some time and then increased only after the intervention, this also improves the credibility of the argument that the intervention contributed to the result. With a stable time series and attention to ruling out other possible explanations, it might be reasonable to predict that if the intervention had not occurred, the stable state would have continued. Program theory can help to identify the various points at which measurements should be made.

CHANGE TRAJECTORY COUNTERFACTUAL FOR EVALUATION OF NEONATAL TREATMENT

A neonatal condition, persistent pulmonary hypertension, had had an 80 percent mortality rate. Physicians tried using extracorporeal membranous oxygenation, a treatment used for other conditions, which involves running the baby's blood through a machine to reoxygenate it. In their trial of forty-five babies, the fatality rate was less than 20 percent. There were no plausible alternative explanations for the improvement in the outcomes, and the new treatment was recommended for wider use (Bartlett and others, 1982). At the time, this evidence was not seen as sufficiently credible, and an RCT was required in order to demonstrate the effectiveness of the treatment (Bartlett and others,

1985; Ware and Epstein, 1985; Ware, 1989; Worrall, 2002). Subsequently these types of designs have been seen to be appropriate for rare conditions such as this.

Asking Participants

Simply having participants describe what would have happened in the absence of the intervention does not provide compelling evidence. However, they may be able to provide a plausible narrative showing the likely interplay between their choices and resources in the absence of the program.

In the criminal justice program we discussed earlier (CIRCLE, 2003), offenders were asked what they thought would have happened in the absence of the program. Some of them with considerable experience with the criminal justice system were able to make credible comparisons with previous experiences. One offender made the startling claim that the program had saved his life, but then went on to make a credible explanation of why this was so: "I wouldn't be here today at all. I'd honestly be dead. Because I saw the way my life was heading and I knew that I was heading back to jail . . . and I couldn't handle that again, no way. I'd rather die . . . this program has not only saved my life, but has started to give me one as well" (CIRCLE, 2003).

Addressing Complications When Considering Counterfactuals

Causal inference should pay particular attention to issues of necessariness and sufficiency when there are complicated aspects of the intervention or the situation.

Necessariness   With-or-without comparisons are appropriate for simple interventions that are both necessary and sufficient to produce the impacts of interest. When this is not the case, the evaluation needs to investigate alternate paths to achieving the impacts that might have been used by nonparticipants.

The Comprehensive Child Development Program (CCDP) provides an example of considering necessariness when undertaking causal inference. The CCDP was a federal government program implemented at twenty-four sites across the United States in the 1990s at an annual cost of $25 million (St. Pierre, Layzer, Goodson, and Bernstein, 1996; St. Pierre and Rossi, 2006). It focused on ensuring the delivery of early and comprehensive services to children and families with the aim of enhancing child development and helping low-income families achieve economic self-sufficiency.

The CCDP design relied on a case manager who was responsible for coordinating the service needs of a group of families. Results, both intermediate and final, were congruent with program theory. Families in the program had increased their use of services and had improved their level of functioning across a number of domains. However, a comparison with results for a control group showed that its members had also increased their use of services and improved their level of functioning. Some interpretations of this example have argued this shows that the program was ineffective since control group families achieved similar results (Goodson and others, 2000; Savedoff, 2005).

Clearly the program was not necessary, in the philosophical sense, since it was possible to access services without it. However, this is not the same as saying it had no value for participants. To determine the value of the program, a comparison would need to be made between the costs incurred in accessing these services without the program.

Sufficiency   It is also important to consider the issue of sufficiency when making comparisons—in particular, considering whether interventions might have different outcomes in different contexts.

The Minneapolis Domestic Violence Experiment in the 1980s evaluated a proposed new policy of mandatory arrest in cases of domestic violence (Sherman and Berk, 1984). Cases were randomly assigned to either the proposed new policy or the existing policy, which left the decision about arrest to the responding police officer. Offenders who were treated under the new policy were found to be significantly less likely to be subsequently reported for reoffending.

On the basis of these results, the policy was introduced in many other jurisdictions, and the study was replicated in six cities. However, the results of these studies were conflicting: in some cities, offenders dealt with under the mandatory arrest policy were less likely to reoffend, but in others they were more likely to reoffend. Subsequent investigation identified a very different response from offenders who were unemployed (Sherman, 1992). A later analysis suggested that the causal mechanism involved in the policy (naming and shaming) was effective only for people who were seeking social acceptance (Pawson, 2006).

Addressing Complexity When Considering Counterfactuals

In complex situations, this can be impossible, since in the absence of the intervention, the environment would have been completely different, and it is not possible to predict what would have happened. To consider this situation, imagine how your life might have been different if you had (or had not) married your first romantic partner. The causal chain is too long, and too uncertain, to be able to make a meaningful prediction. Similarly, in complex situations, it cannot be credibly estimated what would have happened in the absence of a particular intervention. It might have meant that the intervention did not happen at all or that another agency might have initiated something to address the same issues. In complex situations, speaking of a single counterfactual is not meaningful; it is more useful to discuss a number of plausible counterfactuals.

CRITICAL REVIEW

One of the risks of program theory evaluation is gathering only evidence that is consistent with the theory and analyzing it only in ways that are consistent with the theory. This is why we believe that for every evaluation, there should be some attempt to include the third component of causal analysis: iteratively checking out exceptions to the patterns and other possible explanations.

Methods for Critical Review

As Burt Barnow (2010) has said, "random assignment is not a substitute for thinking" (p. x):

- *Identifying alternative explanations.* Alternative explanations might come from insiders (participants, key informants), previous research, or speculation.

- *Identifying and explaining exceptions.* Exceptional cases might be successes that were expected to be failures or vice versa. Ideally an evaluation can explain these, or at least document that they exist, and not lose this information by focusing only on the overall pattern.

- *Comparing expert predictions with actual results.* Evidence about policies and procedures and other opportunities in some cases can be used to develop a realistic scenario of the chain of events in the absence of an intervention.

- *Asking participants.* Participants can be a valuable source of alternative explanations for observed outcomes or exceptions that do not fit the overall pattern.

- *Asking other key informants.* Experts, including experienced practitioners, may be able to suggest alternative explanations or identify, and possibly explain, cases that don't fit the overall pattern.

- *Making comparisons across different cases.* Comparisons across cases can readily identify exceptions and might suggest reasons for these.

Investigating Alternative Explanations

More convincing evidence of the role of interventions in producing outcomes can come from systematic investigation of alternative explanations for the observed changes in clinical indicators, drawing on Campbell and Stanley's (1963) classic list of threats to internal validity: history, maturation, repeated testing, instrumentation, regression to the mean, experimental mortality, selection, and interactions between selection and other factors.

Families in the Sustainable Farming Families (SFF) health promotion program, referred to previously in this chapter, had seen improvements in a number of important health indicators consistent with the program theory. There was also evidence of the achievement of specific health-related behavior changes as part of participants' personal action plans, which had been developed during the program. Participants stated as well that the program had helped them to change their health behaviors and could specify how it had done this. The program was implemented in small, rural communities where it was not possible to create a control group or a credible comparison group. Instead the following alternative explanations were systematically investigated (Boymal and others, 2007).

History   This might involve another event during the period that could be the actual cause of the observed changes. A good program theory will include identifying other factors that could potentially affect the outcomes of interest.

Maturation   In some programs, natural development might lead to performance improvement. This is most commonly seen in children who are likely to grow and perform better due to maturation. Identifying factors that might affect outcomes is an important part of developing a good program theory.

Repeated Testing   This can have an effect where knowledge or behavior appears to improve due to familiarity with testing procedures.

Instrumentation   This refers to the impact of changes in the continuity of instruments used to collect data. This is more likely to be an issue where there have been changes in the personnel collecting data, research scales have been used over a long time and need to be reformed, or evidence comes from methods such as observer ratings, which can vary over time.

Regression to the Mean   This can be a threat where participants have been selected on the basis of lower-than-average performance on an indicator with considerable error in its measurement. A number of those whose performance measured just below the average are likely to have higher levels of actual performance than measured performance. The next time they are measured, if the measured performance more accurately reflects their actual performance, the result will be an increase in measured performance, which can be mistaken for an increase in actual performance.

Experimental Mortality   This refers to differential dropout rates among groups that are being compared, as well as to mortality in terms of actual death. For example, if participants for whom a project was not successful dropped out, average results among those who finished would be better than the average of those who started even if there were no actual changes for individual participants.

Selection   This can threaten validity if two groups that are being compared are different in terms of an important variable that by itself could explain the difference in observed performance. Although there is not always an explicit comparison between program participants and another group, there is an implicit comparison with those who chose not to participate. In many cases, participants in voluntary projects might be expected to be more highly motivated to make changes in their behaviors than nonparticipants.

Identifying and Explaining Exceptions

As well as looking for patterns in our data, rigorous analysis requires attention to, and explanation of, exceptions.

One of our evaluations involved around two hundred varied early intervention projects funded under a federal government program in Australia. As well as documenting the outcomes from these projects, we were asked to analyze what factors helped and hindered projects. The program theory had hypothesized that one of the factors that would affect the success of projects was whether they had been supported by their agency. We had found that many projects had identified support from the agency (which had received the funding) both before and during the project as particularly important. To check out this statement, we looked for exceptions to the rule.

We had independently rated the overall achievements of projects and gathered the projects' assessment of the quality of support from the auspice. If the theory held, we would expect that high-performing projects would have rated their agency highly, and low-performing projects would have rated it low. This pattern was generally true. Two of the four projects that had been rated as having achieved little success or had been terminated appeared to have experienced significant difficulties with their agency. In one case, the project had been terminated because the agency had become insolvent. When we looked at the two projects rated as having only moderate or mixed success which had rated their agency as having been unhelpful or very unhelpful, we found that in both cases, the difficulties with their agency had substantially contributed to their level of success. Lack of support, lack of linkages, and lack of referrals were factors in both cases.

We then looked for exceptions. Conversely, factors other than the auspice had contributed to the difficulties that remaining projects of moderate or mixed success had experienced.

We identified one project that had been rated as outstanding despite having rated its agency as very unhelpful. We examined the data to see how it was that it appeared to have been successful in spite of the unhelpful agency. We found that it was a relatively small project located within a very large agency, which may have contributed to its sense of isolation and the difficulties of getting things done that can occur in large organizations. What seemed to make it work was a combination of the determination and optimism of the project leader, the mutual support among the staff in the context of working with a highly traumatized group of people, and ultimately the desire of those people to make a difference to their own lives and the lives of others. It had persevered in the face of considerable resistance or apathy, or both, on behalf of the agency and perhaps some other potential partners as well.

Checking out the exceptions helped strengthen the credibility of the conclusion that support from their organization was an important factor in project success but that it was not sufficient and that some projects were able to find ways to overcome or circumvent the negative effects of lack of support. They pursued alternative paths to success.

SUMMARY

Causal inference is an important part of a program theory evaluation. Paying attention to congruency, counterfactual comparisons, and critical review can significantly improve the quality of causal inference in small or large evaluations.

EXERCISES

Read an evaluation report that has used program theory, and identify the strategies used for causal inference.

1. To what extent does it analyze and report congruence of results with the program theory along the outcomes chain? Does it compare results for cases with and without a complete causal package, as predicted by the program theory?

2. What counterfactual comparisons have been used? How credible are they?

3. To what extent have exceptions and alternative explanations been identified and investigated?

16

Synthesis and Reporting

EVALUATIONS GENERATE information about multiple aspects of an intervention. This information needs to be synthesized and reported in ways that provide coherent and clear messages while retaining the details of important patterns. This chapter discusses how program theory can provide a framework for synthesizing and reporting evidence for a single evaluation and across multiple evaluations.

SYNTHESIS AND REPORTING FOR A SINGLE EVALUATION

Evaluation reports are sometimes structured around methods of data collection with, for example, a chapter on survey results, another chapter on case studies, and a chapter on analysis of documentation. This is generally a poor practice as it often fails to undertake the synthesis and triangulation needed to produce coherent messages. Instead we advocate following Davidson's advice (2009) to organize evaluation reports around the key evaluation questions. Program theory can help to structure these.

Reports can be organized around outcomes in the outcomes chain together with additional chapters about the outcomes chain as a whole (the whole as more than the sum of its parts) and other chapters that may be about specific issues of interest. This approach encourages synthesis of a range of different types of quantitative and qualitative data around evaluation questions that relate to outcomes. In this way, the methods become the servants of the evaluation process rather than dictating the form of the evaluation report.

Program theory can provide a structure for bringing together diverse evidence into a coherent narrative of how an intervention contributed to bringing about change. Although the examples that follow all provide a single narrative, it is possible to develop multiple narratives that explain how the intervention worked differently for different people or in different situations.

Synthesizing Diverse Evidence Along the Outcomes Chain

The outcomes chain can provide a framework for assembling evidence about an intervention. It can be done before an evaluation to identify what is already known from existing data and what the priorities should be for additional data collection. During an evaluation, it can be used to provide updates on what is being learned from the evaluation. And after an evaluation, it can be used to bring together diverse evidence collected and retrieved throughout the evaluation (Cooksy, Gill, and Kelly, 2001).

For example, Bron McDonald, an Australian evaluator, used Bennett's hierarchy (discussed in Chapter Thirteen) as a framework for presenting information about the impact of Target 10, a program that helped dairy farmers improve their productivity (McDonald, Rogers, and Kefford, 2003), drawing on existing data. Although the program had undergone various forms of evaluation, these had not adequately met the information needs of the stakeholders who were making decisions about the future of the program. These existing data, together with results from a questionnaire designed to fill in data gaps, were used to present a coherent story of impact along the seven levels of Bennett's hierarchy: with the resources expended by the project (1), these activities were undertaken (2), involving people with particular characteristics (3), who had reactions to their experience (4) that changed their knowledge, attitudes, skills, and aspirations (5), leading to changes in their behaviors (6) that achieved end results of economic, social, and/or environmental value (7).

This approach has been further developed by Jessica Dart, an Australia-based evaluator and community development facilitator, as the collaborative outcomes reporting technique (CORT) (Dart, 2010). CORT begins by developing a program theory that is then used as the framework for synthesis. Available data are retrieved and mapped onto the program theory. Additional data are gathered in a data trawl to fill gaps or answer additional emerging questions, often involving stakeholders in the process of data gathering. These are also mapped onto the program theory. People with relevant scientific, technical, or sectoral knowledge are brought together in an outcomes panel to review the evidence in terms of the descriptive claims (the outcomes happened) and the causal claims (the intervention contributed to producing these outcomes). Ideally the experts achieve a consensus; if this is not possible, they identify where they are unable to achieve agreement. The final stage is an evaluation summit workshop, where key stakeholders convene to review all the material that has been produced and develop recommendations.

Those who are reporting findings using the outcomes chain must ensure that, in addition to reporting results at each level, there is sufficient attention paid to the links among the results. For example, in addition to reporting participation and practice change, what evidence is there of the extent to which practice change was influenced by participation in the program? Attention to patterns between the levels and systematic causal inference are also needed.

Rapid Reporting

Outcomes chains can be particularly important for rapid reporting of evaluation findings. Imagine you have been allocated half an hour in a meeting to get the main messages of your evaluation across to senior management in an organization for which you have conducted the evaluation. Condensing all that wonderfully detailed work to produce a credible robust evaluation is a challenge that frequently confronts evaluators.

One way to address the issue is to use an outcomes chain, simplified if needed, as the framework for addressing the main findings of your report. Of course, there is always a need for a written report and executive summaries of it. The oral report is typically an additional rather than an alternative means of communication.

Exhibit 16.1 shows an example of a layout used for an oral report of an evaluation of a road safety education program for high school students that involved producing an education kit to be used by trained police and teachers. This one-page document could not stand alone. It needs the evaluator to be able to explain the overall concept, the bulleted points, and the links among the outcomes achieved. In the meeting in which this oral reporting framework was used, the stakeholders from several partner agencies used the framework to map out their future directions, develop recommended actions, and so on. The evaluation reporting fed directly into an action plan that was developed not by the evaluator but by the stakeholders to ensure joint ownership across the agencies.

Exhibit 16.1 shows that no information was collected in relation to the top three outcomes. This related to the timing of the evaluation (early implementation) in that most of the students who had participated in the program were not yet driving. Information about claimed behavioral causes of accidents among young drivers in the general population (information about the current situation) was, however, used as part of the assessment of the relevance of the kit.

Because the program was in its early stages of implementation, many of the early objectives and targets that had been agreed upon among the various partner agencies related to program initiation and implementation. The stakeholders wished these outcomes be given prominence through inclusion in the outcomes chain rather than buried in other parts of the description of the program theory (for example, the description of program activities and resources). Hence, the lower levels of the outcomes chain are more about program implementation than about program outcomes in terms of impacts on target audiences. The oral presentation also explored links among the various levels of outcomes but at a very basic level (for example, whether there was a strong, positive relationship between student interest

EXHIBIT 16.1 USING AN OUTCOMES CHAIN AS A STRUCTURE FOR AN ORAL REPORT: AN EVALUATION OF ROAD SAFETY PRESENTATIONS BY POLICE TO HIGH SCHOOL STUDENTS

9. Reduction in mortality, Not assessed
   injuries, costs

↑

8. Reduction in accidents Not assessed

↑

7. Improvements in Not assessed except in relation to behavioral inten-
   young driver behavior tions as both drivers and passengers

↑

6. Improvements in ✓ Concerning:
   student road safety
   knowledge and • Causes of accidents
   attitudes
                          • What actions to take

                          • Personal responsibility
↑
                          • Drinking and driving relationship

5. Student interest in ✓ Especially
   road safety presenta-
   tions • Videos

                          • Opportunity to talk to police but want more oppor-
↑   tunity

                          ✓ Impediments:

                          • Large class sizes

                          • Police teaching skills

   Effective implementation below. Student and road safety outcomes above.

4. Trained police and Relative to targets
   teachers make
   maximum use of the ? Questionable relative to potential use—
   kit with the target   considerable opportunity for improvement.
   student population   Use was affected by:

                          • Logistics
↑
                          • Competing priorities (police and schools)

                          • Marketing (information base and responsibilities)

                          • Teacher access to kits

                                                        (Continued)

EXHIBIT 16.1 *(Continued)*

3. Sufficient and appro-
   priate police are
   effectively trained and
   efficiently used

   ↑

✓ Relative to target

X Number trained relative to number used to
   deliver kit

? Questionable whether appropriate police are
   always selected

✓ Quality and effectiveness of training is questionable

2. Kit is available to
   those who require it

   ↑

✓ Relative to targets set for distribution

? Continuing availability of complete kits in patrols

X Accessibility to teachers is a problem; police are
   sometimes too possessive of kit

1. High-quality road
   safety education kit
   relevant to needs is
   produced

✓ Coverage of behavioral causes of young driver
   accidents

✓ Police make most use of modules on alcohol,
   speed, seat belts, general safety, videos, and
   posters

X Not all parts are equally suited to presentation by
   police

X Not well adapted to different student subpopula-
   tions (for example, rural)

? Some better linkages to national curriculum
   statements and young driver education programs is
   desirable

Note: ✓ = in general, this outcome had been well achieved.
X = this outcome was not well achieved.
? = either performance was questionable or variable (for example, across different parts of the state).

and improved knowledge). It also addressed issues concerning the overall
management framework for the program, the roles and responsibilities of
the three agencies involved in the program, and the positioning of the program relative to other strategies.

Brief reporting using the outcomes chain can also be used to produce an "elevator pitch"—a brief account of how the intervention has been shown to contribute to important outcomes, which can be prepared and used at opportunities to speak to important stakeholders. John Davidson (2007, p. x), in an address to county extension directors, has outlined the advantages of succinct success stories that clearly explain the value added by the investment in a program, including this story from Clinton McRae:

> According to the [North Carolina] Department of Juvenile Justice, Hoke County is a "target county" because a large number of youth are sent to training school. Between January and May 2004, it has cost the county $25,272 to detain youth. To address the character and decision making skills of youth, Cooperative Extension offers a prevention program called 4H Life Skills. Cooperative Extension collaborates with Hoke County Schools to offer the program to elementary and middle school students. The program has reached 158 students this year and none of the students who have graduated from the program have been referred to juvenile justice. For each child not referred to juvenile justice, the county saves $72.00 a day and $26,280 a year.

In response to demands from policymakers and funders for succinct statements of impact, Richardson has suggested structuring reports around the following elements: problem, program, people, partners, impact, and conclusion. The following example shows how a report can combine evidence from a program evaluation and previous research to tell a coherent story:

> Heart Healthy Program Successful. Heart disease is the number 1 killer in Robeson County [Problem]. Cooperative Extension in cooperation with local hospital and Health Dept. [Partners] sponsors quarterly "heart healthy" cooking workshops for the general public. Emphasis is on health recipes and cooking techniques participants can use in their daily lives [Program]. 49 consumers participated in the four workshops [People]. Participants were surveyed several weeks following classes, and 36 indicated they had used class recipes or modified their own recipes to reduce fat, sodium, or sugar.

According to the JADA [*Journal of the America Dental Association*], nutrition intervention saves on average $8000 in medical costs per patient. Thus, for the small group surveyed there was an estimated total savings of $288,000 [Impact]. Due to interest of participants and program value, workshops will continue quarterly [Conclusion] [Richardson, 1999, p. x].

Software to Report Evidence in Terms of a Logic Model

Computer-based reporting makes it possible to link to multiple levels of information from an overall logic model.

DoView, software developed by New Zealand evaluator Paul Duignan, can embed summaries of data relating to specific outcomes in the outcomes chain and provide a link to another page with more detail. Figure 16.1 shows a logic model for a smoking reduction program with indicators listed under various outcomes (clearly showing those where there are no data being collected) and a thumbnail of a graph of actual results.

Stewart Donaldson and Tarek Azzam have developed a Web-based presentation of logic models using Shockwave, which can easily link to pages with evidence. Figure 16.2 shows a logic model for a computer education and employment program, Computers in Our Future (CIOF), and then a linked page, which presents the major findings in terms of one of the intermediate outcomes and the supporting evidence for these.

SYNTHESIS AND REPORTING ACROSS EVALUATIONS

There is increasing interest in learning from multiple evaluations to make better use of the significant investment that has been made and to provide better information on which to base future interventions and evaluations. Program theory can be extremely useful for evaluations of programs that are made up of many separate and diverse projects and for syntheses of evidence across different evaluations.

Figure 16.1   **Reporting Findings Using DoView**

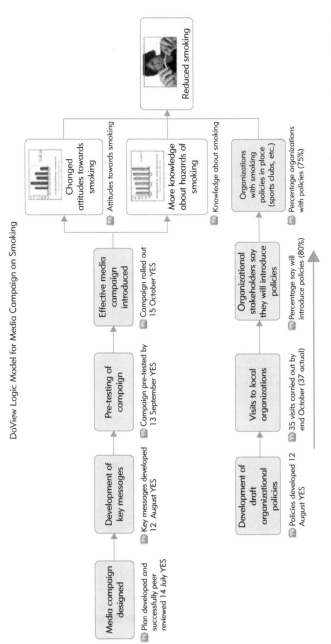

DoView Logic Model for Media Campaign on Smoking

Source: From Duignan (2009b)

Figure 16.2   **Reporting Findings Using Shockwave**

Overall Logic Model

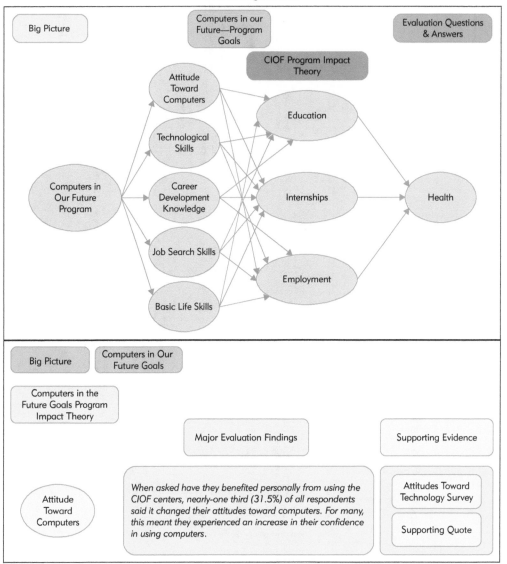

Source: From Donaldson (2009); Donaldson and Azzam (forthcoming)

Figure 16.2   *(Continued)*

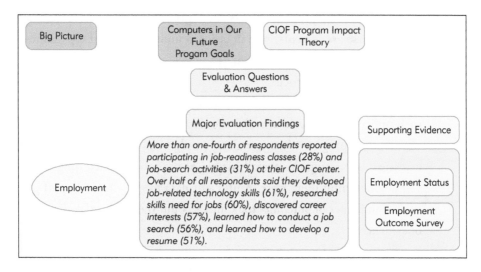

Synthesis Across Linked Projects

If the projects funded under a larger program are very similar, with a common program theory and performance indicators, aggregating data arithmetically and reporting in terms of both the overall program and individual projects is a simple matter. For example, a fast food chain with standardized products can use the same reports for individual stores and the entire company. However, when diverse projects are funded under a program and outcome measures are necessarily different, different synthesis methods are needed.

In our evaluation of the Stronger Families and Communities Strategy 2000–2004, providing an overall synthesis to the government department that had funded the projects was important. The projects were extremely diverse, from small ones to provide training and support for volunteers to large-scale community development and family support projects. Although they had been organized around a common program theory of capacity building (as outlined in Chapter Twelve), they had collected different data that were relevant to the particular type of project and reported these in progress and final reports and, in some cases, in separate evaluation reports. In order to look for relationships and patterns, we needed to make the data from different sources manageable. In order to be confident about the different sources of data we were using, we needed to assess the quality of data first. In order to be able to judge how successful projects had been, we needed criteria for judging success and a process for demonstrating consistency in these assessments. And in order to understand what types of results that the projects and the program as a whole were achieving, we needed to preserve an outcomes focus.

Our evaluation team worked through the project documentation for each project, coding evidence about outcomes into one of the outcomes in the common program theory, making a judgment about the quality of that evidence, and rating the overall success of the project (including, in some cases, concluding there were insufficient data to do so). Coding of global success and types of outcomes took into account the results of coding quality of evidence.

The coding of outcomes provided many specific examples for inclusion in the report showing what the program theory looked like when translated into actual projects. The coded outcomes were aggregated to show the overall success of the program in terms of the different levels of its results chain.

Outcomes that were rated as having verifiable evidence were those where the evidence was both plausible and relatively easily verifiable—for example, participant feedback had been documented and in principle could be requested—or surveys, measures, and other records had been used—for example, a recorded response from a participant in an interview or questionnaire. Some used relatively strong research designs such as those involving measurement over several occasions (before and after an intervention) and with comparison groups using standardized instruments. These types of studies were rare, however, and in most cases neither appropriate nor feasible given the nature of the projects and the populations they were serving. Not surprisingly, it was easier to find verifiable evidence for lower-level outcomes such as effective engagement with the target group than for higher-level outcomes such as improved community well-being.

Outcomes that were rated as having plausible evidence were those whose reports included plausible claims concerning specific outcomes for specific individuals, groups, or the community as a whole—that participants had gone on to do particular things as a result of what they had learned and there was a clear logic to the connection, that they had articulated what they had learned to the project officer, that others had observed changes in behavior. For example, a project where marginalized youths worked with older mentors to build a ramp for people with disabilities reported feedback from the young people that they had gained skills and improved self-esteem through making their contribution. Outcomes that were rated as having no or negligible evidence were those where the outcome was simply claimed, typically in general terms without examples or other supporting evidence—for example, a statement that "the community is now more cohesive."

The overall success of each project was coded as outstanding, generally successful, mixed or moderate success, little or no success, or unknown (insufficient evidence). All categories were carefully defined with descriptors. To illustrate the categories, the report included descriptions of projects at different levels of success and different size budgets and explained why they had been coded in this way.

This coding allowed us to report on the overall achievements of the program. Although the evidence had gaps, we were able to make a conservative estimate of what had been achieved and confidently report significant achievements had been made in many areas. It allowed us to look for patterns across projects, including multiple regression, to identify predictors of overall project success, such as effective support from a project's organization and adherence to the principle of evidence-based practice. It also made possible an exploration of the exceptions to these patterns by doing further disaggregation and qualitative analysis.

Synthesis Across Different Evaluations

Does home visiting of new mothers improve outcomes for babies and families? Is hormone replacement therapy effective in reducing cardiovascular disease? What are the most effective ways of teaching children to read? Synthesis of evidence from multiple evaluations is needed to answer these questions.

Synthesis can be entirely quantitative following the application of rules for inclusion of studies to be analyzed as in the case of statistical meta-analysis. It can also apply a blend of quantitative analysis and reasoning, explanation, and argument, as in the case of realist synthesis.

Statistical Meta-Analysis   Meta-analysis statistically summarizes the results from a number of evaluations or research studies—not by simply averaging across studies, but taking into account the quality of the studies (excluding those of insufficient quality or which do not provide an effect size) and their sample size (Lipsey and Wilson, 2001). Two international initiatives, the Cochrane Collaboration and the Campbell Collaboration, have been developed to undertake systematic reviews of interventions in health (Cochrane) and in education, crime and justice, and social welfare (Campbell), particularly using meta-analysis of evidence from randomized controlled trials (RCTs).

A systematic review using meta-analysis seeks to produce a summary statement of knowledge such as the following summary of a systematic review of the use of hormone replacement therapy to reduce cardiovascular disease in postmenopausal women: "Hormone therapy has no heart-protective bene-fit to healthy postmenopausal women or to those with heart disease" (Gabriel Sanchez and others, 2005).

A number of authors have outlined the advantages of combining pro-gram theory with meta-analysis (for example, Petrosino, 2000), includ-ing Mark Lipsey's "What Can You Build with Thousands of Bricks? Musing on the Cumulation of Knowledge in Program Evaluation" (1997). Using program theory in these reviews can help to identify interventions that work in quite different ways, despite their common label, and to summa-rize these separately. For example, the analysis of different types of HRT on women of different ages can be separated to test the timing theory of HRT that suggests early initiation might have quite different results from late ini-tiation (Mendelsohn and Karas, 2007; Hodis, 2008; Grodstein, Manson, Stampfer, and Rexrode, 2008). In practice, the ability of systematic reviews to explore these types of patterns is limited by the disaggregation of data and use of program theory in the original studies.

Realist Synthesis   If meta-analysis is about assembling bricks, that is, standard-ized pieces of knowledge, realist synthesis might best be thought of as a way of assembling rocks, or nonstandardized pieces of knowledge. Devel-oped by British sociologist Ray Pawson, realist synthesis can use a range of evidence, including results from RCTs, quasi-experimental studies, and nonexperimental studies, including single case studies, as long as they are of sufficient quality. Quality is not assessed with reference to a hierarchy of research designs for the entire study but by assessing whether threats to validity have been adequately addressed in terms of the specific piece of evidence being used (Pawson, 2008).

Realist synthesis uses purposive and iterative sampling of a wide variety of evidence to develop, refine, and test theories about how, for whom, and in what contexts policies will be effective. Like realist evaluation, to which we referred in our discussion of realist matrices in Chapter Nine, it is based on scientific realist philosophy that outcomes arise from a combination of structure and agency and that evaluation should therefore investigate both "the resources offered by programs and the choices made by their subjects" (Pawson, 2006b, p. 45), and identify causal mechanisms that operate only in particular contexts. Realist synthesis is inherently a process of building, testing, and refining program theory, as Australian realist evaluator Gill Westhorp (2008) explains:

> Realist synthesis seeks to provide, not just description of relationships between variables, but "explanatory synthesis" (Pawson, 2006, p 69) and to avoid the gross over-simplifications of the "what works" literature (Pawson, 2006, pp 70–71). Consequently, it engages in active theory-building rather than just "aggregation" or "reportage" (Pawson, 2006, p 73–74). It begins from a clear identification of a topic and an "embryonic" realist theory and uses primary data as "case studies . . . to test, revise and refine the preliminary theory" (Pawson, 2006, p 74). It proceeds by way of juxtaposing ("for instance, when one study provides the process data to make sense of the outcome pattern noted in another", Pawson, 2006, p 74), reconciling (by identifying contextual or implementation differences which explain apparently contradictory sets of findings), and if necessary adjudicating between studies; and by "consolidating" and "situating" sets of studies to comprise different CMOC's [Context-Mechanism-Outcome-Configurations]. The product of this process is intended to be "not an arithmetic verdict on a family of programmes

but a refinement of its underlying theory" (Pawson, 2006, p 76), with reasoning and abstraction the tools of the trade [Westhorp, 2008, p. 189].

A realist synthesis will seek to identify and explain differences in patterns of outcomes. For example, Ray Pawson's realist synthesis of the naming-and-shaming theory of change (Pawson, 2006a, 2006b) identified different outcomes when this was used for a car theft index naming automobile manufacturers, tables of school educational results, registers of sex offenders, and naming of poll tax protesters in local newspapers and explained them with reference to Merton's typology of aspirations to group membership (Merton, 1938).

SUMMARY

Program theory can provide a robust framework for synthesis and reporting of evidence from a single evaluation or from multiple evaluations, even when these involve diverse projects and diverse evidence.

EXERCISES

1. For an evaluation report with which you are familiar, identify ways in which it might have been presented differently if it had used program theory as the organizing framework.

2. Read a systematic review in an area where you have some experience and knowledge. What might be a relevant program theory that could have been used to structure the synthesis?

As William James (1907) once famously said ". . . no theory is absolutely a transcript of reality but that any one of them may from some point of view be useful" (p. 26). Purposeful program theory is about developing program theories that serve useful purposes. It is not about developing perfect theories. We see program theory development as an iterative process and a process that is as much about stimulating important questions as it is about answering those questions.

We hope that this book will excite others to undertake research and reflect on practices in ways that will push the frontiers of program theory and further develop some of the approaches that we have identified. In particular, we see significant opportunities to further improve our processes for developing and using program theories, in terms of:

- *Further exploration of how complicated and complex aspects of programs can be understood* and represented in program theories and the implications for evaluation and causal inference, for example, in programs where multiple agencies contribute to the results.

- *Improving the usefulness of program archetypes and policy tools that are currently available as resources for developing program theories.* To this end, further research is needed into both existing and new program archetypes to better understand their important features and their many variations. We see potential for collaborative work between program theorists, realist evaluators, and meta-analysts in exploring mechanisms and conditions under which those mechanisms do and do not apply.

- *Exploring and documenting how a wide range of recognized theories of change from various disciplines and policy areas could contribute to program theory development.* In this book we have selected just a few examples of recognized theories of change (such as diffusion theory) to demonstrate how these types of theories can be useful. As our appreciation of a wider range of theories evolves, we will need to become more sophisticated about how we choose relevant theories.

- *More effective use of program theory to synthesize evidence from multiple evaluations* to build better knowledge about what works and how it works, including systematic analysis of the conditions under which they do and do not work.

- *Better use and adaptation of new technologies* to develop and record program theories, to synthesize and report evaluation findings in relation to program theories, and to engage stakeholders more widely in conversations about program theories.

REFERENCES

ActKnowledge. "Guided Example: Project Superwoman," 2003. http:// www.theoryofchange .org/pdf/Superwomen_Example.pdf.

Adler, M. A. "The Utility of Modeling in Evaluation Planning: The Case of the Coordination of Domestic Violence Services in Maryland." *Evaluation and Program Planning,* 2002, *25*(3), 203–213.

Agency for Healthcare Research and Quality. *Request for Proposal Number AHRQ-2009–10001 Assessing the Evidence Base for Context-Sensitive Effectiveness and Safety of Patient Safety Practices: Developing Criteria.* Washington, D.C.: U.S. Government Printing Office, 2008.

Ajzen, I. "From Intentions to Actions: A Theory of Planned Behavior." In J. Kuhl and J. Beckmann (eds.), *Action Control: From Cognition to Behavior.* New York: Springer-Verlag, 1985.

Ajzen, I. *Attitudes, Personality, and Behavior.* Bristol, Pa.: Open University Press, 1988.

Ajzen, I., and Fishbein, M. *Understanding Attitudes and Predicting Social Behavior.* Englewood Cliffs, N.J.: Prentice Hall, 1980.

Anderson, A. *The Community Builder's Approach to Theory of Change: A Practical Guide to Theory and Development.* New York: Aspen Institute Roundtable on Community Change, 2005. http://www.aspeninstitute .org/sites/default/files/content/docs/roundtable%20on% 20community%20change/rcccommbuildersapproach.pdf.

Anderson, L. A., and others. "Using Concept Mapping to Develop a Logic Model for the Prevention Research Centre's Program." *Preventing Chronic Disease,* 2006, *3*(1), A6.

Angell, M. *The truth about the drug companies.* New York: Random House, 2004.

Argyris, C. *Intervention Theory and Method: A Behavioral Science View.* Reading, Mass.: Addison Wesley, 1970.

Argyris, C., and Schön, D. *Theory in Practice: Increasing Professional Effectiveness.* San Francisco: Jossey-Bass, 1974.

Argyris, C., and Schön, D. *Organizational Learning: A Theory of Action Perspective.* Reading, Mass.: Addison-Wesley, 1978.

Armstrong, D. "How the New England Journal Missed Warning Signs on Vioxx." *Wall Street Journal,* May 15, 2006, pp. A1–A2.

Arnkil, R., and others. "Emergent Evaluation and Learning in Multi-Stakeholder Situations." Paper presented at the Conference of the European Evaluation Society, Seville, 2002. http://www.evaluationcanada.ca /distribution/20021010_arnkil_robert.pdf.

Aspen Institute for Rural Economic Policy Program. *Measuring Community Capacity Building: A Workbook-in-Progress for Rural Communities.* Aspen, Colo.: Aspen Institute, 1996.

The Aspen Institute. "The Community Builder's Approach to Theory of Change: A Practical Guide to Theory Development." New York: The Aspen Institute Roundtable on Community Change, 2005. http://www .aspeninstitute.org/sites/default/files/content/ docs/roundtable%20on%20community%20change/rcccommbuildersapproach.pdf.

Astbury, B. "Social Mechanisms, Theory-Building and Evaluation Research." Paper presented at the 2009 conference of the American Evaluation Association, Orlando, Fla., 2009.

Astbury, B., and Leeuw, F. "Unpacking Black Boxes: Mechanisms and Theory Building in Evaluation." *American Journal of Evaluation,* 2010, *31*(3), 363–381.

Australian Public Service Commission. *Changing Behaviour: A Public Policy Perspective.* Canberra: Commonwealth of Australia, 2007a. http://www .apsc.gov.au/publications07/ changingbehaviour.pdf.

Australian Public Service Commission. *Tackling Wicked Problems: A Public Policy Perspective.* Canberra: Commonwealth of Australia, 2007b.

Bakewell, O., and Garbutt, A. *The Use and Abuse of the Logical Framework Approach.* Stockholm: Swedish International Development Agency, 2005.

Bandura, A. "Self-Efficacy: Toward a Unifying Theory of Behavioral Change." *Psychological Review*, 1977, *84*, 191–215.

Bandura, A. *Social Learning Theory.* New York: General Learning Press, 1977.

Bandura, A. *Self Efficacy: The Exercise of Control.* New York: Freeman, 1997.

Barnes, M., Matka, E., and Sullivan, H. "Evidence, Understanding and Complexity: Evaluation in Non-Linear Systems." *Evaluation*, 2003, *9*(3), 265–284.

Barnes, M., Sullivan, H., and Matke, E. *"The Development of Collaborative Capacity in Health Action Zones: A Final Report from the National Evaluation."* Birmingham: University of Birmingham, 2004. http://www.socialresearch.bham.ac.uk/downloads/HAZ_FINAL _REPORT_JAN_11.pdf.

Bartlett, R. H., and others. "Extracorporeal Membrane Oxygenation for Newborn Respiratory Failure: 45 Cases." *Surgery*, 1982, *92*, 425–433.

Bartlett, R. H., and others. "Extracorporeal Circulation in Neonatal Respiratory Failure: A Prospective Randomized Study." *Pediatrics*, 1985, *76*, 479–487.

Batterham, R., Dunt, D., and Disler, P. "Can We Achieve Accountability for Long-Term Outcomes?" *Archives of Physical Medicine and Rehabilitation*, 1996, *77*, 1219–1225.

Baumgartner, F., and Jones, B. D. *Agendas and Instability in American Politics.* Chicago: University of Chicago Press, 1993.

Bemelmans-Videc, M., Rist, R. C., and Vedung, E. (eds.). *Carrots, Sticks and Sermons: Policy Instruments and Their Evaluation.* New Brunswick, N.J.: Transaction Publishers, 1998.

Bennett, C. F. "Up the Hierarchy." *Journal of Extension*, 1975, *75*(2). http://www.joe.org/joe/ 1975march/1975-2-a1.pdf.

Bennett, C. F., and Rockwell, K. "Targeting Outcomes of Programs (TOP): A Hierarchy for Targeting Outcomes and Evaluating Their Achievement." 1999. http://citnews.unl.edu/ TOP/index.html.

Berry, F. S., and Berry, W. D. "Innovation and Diffusion Models in Policy Research." In P. A. Sabatier (ed.), *Theories of the Policy Process.* Boulder, Colo.: Westview Press, 1999.

Bickman, L. (ed.). *Using Program Theory in Evaluation.* New Directions for Evaluation, no. 33. San Francisco: Jossey-Bass, 1987a.

Bickman, L. "The Function of Program Theory." In L. Bickman (ed.), *Using Program Theory in Evaluation.* New Directions for Evaluation, no. 33. San Francisco: Jossey-Bass, 1987b.

Bickman, L. "Barriers to the Use of Program Theory." *Evaluation and Program Planning*, 1989, *12*(4), 387–390.

Bickman, L. (ed). *Advances in Program Theory.* New Directions for Evaluation, no. 47. San Francisco: Jossey-Bass, 1990.

Bickman, L., and Peterson, K. A. "Using Program Theory to Describe and Measure Program Quality." In L. Bickman (ed.), *Advances in Program Theory.* New Directions for Evaluation, no. 47. San Francisco: Jossey-Bass, 1990.

Biggs, S., and Matsaert, H. "An Actor-Oriented Approach for Strengthening Research and Development Capabilities in Natural Resource Systems." *Public Administration and Development*, 1999, *19*(3), 231–262.

Bikhchandani, S., Hirshleifer, D., and Welch, I. "A Theory of Fads, Fashion, Custom, and Cultural Change as Informational Cascades." *Journal of Political Economy*, 1992, *100*(5), 992–1026.

Bodenheimer, T. "Uneasy Alliance: Clinical Investigators and the Pharmaceutical Industry." *New England Journal of Medicine*, 2000, *342*, 1539–1544.

Borgatti, S. P., and Molina, J.-L. "Towards Ethical Guidelines for Network Research in Organizations." *Social Networks*, 2005, *27*(2), 107–117.

Boymal, J., and others. "Living Longer on the Land: An Economic Evaluation of a Health Program That Works for Rural Australians." Kingston, ACT, Australia: Rural Industries Research and Development Corporation, 2007. http://www.rirdc.gov.au/reports/ HCC/07–094sum.html

Briggs, B. "Re: Assessing Program Theory." Post to discussion list EVALTALK, Apr. 2, 1998.

Bronfenbrenner, U. *The Ecology of Human Development.* Cambridge, Mass.: Harvard University Press, 1979.

Bronfenbrenner, U. "Ecological Models of Human Development." In *International Encyclopedia of Education.* (2nd ed.) London: Pergamon Press, 1994.

Bullen, P., and Weller, S. "Applying an Expanded Outcomes Hierarchy Model in the Evaluation of the National Migrant Access Projects Scheme." In *Proceedings of the Australasian Evaluation Society International Conference.* Melbourne, 1992.

Campbell, D. T., and Stanley, J. C. *Experimental and Quasi-Experimental Designs for Research.* Skokie, Ill.: Rand McNally, 1963.

Cancer Prevention Research Center. "Detailed Overview of the Transtheoretical Model." 2010. http//www.uri.edu/research/cprc/.

Case Management Society of America. *Standards of Practice for Case Management.* Little Rock, Ark.: Case Management Society of America, 2010.

CC PACE. *Agile Project Management.* White Paper, 2003–2009. http://www .ccpace.com/ Resources/documents/AgileProjectManagement.pdf.

Centre for Community Research. "Using Microsoft PowerPoint to Create Logic Models." 2008. http://www.communitybasedresearch.ca/resources/How%20to%20make%20logic %20models%20in%20Microsoft%20 Powerpoint.pdf.

Chambers, R. *Whose Reality Counts? Putting the First Last.* London: ITDG Publications, 1997.

Chen, H. "Special Issue: The Theory-Driven Perspective." *Evaluation and Program Planning,* 1989, *12*(2).

Chen, H. *Theory Driven Evaluation.* Thousand Oaks, Calif.: Sage, 1990a.

Chen, H. "Issues in Constructing Program Theory." In L. Bickman (ed.), *Advances in Program Theory.* New Directions for Evaluation, no. 47. San Francisco: Jossey-Bass, 1990b.

Chen, H. "Current Trends and Future Directions in Program Evaluation." *Evaluation Practice,* 1994a, *15*(3), 229–237.

Chen, H. "Theory-Driven Evaluations: Need, Difficulties, and Options." *Evaluation Practice,* 1994b, *15*(1), 79–82.

Chen, H., Quane, J., Garland, T. N., and Marcin, P. "Evaluating an Antismoking Program: Diagnostics of Underlying Causal Mechanisms." *Evaluation and the Health Professions,* 1988, *11*(4), 441–464.

Chen, H., and Rossi, P. "The Multi-Goal, Theory-Driven Approach to Evaluation: A Model Linking Basic and Applied Social Science." *Social Forces,* 1980, *51*(1), 106–122.

Chen, H., and Rossi, P. "Evaluating with Sense: The Theory-Driven Approach." *Evaluation Review,* 1983, *7*(3), 283–302.

Chen, H., and Rossi, P. "Issues in the Theory-Driven Perspective." *Evaluation and Program Planning,* 1989, *12,* 299–306.

Chen, H., and Rossi, P. H. "The Theory Driven Approach to Validity." *Evaluation and Program Planning,* 1987, *10*(1), 95–103.

CIRCLE. "Evaluation of the Pilot Bail Advocacy Program." Melbourne: Department of Justice, Victoria, Jan. 2003. http://www.justice.vic.gov.au/wps/wcm/connect/ 9f37f580404a6b4896f4fff5f2791d4a/Pilot_Bail_Advocacy_and_Support+Services _Program_Evaluation.pdf?MOD=AJPERES.

CIRCLE. "Third Newsletter, Evaluation of the Stronger Families and Communities Strategy 2000–2004." Melbourne: RMIT University, July 2006. http://www.rmit.edu.au/casr/ sfcse.

CIRCLE. "Final Report. Evaluation of the Stronger Families and Communities Strategy, 2000–2004." Melbourne: RMIT University, 2008. http://www.rmit.edu.au/casr/sfcse.

Clark, H., and Anderson, A. A. "Theories of Change and Logic Models: Telling Them Apart." Paper presented at the annual meeting of the American Evaluation Association, Atlanta, Ga., Nov. 2004.

Cockburn, A. *Agile Software Development.* Reading, Mass.: Addison-Wesley, 2001.

Coffman, J. "Evaluation Based on Theories of the Policy Process." *Evaluation Exchange,* 2007, *13*(1), 2–4. www.gse.harvard.edu/hfrp/eval/issue34/eval2.html.

Cohen, M. D., March, J. G., and Olsen, J. P. "A Garbage Can Model of Organizational Choice." *Administrative Science Quarterly*, 1972, *17*(1), 1–25.

Cole, G. E. "Advancing the Development and Application of Theory-Based Evaluation in the Practice of Public Health." *American Journal of Evaluation*, 1999, *20*(3), 453–470.

Connell, J. P., and Kubisch, A. C. "Applying a Theory of Change Approach to the Evaluation of Comprehensive Community Initiatives: Progress, Prospects, and Problems." In K. Fulbright-Anderson, A. C. Kubisch, and J. P. Connell (eds.), *New Approaches to Evaluating Community Initiatives*. Washington, D.C.: Aspen Institute, 1998.

Conrad, K. "Measuring and Testing Program Philosophy." In L. Bickman (ed.), *Using Program Theory in Evaluation*. New Directions for Evaluation, no. 33. San Francisco: Jossey-Bass, 1987.

Conrad, K., and Buelow, J, "Developing and Testing Program Classification and Function Theories." In L. Bickman (ed.), *Using Program Theory in Evaluation*. New Directions for Evaluation, no. 33. San Francisco: Jossey-Bass, 1987.

Conrad, K., and Miller, T. "From Program Theory to Tests of Program Theory." In L. Bickman (ed.), *New Directions for Program Evaluation, 33*, 19–42, 1987.

Cook, T. D. "The False Choice Between Theory-Based Evaluation and Experimentation." In P. J. Rogers, T. A. Hasci, A. Petrosino, and T. A. Huebner (eds.), *Challenges and Opportunities in Program Theory Evaluation*. New Directions for Evaluation, no. 87. San Francisco: Jossey-Bass, 2000.

Cooksy, L. "The Meta-Evaluand: The Evaluation of Project TEAMS." *American Journal of Evaluation*, 1999, *20*(1), 123–136.

Cooksy, L., Gill, P., and Kelly, P. "The Program Logic Model as an Integrative Framework for a Multimethod Evaluation." *Evaluation and Program Planning*, 2001, *24*(2), 119–128.

Cooperrider, D. L., Barrett, F., and Srivastva, S. "Social Construction and Appreciative Inquiry: A Journey in Organizational Theory." In D. Hosking, P. Dachler, and K. Gergen (eds.), *Management and Organization: Relational Alternatives to Individualism*. Aldershot, U.K.: Avebury Press, 1995.

Cooperrider, D. L., Whitney, D., and Stavros, J. M. *Appreciative Inquiry: The Handbook*. (2nd ed.). Brunswick, Ohio: Crown Custom Publishers, 2002.

Cordray, D. S. "Optimizing Validity in Program Research: An Elaboration of Chen and Rossi's Theory-Driven Approach." *Evaluation and Program Planning*, 1990, *12*, 4: 379–385.

Coryn, C.L.S., Noakes, L. A., Westine, C. D., & Schröter, D. C. "A Systematic Review of Theory-Driven Evaluation Practice from 1990–2009." *American Journal of Evaluation*, forthcoming.

Costner, H. L. "The Validity of Conclusions in Evaluation Research: A Further Development of Chen and Rossi's Theory-Driven Approach." *Evaluation and Program Planning*, 1990, *12*, 4: 345–353.

Cummings, R. "'What If': The Counterfactual in Program Evaluation." *Evaluation Journal of Australasia*, 2006, *6*(2), 6–15. http://www.aes.asn .au/publications/Vol6No2/What_if _the_counterfactual_in_program _evaluation.pdf.

Cummings, R., Stephenson, K., and Hale, L. "Using Program Theory in an Educational Setting: What Can We Learn from It?" *Evaluation Journal of Australasia*, 2001, n.s. *1*(1), 29–39.

Dahler-Larsen, P. "From Programme Theory to Constructivism: On Tragic, Magic and Competing Programmes." *Evaluation*, 2001, *7*(3), 331–349.

Dart, J. J. "Report on Outcomes and Get Everyone Involved: The Participatory Performance Story Reporting Technique." Presentation at the 2008 International Conference of the Australasian Evaluation Society, Perth, Sept. 2008. http://www.aes.asn.au/ conferences/2008/papers/p100.pdf.

Dart, J. J. "Participatory Performance Story Reporting Technique." Paper presented at the 2009 Impact Evaluation conference, International Initiative for Impact Evaluation/African Evaluation Association, Cairo, Mar. 2009.

Dart, J. J. "Collaborative Outcomes Reporting Technique." Paper presented at Evaluation Revisited: Improving the Quality of Evaluative Practice by Embracing Complexity.

Utrecht, May 2010. http://evaluationrevisited .files.wordpress.com/2010/05/overview-of
-cort.pdf.

Dart J. J., and Mayne, J. "Performance Story." In S. Mathison (ed.), *Encyclopedia of
Evaluation.* Thousand Oaks, Calif.: Sage, 2005.

Dart, J. J., and McGarry, P. "People-Centred Evaluation." Presentation at the 2006 International
Conference of the Australasian Evaluation Society, Darwin, Sept. 2006. http://www.aes.asn
.au/conferences/2006/papers/099%20People-Focused%20Evaluation.pdf.

Davidson, E. J. "Ascertaining Causality in Theory-Based Evaluation." In P. J. Rogers, A. J.
Petrosino, T. Hacsi, and T. Huebner (eds.), *Program Theory in Evaluation: Challenges and
Opportunities.* New Directions for Evaluation, no. 87. San Francisco: Jossey-Bass, 2000.

Davidson, E. J. *Evaluation Methodology Basics: The Nuts and Bolts of Sound Evaluation.*
Thousand Oaks, Calif.: Sage, 2005.

Davidson, E. J. "Causal Inference Nuts and Bolts." Demonstration session at the 2006
Conference of the American Evaluation Association, Portland, Ore., Nov. 2006. http://
realevaluation.co.nz/pres/causation-aea06.pdf.

Davidson, E. J. "Editorial: Unlearning Some of Our Social Scientist Habits." *Journal of
Multidisciplinary Evaluation,* 2007, *4*(8), iii-vi.

Davidson, E. J. "Improving Evaluation Questions and Answers: Getting Actionable Answers
for Real-World Decision Makers." Demonstration session at the 2009 Conference of the
American Evaluation Association, Orlando, Fla., Nov. 2009. http://realevaluation.co.nz/
past-presentations/.

Davidson, E. J. "Why Genuine Evaluation Must Include Causal Inference." Feb. 20, 2010a.
http://genuineevaluation.com/why-genuine-evaluation -must-include-causal-inference/.

Davidson, E. J. "Outcomes, Impacts and Causal Attribution: Affordable, Practical, Feasible
Options for the Real World." Interactive symposium presented at the regional symposium
of ANZEA (Aoterea/New Zealand Evaluation Association), May 2010b, Auckland, New
Zealand. http://realevaluation.com/outcomes-impacts-causal-attribution-handouts-from
-auckland-anzea-symposium/.

Davidson, John. "Impacts and Their Values in Extension Accountability." Presentation to
Florida County Extension Directors In-service Training, Gainesville, Florida, November
8, 2007. http://ded.ifas.ufl.edu/files /Florida_Accountability_valuing_in_Extension.ppt.

Davies, R. "Improved Representations of Change Processes: Improved Theories of Change."
Paper presented at the Fifth Biennial Conference of the European Evaluation Society,
Seville, Spain, Oct. 2002.

Davies, R. "Scale, Complexity and the Representation of Theories of Change (Part I)."
Evaluation, 2004, *10*, 101–121.

Davies, R. "Scale, Complexity and the Representation of Theories of Change (Part II)."
Evaluation, 2005, *11*, 133–149.

Dawkins, R. "Eulogy for Douglas Adams." Church of Saint Martin in the Fields, London,
Sept. 7. 2001. http://www.edge.org/documents/adams _index.html.

Department of Labour, New Zealand. *Final evaluation report of the Recognised Seasonal
Employer policy (2007–2009),* Wellington, Department of Labour. 2010.

Deprez, S., & Van Steenkiste, C. *Planning, Learning and Accountability for Sustainable
Agricultural Chain Development: The case of VECO Indonesia.* Case presented at the
Evaluation Revisited Conference, Utrecht, May 2010. http://evaluationrevisited.files.
wordpress.com/2010/05/case -presentation_deprez-fnal.pdf.

Dhamotharan, M. *Handbook on Integrated Community Development: Seven D Approach to
Community Capacity Development.* Tokyo: Asian Productivity Organization, 2009.
http://www.apo-tokyo.org/00e-books/AG -21_APO_ICD_Manual/AG-21_APO_ICD
_Manual.pdf

Dhamorathan, M. "Effective Behaviour Through Genuine Interactions." *Capacity,* May 6–7,
2010. http://www.capacity.org/en/journal/practice _reports/effective_behaviour_through
_genuine_interactions.

DiClemente, R. J., Crosby, R. A., and Kegler, M. *Emerging Theories in Health Promotion
Practice and Research.* San Francisco: Jossey-Bass, 2009.

Donaldson, S. I. "Using Program Theory-Driven Evaluation Science to Crack the Da Vinci Code. In M. C. Alkin and C. A. Christie (eds.), *Theorists' Models in Action.* New Directions for Evaluation, no. 106. San Francisco: Jossey-Bass, 2005.

Donaldson, S. I. *Program Theory-Driven Evaluation Science: Strategies and Applications.* Mahwah, N.J.: Erlbaum, 2007.

Donaldson, S. I., and Azzam, T. "Interactive Evaluation Report." 2009. http://web.cgu.edu/media/sbos/Donaldson.swf.

Donaldson, S. I., and Azzam, T. "Developing Complex Theories of Change: The Promise of Technology-Enhanced Interactive Conceptual Frameworks." Forthcoming.

Douthwaite, B., and others. "Participatory Impact Pathways Analysis: A Practical Application of Program Theory in Research for Development." Canadian Journal of Program Evaluation, 2007, *22,* 127–159.

Douthwaite, B., Alvarez, S., Theile, G., and Mackay, R. Participatory Impact Pathways Analysis: A Practical Method for Project Planning and Evaluation. May 2008. http://www.cgiar-ilac.org/content/participatory-impact -pathways-analysis.

Douthwaite, B., Delve, R., Ekboir, J., and Twomlow, S. "Contending with Complexity: The Role of Evaluation in Implementing Sustainable Natural Resource Management." *International Journal of Agricultural Sustainability,* 2003, *1*(1), 51–66.

Douthwaite, B., Kuby, T., van de Fliert, E., and Schulz, S. "Impact Pathway Evaluation: An Approach for Achieving and Attributing Impact in Complex Systems." *Agricultural Systems,* 2003, *78,* 243–265.

Duignan, P. "Advocacy Group High Level Outcomes Model." 2009a. http://www.outcomesmodels.org/models/advocacy4.html.

Duignan, P. "Painless Performance Indicators." 2009b. http://youtube.com/watch?v=2cvJasO_gdcandfeature=related.

Durland, M. "Exploring and Understanding Relationships." In M. M. Durland and K. A. Fredericks (eds.), *Social Network Analysis in Program Evaluation.* New Directions for Evaluation, no. 107. San Francisco: Jossey-Bass, 2005.

Durland, M. M., and Fredericks, K. A. (eds.). *Social Network Analysis.* New Directions for Evaluation, no. 107. San Francisco: Jossey-Bass, 2005.

Dyehouse, M., and others. "A Comparison of Linear and Systems Thinking Approaches for Program Evaluation Illustrated Using the Indiana Interdisciplinary GK12." *Evaluation and Program Planning,* 2009, *32,* 187–196.

Earl, S., Carden, F., and Smutylo, T. *Outcome Mapping: Building Learning and Reflection into Development Programs.* Ottawa: International Development Research Centre, 2001.

Einstein, "On the Method of Theoretical Physics," The Herbert Spencer Lecture, delivered at Oxford (10 June 1933); also published in *Philosophy of Science,* Vol. 1, No. 2 (April 1934), pp. 163–169, p. 165.

Elsworth, G. "Theory-Informed Quantitative Data Analysis in Realist Program Evaluation." Paper presented at the Joint Conference of the European Evaluation Society and the United Kingdom Evaluation Society, London, Oct. 2006.

Eng, E., and Young, R. "Lay Health Advisors as Community Change Agents." *Journal of Family and Community Health,* 1992, *15,* 24–40.

Eoyang, G., and Berkas, T. "Evaluation in Complex Adaptive Systems." In M. Lissack and H. Gunz (eds.), *Managing Complexity in Organizations.* Westport, Conn.: Quorum Books, 1998.

Evaluation Support Scotland. "Developing a Logic Model." 2009. http://www.evaluationsupportscotland.org.uk/downloads/Supportguide1.2logicmodelsJul09.pdf.

European Commission. *Evaluating Socio Economic Development, Sourcebook 2: Methods and Techniques Evaluability Assessment.* Brussels: European Commission, 2009. http://ec.europa.eu/regional_policy/sources/ docgener/evaluation/evalsed/sourcebooks/method _techniques/ structuring_evaluations/evaluability/index_en.htm.

Eyben, R., Kidder, T., Rowlands, J., and Bronstein, A. "Thinking About Change for Development Practice: A Case Study from Oxfam GB." *Development in Practice,* 2008, *18*(2), 201–212.

Festinger, L. *A Theory of Cognitive Dissonance.* Stanford, Calif.: Stanford University, 1957.

Fishbein, M., and Ajzen, I. *Belief, Attitude, Intention, and Behavior: An Introduction to Theory and Research.* Reading, Mass.: Addison-Wesley, 1975.

Fishbein, M., and Ajzen, I. *Predicting and Changing Behavior: The Reasoned Action Approach.* New York: Psychology Press, 2009.

Fishbein, M., Hennessy, M., Kamb, M., Bolan, G. A., Hoxworth, T., Iatesta, M., et al. "Using Intervention Theory to Model Factors Influencing Behavior Change: Project RESPECT." *Evaluation and the Health Professions,* 2001, *24,* 363–384.

Fitz-Gibbon, C. T., and Morris, L. L. "Theory-Based Evaluation." *Evaluation Comment,* 1975, *5*(1), 1–4.

Fitzpatrick, J. "Dialog with Stewart Donaldson About the Theory-Driven Evaluation of the Work and Health Initiative." *American Journal of Evaluation,* 2002, *23*(3), 347–365.

Flannery, T. P., Hofrichter, D. A., and Platten, P. E. *People, Performance and Pay.* New York: Free Press, 1996.

Foster, J., and Hope, T. *Housing, Community and Crime: The Impact of the Priority Estates Project.* London: Her Majesty's Stationery Office, 1993.

Fraser, D. "Non-logics and Non-solutions: A Typology of Irrational Program Designs." *Evaluation Journal of Australasia,* 2001, *1*(1), 20–26.

Frechtling, J. A. *Logic Modeling Methods in Program Evaluation.* San Francisco: Jossey-Bass, 2007.

Freedman, D. A. "On Types of Scientific Enquiry: Nine Success Stories in Medical Research." In J. M. Box-Steffensmeier, H. E. Brady, and D. Collier (eds.), *The Oxford Handbook of Political Methodology.* New York: Oxford University Press, 2008. http://www.stat.berkeley.edu/~census/anomaly.pdf.

Friedman, V. "Designed Blindness: An Action Science Perspective on Program Theory Evaluation." *American Journal of Evaluation,* 2001, *22*(2), 2001, 161–181.

Funnell, S. C. "Developments in the Use of the NSW Approach to Analysing Program Logic." In *Proceedings of the 1990 National Evaluation Conference of the Australasian Evaluation Society.* Sydney, 1990.

Funnell, S. C. "Program Logic: An Adaptable Tool for Designing and Evaluating Programs." *Evaluation News and Comment,* 1997, *6*(1), 5–12.

Funnell, S. C. "Developing and Using a Program Theory Matrix for Program Evaluation and Performance Monitoring." In P. J. Rogers, A. J. Petrosino, T. Hacsi, and T. A. Huebner (eds.), *Program Theory in Evaluation: Challenges and Opportunities.* New Directions for Evaluation, no. 87. San Francisco: Jossey-Bass, 2000.

Funnell, S. C. (2000). *Applications of program logic to evaluation, monitoring and program design.* Demonstration session presented at the 2000 AEA Conference, Hawaii.

Funnell, S. C. "Evaluating Partnership Programs: Challenges and Approaches." In *Proceedings of the 2006 International Conference of the Australasian Evaluation Society.* Sept. 2006. http://www.aes.asn.au/ conferences/2006/papers/031%20Sue%20Funnell.pdf.

Funnell, S.C., and Lenne, B. "A Typology of Public Sector Programs." Paper presented at the Annual Meeting of the American Evaluation Association, San Francisco. November 1989.

Funnell, S. C., and Lenne, B. "Clarifying Program Objectives for Program Evaluation." *Program Evaluation Bulletin,* Jan. 1990.

Funnell, S. C., Rogers, P. J., and Scougall, J. "Issues Paper on Community Capacity Building for the Evaluation of the Stronger Families and Communities Strategy." Canberra: Department of Family and Community Services, 2004. http://www.rmit.edu.au/casr/sfcse.

Gabriel Sanchez, R., and others. "Hormone Replacement Therapy for Preventing Cardiovascular Disease in Post-Menopausal Women." *Cochrane Database of Systematic Reviews,* Apr. 2005, *2,* art. CD002229. http://www2.cochrane.org/reviews/en/ab002229.html.

Generon Consulting. *The U-Process: A Social Technology for Addressing Highly Complex Challenges.* Version 4.1, May 14, 2005. http://www .generonconsulting.com/publications/papers/pdfs/U-Process_Social _Technology.pdf.

George, A. L., and McKeown, T. J. "Case Studies and Theories of Organizational Decision Making." In R. Coulam and R. Smith (eds.), *Advances in Information Processing in Organizations.* Stamford, Conn.: JAI Press, 1985.

Gladwell, M. *The Tipping Point: How Little Things Can Make a Big Difference.* New York: Little, Brown, 2000.

Glanz, K., Rimer, B. K., and Lewis, F. M. *Health Behavior and Health Education. Theory, Research, and Practice.* San Francisco: Wiley & Sons, 2002.

Glasbergen, P. "Global Action Networks: Agents for Collective Action." *Global Environment Change,* 2010, *20*(1), 130–141.

Glass, G. "Interrupted Time Series Quasi-Experiments." In R. M. Jaeger (ed.), *Complementary Methods for Research in Education.* (2nd ed.) Washington, D.C.: American Educational Research Association, 1997.

Glouberman, S. *Towards a New Perspective on Health Policy.* Ottawa: Canadian Policy Research Networks 2001.

Glouberman, S., and Zimmerman, B. *Complicated and Complex Systems: What Would Successful Reform of Medicare Look Like?* Ottawa: Commission on the Future of Health Care in Canada, 2002. http://www .healthandeverything.org/files/Glouberman_E.pdf.

Goertzen, J. R., Fahlman, S. A., Hamption, M. R., and Jeffery, B. L. "Creating Logic Models Using Grounded Theory: A Case Example Demonstrating a Unique Approach to Logic Model Development." *Canadian Journal of Program Evaluation,* 2003, *18*(2), 115–138.

Goodson, B. D., and others. "Effectiveness of a Comprehensive, Five-Year Family Support Program for Low-Income Children and Their Families: Findings from the Comprehensive Child Development Program." *Early Childhood Research Quarterly,* 2000, *15*(1), 5–39.

Government Chief Social Researcher's Office. *The Magenta Book: Guidance Notes for Policy Evaluation and Analysis.* London: Prime Minister's Strategy Unit, Cabinet Office, 2003.

Granovetter, M. S. "The Strength of Weak Ties." *American Journal of Sociology,* 1973, *78*(6), 1360–1380.

Green, L. W., and Kreuter, M. W. *Health Program Planning: An Educational and Ecological Approach.* New York: McGraw-Hill, 2005.

Grizzell, J. "Behavior Change Theories and Models." 2003. http://www .csupomona.edu/ ~jvgrizzell/best_practices/bctheory.html.

Grodstein, F., Manson, J. E., Stampfer, M. J., and Rexrode, K. "Postmenopausal Hormone Therapy and Stroke: Role of Time Since Menopause and Age at Initiation of Hormone Therapy." *Archives of Internal Medicine,* 2008, *168,* 861–868.

Gugiu, P. C., and Rodríguez-Campos, L. "Semi-Structured Interview Protocol for Constructing Logic Models." *Evaluation and Program Planning,* 2007, *30*(4), 339–350.

Guijt, I. *Seeking Surprise: Rethinking Monitoring for Collective Learning in Rural Resource Management.* Wageningen, Netherlands: Wageningen Agricultural University, 2008.

Guijt, I., and Woodhill, J. "A Guide for Project M and E." Rome: International Fund for Agricultural Development, 2002. http://www.ifad.org/evaluation/guide/.

Hacsi, T. A., "Using Program Theory to Replicate Successful Programs." In P. J. Rogers, T. Hacsi, A. J. Petrosino, and T. A. Huebner (eds.), *Program Theory in Evaluation: Challenges and Opportunities.* New Directions for Evaluation, no. 87, San Francisco: Jossey-Bass, 2000.

Hall. W. K. and O'Day, J. "Causal Chain Approaches to the Evaluation of Highway Safety Countermeasures." *Journal of Safety Research,* 1971, *3,* 1: 9–20.

Halpern, D., Bates, C., Mulgan, G., and Aldridge, S. *Personal Responsibility and Changing Behaviour: The State of Knowledge and Its Implications for Public Policy.* 2004. http://www .cabinetoffice.gov.uk/media/cabinetoffice/strategy/.

Hardin, G. "The Tragedy of the Commons." *Science,* 1968, *162*(3859), 1243–1248.

Harlem Children's Zone. "Whatever It Takes: A White Paper on Harlem Children's Zone." New York: Harlem Children's Zone, 2009. http://www .urbanstrategies.org/programs/ schools/documents/HCZWhitePaper.pdf.

Harris, G. "F.D.A. Says Bayer Failed to Reveal Drug Risk Study." *New York Times,* Sept. 29, 2006, pp. A1, A9.

Hastings, J. T. "Curriculum Evaluation: The Why of the Outcomes." *Journal of Educational Measurement,* 1966, *3*(1), 27–32.

Hatry, H. P., Greiner, J. M., and Gollub, R. J. *An Assessment of Local Government Management Motivational Programs: Performance Targeting with and Without Monetary Incentives.* Washington, D.C.: Urban Institute, 1981.

Hatry, H. P., Winnie, R. E., and Fisk, D. M. *Practical Program Evaluation for State and Local Governments.* Washington, D.C.: Urban Institute Press, 1981.

Health Scotland. "Multiple Results Chains Showing Partner Contributions to Shared Health Outcomes." In *Health Improvement Performance Management Tools and Intended Users.* Edinburgh: Health Scotland, 2009. http://www.healthscotland.com/uploads/documents/8927 -HIPM%20SOA%20Guidance-%20Use%20of%20tools.pdf.

Hecksher Foundation. "The Heckscher Foundation for Children Logic Model Guidelines." 2009. http://www.heckscherfoundation.org/guidelines/logicmodel.html/_res/id=sa _File1/Logic%20Model%20 Guidelines.pdf.

Heider, F. *The Psychology of Interpersonal Relations.* Mahwah, N.J.: Erlbaum, 1958.

Hendricks, M., Plantz, M. C., and Pritchard, K. J. "Measuring Outcomes of United Way– Funded Programs: Expectations and Reality." In J. G. Carman and K. A. Fredericks (eds.), *Nonprofits and Evaluation.* New Directions for Evaluation, no. 199. San Francisco: Jossey-Bass, 2008.

Hodis, H. N. "Assessing Benefits and Risks of Hormone Therapy in 2008: New Evidence, Especially with Regard to the Heart." *Cleveland Clinic Journal of Medicine.* 2008, *75*(4), 3–12.

Holla, A., and Kremer, M. "Pricing and Access: Lessons from Randomized Evaluations in Education and Health." Washington, D.C.: Center for Global Development, 2009. http://www.cgdev.org/content/publications/detail/1420826.

Homel, R. *Policing and Punishing the Drinking Driver: A Study of General and Specific Deterrence.* New York: Springer-Verlag, 1988.

Homel, R. "Random Breath Testing in NSW: The Evaluation of a Successful Social Experiment." In *Proceedings of the 1990 National Evaluation Conference of the Australasian Evaluation Society.* Sydney, 1990.

Homel, R., Lenne, B., and Walker, R. *Guidelines for Analysing the Deterrent Effect of Regulatory Programs.* Sydney: NSW Office of Public Management, 1989.

House, E. "Blowback: Consequences of Evaluation for Evaluation." *American Journal of Evaluation,* 2008, *29*(4), 416–426.

Huebner, T. A. "Gaining a Shared Understanding Between School Staff and Evaluators." In P. J. Rogers, T. Hacsi, A. J. Petrosino, and T. A. Huebner (eds.), *Program Theory in Evaluation: Challenges and Opportunities.* New Directions for Evaluation, no. 87, San Francisco: Jossey-Bass, 2000.

Intergovernmental Panel on Climate Change. "The Scientific Basis." Third Assessment Report Climate Change Working Group 1. 2001. http://www.grida.no/publications/other/ipcc_tar.

James, W. *Pragmatism: A New Name for Some Old Ways of Thinking.* New York: Longman Green, 1907.

Janis, I. L. *Victims of Groupthink.* Boston: Houghton Mifflin, 1972

Johnston, A. "Andrea Johnston on the Waawiyeyaa Evaluation Tool." Oct. 1, 2010. http://aea365.org/blog/?p=1781.

Johnston Research. *The Waawiyeyaa Evaluation Tool: Application Manual: Facilitation to Analysis/Reporting.* Toronto: Johnston Research, 2010.

Joint Committee on Standards for Educational Evaluation. *The Program Evaluation Standards.* (2nd ed.) Thousand Oaks, Calif.: Sage, 1994.

Judge, K., and Bauld, L. "Strong Theory, Flexible Methods: Evaluating Complex Community-Based Initiatives." *Critical Public Health,* 2001, *11*(1), 19–38.

Julian, J. A., Jones, A., and Dey, D. "Open Systems Evaluation and the Logic Model: Program Planning and Evaluation Tools." *Evaluation and Program Planning,* 1995, *18*(4), 333–341.

Kalamazoo Wraps. "Evaluation Report to Community Members." Kalamazoo, Mich.: Kalamazoo Wraps, 2010. http://www.kalamazoowrapsevaluation .org/storage/ Final%20Annual%20Report.pdf.

Karau, S. J., and Williams, K. D. "Social Loafing: A Meta-Analytic Review and Theoretical Integration." *Journal of Personality and Social Psychology*, 1993, *65*, 681–706.

Keast, R., and Brown, K. "The Network Approach to Evaluation: Uncovering Patterns, Possibilities and Pitfalls." Paper presented at the 2005 Australasian Evaluation Society International Conference, Brisbane, 2005.

Kelly, J. D., and others. "Performance Management in a Commercial Business. A Case Study: Taronga Zoo." 1989.

Kelly, K., and others. "Tsha Tsha: Key Findings of the Evaluation of Episodes." Grahamstown: Centre for AIDS Development, Research and Evaluation, 2004. http://www.cadre.org .za/files/CADRE_KK_WP _TshaEval_2004.pdf.

Kilduff, M., and Tsai, W. *Social Networks and Organizations.* Thousand Oaks, Calif.: Sage, 2003.

Kipling, R. *Just So Stories.* London: Macmillan, 1902.

Kirkpatrick, D. L. "Techniques for Evaluating Training Programs." *Journal of the American Society for Training and Development,* 1959a, *11,* 1–13.

Kirkpatrick, D. L. "Techniques for Evaluating Training Programs." *Journal of the American Society for Training and Development,* 1959b, *12.*

Kirkpatrick, D. L. "Techniques for Evaluating Training Programs." *Journal of the American Society for Training and Development,* 1960a, *12.*

Kirkpatrick, D. L. "Techniques for Evaluating Training Programs." *Journal of the American Society for Training and Development,* 1960b, *13.*

Kirkpatrick, D. L *Evaluating Training Programs: The Four Levels.* (2nd ed.) San Francisco: Berrett-Koehler, 1998.

Kirkwood, B., and others, "Effect of Vitamin A Supplementation in Women of Reproductive Age on Maternal Survival in Ghana (ObaapaVitA): A Cluster-Randomised, Placebo-Controlled Trial." *Lancet,* May 8, 2010, pp. 1640–1649.

Knoke, D., and Kuklinski, J. H. *Network Analysis.* Thousand Oaks, Calif.: Sage, 1982.

Knowlton L., and Phillips, C. *The Logic Model Guidebook: Better Strategies for Great Results.* Thousand Oaks, Calif.: Sage, 2008.

Kotter, J., and Schlesinger, L. "Choosing Strategies for Change." *Harvard Business Review,* 1979, *57,* 106–114.

Kotvojs, F., and Shrimpton, B. "Contribution Analysis: A New Approach to Evaluation in International Development." *Evaluation Journal of Australasia,* 2007, *7*(1), 27–35. http:// www.aes.asn.au/publications/Vol7No1/Contribution_Analysis.pdf.

Kretzmann, J. P. "Building Communities from the Inside Out." *Shelterforce,* Sept.–Oct. 1995, pp. 8–11. http://www.nhi.org/online/issues/83/ buildcomm.html.

Kretzmann, J. P., and McKnight, J. L. *Building Communities from the Inside Out: A Path Toward Finding and Mobilizing a Community's Assets.* Evanston, Ill.: Institute for Policy Research, 1993.

Kurtz, C. F., and Snowden, D. F. "The New Dynamics of Strategy: Sense-Making in a Complex and Complicated World." *IBM Systems Journal,* 2003, *42*(3), 462–483.

Laumann, E., Marsden, P., and Prensky, D. "The Boundary Specification Problem in Network Analysis." In R. Burt and M. Minor (eds.), *Applied Network Analysis: A Methodological Introduction.* Thousand Oaks, Calif.: Sage, 1983.

Laycock, J. "The Development of a Logic Model for Protection Against Family Violence Act." *Evaluation Journal of Australasia,* 2005, *5*(1), 18–24

Lee, N., and Jacobson, M. "King County's Environmental Behavior Index: Using Social Marketing Research to Monitor and Evaluate Outcomes." Paper presented at Evaluation 2006, Annual Conference of the American Evaluation Association, Portland, Ore., 2006.

Leeuw, F. "Reconstructing Program Theories: Methods Available and Problems to Be Solved." *American Journal of Evaluation,* 2003, *24*(1), 5–20.

Leeuw, F., Gils, G., and Kreft, C. "Evaluating Anti-Corruption Initiatives." *Evaluation,* 1992, *5*(2), 194–219.

Leeuw, F., and Vaessen, J. "Impact Evaluation and Development: NONIE Guidance on Impact Evaluation." Washington, D.C.: Network of Networks on Impact Evaluation, 2009. http://www.worldbank.org/ieg/nonie/guidance.html.

Lenne, B., and Cleland, H. "Describing Program Logic." Sydney: New South Wales Public Service Board, 1987.

Lin, N., Cook, K., and Burt, R. *Social Capital: Theory and Research*. New Brunswick, N.J.: Transaction Publishers, 2001.

Lindblom, C. E. "The Science of Muddling Through." *Public Administration*, 1959, *19*(1), 79–88.

Lipsey, M. W. "Theory as Method: Small Theories of Treatments." In L. Sechrest and A. Scott (eds.), *Understanding Causes and Generalizing About Them*. New Directions for Evaluation, no. 57. San Francisco: Jossey-Bass, 1993.

Lipsey, M. W. "What Can You Build with Thousands of Bricks? Musings on the Cumulation of Knowledge in Program Evaluation." In D. Rog and D. Fournier (eds.), *Progress and Future Directions in Evaluation: Perspectives on Theory, Practice and Methods*. New Directions for Evaluation, no. 76. San Francisco: Jossey-Bass, 1997.

Lipsey, M. W. "Meta-Analysis and the Learning Curve in Evaluation Practice." *American Journal of Evaluation*, 2000, *21*(2), 207–212.

Lipsey, M. W., and Pollard, J. A. "Driving Toward Theory in Program Evaluation: More Models to Choose From." *Evaluation and Program Planning*, 1989, *12*, 317–328.

Lipsey, M. W., and Wilson, D. B. *Practical Meta-Analysis*. Thousand Oaks, Calif.: Sage, 2001.

List, D. "Combining Program Logic with Scenario Networks." In *Proceedings of the 2004 International Conference of the Australasian Evaluation Society, Adelaide*. Oct. 2004. http://www.aes.asn.au/conferences/2004/TH12%20List,%20D.pdf.

London Regeneration Network. *Capacity Building: The Way Forward*. London: London Regeneration Consortium, 1999.

Mackenzie, M., and Blamey, A. "The Practice and the Theory: Lessons from the Application of a Theories of Change Approach." *Evaluation*, 2005, *11*(2), 151–168.

Margoluis, R., Stem, C., Salafsky, N., and Brown, M. "Using Conceptual Models as a Planning and Evaluation Tool in Conservation." *Evaluation and Program Planning*, 2009, *32*, 138–147.

Mark, M. M. "From Program Theory to Tests of Program Theory." In L. Bickman (ed.), *Advances in Program Theory*. New Directions for Evaluation, no. 47. San Francisco: Jossey-Bass, 1990.

Marquart, J. "A Pattern-Matching Approach to Link Program Theory and Evaluation Data." In L. Bickman (ed.), *Advances in Program Theory*. New Directions for Evaluation, no. 47. San Francisco: Jossey-Bass, 1990.

Marsh, D. "The Utility and Future of Policy Network Analysis." In D. Marsh (ed.), *Comparing Policy Networks*. Bristol, Pa.: Open University Press, 1998.

Mathematica Policy Research. *Making a Difference in the Lives of Infants and Toddlers and Their Families: The Impacts of Early Head Start*. Washington, D.C.: U.S. Department of Health and Human Services, 2002.

Mathews, A. W. "Worrisome Ailment in Medicine: Misleading Journal Articles." *Wall Street Journal*, May 10, 2005, pp. A1–A2.

Matthews, A., and Funnell, S. C. "Evaluability Assessment of Drug and Alcohol Programs for Prisoners in New South Wales." Paper presented at the Conference of the Australasian Evaluation Society, Canberra, 1987.

Maxwell, J. "Using Qualitative Methods for Causal Explanation." *Field Methods*, 2004, *16*(3), 243–264.

Mayne, J. "Addressing Attribution Through Contribution Analysis: Using Performance Measures Sensibly." *Canadian Journal of Program Evaluation*, 2001, *16*(1), 1–24.

Mayne, J. "Reporting on Outcomes: Setting Performance Expectations and Telling Performance Stories." Canadian Journal of Program Evaluation, 2004, *19*(1), 31–60.

Mayne, J. "Contribution Analysis: An Approach to Exploring Cause and Effect." Rome: Institutional Learning and Change Initiative, *ILAC Brief No. 16*, May 2008. http://www.cgiar-ilac.org/files/publications/briefs/ILAC_Brief16_Contribution_Analysis.pdf.

McClintock, C. "Conceptual and Action Heuristics: Tools for the Evaluator." In L. Bickman (ed.), *Using Program Theory in Evaluation.* New Directions for Evaluation, no. 33. San Francisco: Jossey-Bass, 1987.

McClintock, C. "Administrators as Applied Theorists." In L. Bickman (ed.), *Advances in Program Theory.* New Directions for Evaluation, no. 47. San Francisco: Jossey-Bass, 1990.

McDonald, B., and Rogers, P. J. "Market Segmentation as an Analogy for Differentiated Program Theory: An Example from the Dairy Industry." Paper presented at the annual meeting of the American Evaluation Association, Orlando, Fla., 1999.

McDonald, B., Rogers, P. J., and Kefford, B. "Teaching People to Fish? Building the Evaluation Capability of Public Sector Organizations." *Evaluation,* 2003, *9*(1), 9–29.

McDonald, R., and Teather, G. "Science and Technology Policy Evaluation Practices in the Government of Canada." In *Policy Evaluation in Innovation and Technology: Towards Best Practices, Proceedings.* Paris: Organization for Economic Co-Operation and Development, 1997.

McLaughlin, J. A., and Jordan, G. B. "Logic Models: A Tool for Telling Your Program's Performance Story." *Evaluation and Program Planning,* 1999, *22*(1), 65–72.

Mendelsohn M., and Karas, R. "Editorial. HRT and the Young at Heart." *New England Journal of Medicine,* June 21, 2007, pp. 2639–2641.

Merton, R. K. "Social Structure and Anomie." *American Sociological Review,* 1938, *3,* 672–682.

Merton, R. K. *Social Theory and Social Structure.* New York: Free Press, 1968.

Miguel, E., and Kremer, M. "Worms: Identifying Impacts on Education and Health in the Presence of Treatment Externalities." *Econometrica,* 2004, *72,* 159–217.

Miles, M. B., and Huberman, A. M. *Qualitative Data Analysis.* (2nd ed.) Thousand Oaks, Calif.: Sage, 1994.

Milne, C. "Outcomes Hierarchies and Program Logic as Conceptual Tools: Five Case Studies." In *Proceedings of the Australasian Evaluation Society Annual Conference.* Brisbane, 1993.

Ministry of Health, New Zealand. Reducing Inequalities in Health. Wellington: Ministry of Health, 2002. http://www.moh.govt .nz/moh.nsf/ 82f4780aa066f8d7cc2570bb006b5d4d/523077 dddeed012dcc256c550003938b/$FILE/ ReducIneqal.pdf.

Montague, S. "Build Reach into Your Logic Model." Ottawa: Performance Management Network, 1998. http://pmn.net/library/build_reach_into _your_logic_model.htm.

Montague, S., Young, G., and Montague, C. "Circles Tell the Performance Story." *Canadian Government Executive,* 2003, *2,* 12–16.

Montibeller Neto G., Benton, V. "Causal Maps and the Evaluation of Decision Options—A Review." *Journal of the Operational Research Society.* 2006, *57,* 7, 779–791.

Morell, J. "Logic Models: Uses, Limitations, Links to Methodology and Data." Presentation to the 2009 annual meeting of the American Evaluation Association, Orlando, Fla., 2009.

Nagarajan, N., and Vanheukelen, M. *Evaluating EU Expenditure Programmes: A Guide. Ex Post and Intermediate Evaluation.* Brussels: Directorate- General for Budgets of the European Union, 1997.

North Carolina State University. Co-Operative Extension. "2002 Program Success Highlights." 2002. http://www.ces.ncsu.edu/AboutCES/ Success/2002/.

Norwegian Agency for Development Co-Operation. *The Logical Framework Approach, Handbook for Objectives-Oriented Planning.* (4th ed.) Oslo: Norwegian Agency for Development Co-Operation, 1999.

Nunns, H., and Roorda, M. "Lifting the Lens: Developing a Logic for a Complicated Policy." *Evaluation Journal of Australasia,* 2009, *9,* 2: 24–32.

Office of the Auditor General, Canada. "Reporting on Outcomes: Setting Performance Expectations and Telling Performance Stories." 2003. http://www.oag-bvg.gc.ca/internet/ English/meth_lp_e_907.html.

Office of the Public Service Commission, South Africa. "Monitoring and Evaluation in the Public Service." Presentation at the Learning Network Session: Transforming the public service for *service excellence,* Johannesburg, Feb. 24, 2006.

Ostrom, E. "Institutional Rational Choice: An Assessment of the IAD Framework." In P. Sabatier (ed.), *Theories of the Policy Process*. Boulder, Colo.: Westview Press, 1999.

Ostrom, E. "Developing a Method for Analyzing Institutional Change." In S. S. Bati and N. Mercuro (eds.), *Alternative Institutional Structures: Evolution and Impact*. New York: Routledge, 2008.

Palumbo, D. J., and Oliviero, A. "Implementation Theory and the Theory-Driven Approach to Validity." *Evaluation and Program Planning*, 1990, *12*, 4: 337–336.

Pascale, R., Sternin, J., and Sternin, M. *The Power of Positive Deviance: How Unlikely Innovators Solve the World's Toughest Problems*. Boston: Harvard Business School Press, 2010.

Patton, M. Q. *Utilization-Focused Evaluation* (1st ed.). Beverly Hills, Calif.: Sage Publications, 1978.

Patton, M. Q. *Utilization-Focused Evaluation* (2nd ed.). Thousand Oaks, Calif.: Sage Publications, 1987.

Patton, M. Q. "A Context and Boundaries for a Theory-Driven Approach to Validity." *Evaluation and Program Planning*, 1990, *12*, 4: 375–377.

Patton, M. Q. "Developmental Evaluation." *American Journal of Evaluation*, 1994, *15*, 311–319.

Patton, M. Q. *Utilization-Focused Evaluation* (3rd ed.). Thousand Oaks, Calif.: Sage Publications, 1997.

Patton, M. Q. "Assessing Program Theory." Post to discussion list EVALTALK, Apr.2, 1998.

Patton, M. Q. "New Dimensions and Meanings of Useful Evaluation." Paper presented at the 2003 conference of the Australasian Evaluation Society, Auckland, 2003.

Patton, M. Q. *Utilization-Focused Evaluation*. (4th ed.) Thousand Oaks, Calif.: Sage, 2008.

Patton, M. Q. *Developmental Evaluation*. New York: Guilford Press, 2010.

Pavlov, I. P. *Conditional Reflexes*. New York: Dover, 1960. (Originally published 1927)

Pawson, R. "Evidence Based Policy: The Promise of Realist Synthesis." *Evaluation*, 2002, *8*(3), 340–358.

Pawson, R. "A Realist Approach to Evidence Based Policy." In B. Carter and C. New (eds.), *Making Realism Work*. London: Routledge, 2003a.

Pawson, R. "Nothing as Practical as a Good Theory." *Evaluation*, 2003b, *9*(3), 309–321.

Pawson, R. "Simple Principles for the Evaluation of Complex Programmes." In A. Killoran, C. Swann, and M. Kelly (eds.), *Public Health Evidence: Tackling Health Inequalities*. New York: Oxford University Press, 2006a.

Pawson, R. *Evidence Based Policy: A Realist Perspective*. Thousand Oaks, Calif.: Sage, 2006b.

Pawson, R., and Tilley, N. "What Works in Evaluation Research?" *British Journal of Criminology* 1994, *34*(3), 291–306.

Pawson, R., and Tilley, N. *Realistic Evaluation*. Thousand Oaks, Calif.: Sage, 1997.

Perkins, D., and Zimmerman, M. "Empowerment Theory, Research and Application." *American Journal of Community Psychology*, 1995, *23*(5), 569–579.

Peters, T. J., and Waterman, R. H. *In Search of Excellence: Lessons from America's Best Run Companies*. New York: Warner Books, 1992.

Petrosino, A. "Whether and Why? The Potential Benefits of Including Program Theory in Meta-Analysis." In P. J. Rogers, A. J. Petrosino, T. Hacsi, and T. Huebner (eds.), *Program Theory Evaluation: Challenges and Opportunities*. New Directions in Evaluation, no. 87. San Francisco: Jossey-Bass, 2000.

Pollack, A., and Abelson, R. "Why the Data Diverge on the Dangers of Vioxx." *New York Times,* May 22, 2006, pp. C1, C5.

Practical Concepts, Inc. *The Logical Framework: A Manager's Guide to a Scientific Approach to Design and Evaluation*. New York: Practical Concepts, 1979.

Preskill, H., and Catsambas, T. *Reframing Evaluation Through Appreciative Inquiry*. Thousand Oaks, Calif.: Sage, 2006.

Preskill, H., and Coghlan, A. T. (eds.). *Using Appreciative Inquiry in Evaluation*. New Directions for Evaluation, no. 100. San Francisco: Jossey-Bass, 2004.

Prochaska, J. O., and DiClemente, C. C. "Stages and Processes of Self-Change of Smoking: Toward an Integrative Model of Change." *Journal of Consulting and Clinical Psychology,* 1983, *51*(3), 390–395.

Prochaska, J. O., DiClemente, C. C., and Norcross, J. "In Search of How People Change: Applications to Addictive Behaviors." *American Psychologist*, 1992, *47*(9), 1102–1114.

Prochaska, J. O., and Velicer, W. F. "The Transtheoretical Model of Health Behavior Change." *American Journal of Health Promotion,* Sep–Oct. 1997, *12*(1):38–48.

Provan, K. G., and Milward, H. B. "A Preliminary Theory of Interorganisational Network Effectiveness: A Comparative Study of Four Community Mental Health Systems." *Administrative Science Quarterly,* 1995, *40*(1), 1–33.

Ragin, C. C. *The Comparative Method: Moving Beyond Qualitative and Quantitative Strategies.* Berkeley: University of California Press, 1987.

Ramalingam, B., Jones, H., Young, J., and Reba, T. *Exploring the Science of Complexity: Ideas and Implications for Development and Humanitarian Efforts.* London: Overseas Development Institute, 2008.

Rappaport, J. "Terms of Empowerment/Exemplars of Prevention: Toward a Theory for Community Psychology." *American Journal of Community Psychology,* 1987, *15*(2), 121–148.

Reed, J. "A Generic Theory Based Logic Model for Creating Scientifically Based Program Logic Models." Paper presented at the American Evaluation Association Conference, Portland, Ore., 2006.

Reeler, D. "A Three-Fold Theory of Social Change and Implications for Practice, Planning, Monitoring and Evaluation." Cape Town: Community Development Resources Association, 2007. http://www .cdra.org.za/articles/A%20Theory%20]of%20Social%20 Change% 20by%20Doug%20Reeler.pdf.

Regine, B., and Lewin, R. *Complexity Science in Human Terms: A Relational Model of Business.* N.d. http://www.harvest-associates.com/pubs/human .html.

Renger, R., and Titcomb, A. "A Three-Step Approach to Teaching Logic Models." *American Journal of Evaluation,* 2002, *23*(4), 493–503.

Richardson, G. P. "How to Anticipate Change in Tobacco Control Systems." In A. Best, P. I. Clark, S. J. Leischow, & W.M.K. Trochim (eds.), *Greater Than the Sum: Systems Thinking in Tobacco Control.* Bethesda, Md.: U.S. Department of Health and Human Services, National Institutes of Health, National Cancer Institute, 2007.

Richardson, J. G. "Developing and Communicating Effective Program Success Stories for Enhanced Accountability." Paper presented at the Southern Association of Agricultural Scientists Conference, Memphis, Tenn., 1999.

Richardson, J. G. "Impacts and Their Values in Extension Accountability." Workshop for Florida County Extension Directors In-Service Training, Gainesville, Fla., Nov. 2007. http://ded.ifas.ufl.edu/files/Florida _Accountability_valuing_in_Extension.ppt.

Riggan, J. M. "Looking Where the Key Is: Surfacing Theories of Change in Complex, Collaborative Initiatives." Unpublished doctoral dissertation, University of Pennsylvania, 2005. http://repository.upenn.edu/ dissertations/AAI3179796.

Riggin, L. "Linking Program Theory and Social Science Theory." In L. Bickman (ed.), *Advances in Program Theory.* New Directions for Evaluation, no. 47. San Francisco: Jossey-Bass, 1990.

Roberts Evaluation. *Bennett's Hierarchy.* 2010. http://www.robertsevaluation .com.au/index .php?option=com_content&task=view&id=48.

Rockwell, K., and Bennett, C. "A Hierarchy for Targeting Outcomes and Evaluating Their Achievement." N.d. http://citnews.unl.edu/TOP/english/.

Rogers, E. "Progress, Problems and Prospects for Network Research: Investigating Relationships in the Age of Electronic Communication Technologies." *Social Networks,* 1987, *9,* 285–310.

Rogers, E. *Diffusion of Innovations.* (4th ed.) New York: Free Press, 1995.

Rogers, P. J. "Program Theory: Not Whether Programs Work, But How They Work." In D. L. Stufflebeam, G. F. Madaus, and T. Kellaghan (eds.), *Evaluation Models: Viewpoints on Educational and Human Services Evaluation.* Norwell Mass.: Kluwer, 2000a.

Rogers, P. J. "Causal Models in Program Theory Evaluation." In P. J. Rogers, A. J. Petrosino, T. Hacsi, and T. Huebner (eds.), *Program Theory Evaluation: Challenges and Opportunities.* New Directions in Evaluation, no. 87. San Francisco: Jossey-Bass, 2000b.

Rogers, P. J. "Logic Models." In S. Mathison (ed.), *Encyclopedia of Evaluation*. Thousand Oaks, Calif.: Sage, 2004.

Rogers, P. J. "Theory-Based Evaluation: Reflection Ten Years On." In S. Mathison (ed.), *Enduring Issues in Evaluation: The 20th Anniversary of the Collaboration Between NDE and AEA*: New Directions for Evaluation, no. 114. San Francisco: Jossey-Bass, 2007.

Rogers, P. J. "Using Programme Theory for Complicated and Complex Programmes." *Evaluation*, 2008, *14*(1), 29–48.

Rogers, P. J., Edgecombe, G., and Kimberley, S. "Early Intervention, Especially in Early Childhood: Issues Paper." Canberra: Department of Families and Community Services, 2004.

Rogers, P. J., Guijt, I., and Williams, B. "Thinking Systemically: Seeing from Simple to Complex in Impact Evaluation." Presented at the 3IE/African Evaluation Association Impact Evaluation Conference, Cairo, Egypt, 2009.

Rogers, P. J., Petrosino, A. J., Hacsi, T., and Huebner, T. (eds.). *Program Theory Evaluation: Challenges and Opportunities*. New Directions for Evaluation, no. 87. San Francisco: Jossey-Bass, 2000a.

Rogers, P. J., Petrosino, A. J., Hacsi, T., and Huebner, T. A. "Program Theory Evaluation: Practice, Promise and Problems." In P. J. Rogers, T. Hacsi, A. J. Petrosino, and T. A. Huebner (eds.), *Program Theory in Evaluation: Challenges and Opportunities*. New Directions for Evaluation, no. 87. San Francisco: Jossey-Bass, 2000b.

Rogers, P. J., Stevens, K., and Boymal, J. "Qualitative Cost-Benefit Evaluation of Complex, Emergent Programs." *Evaluation and Program Planning*, 2009, *32*(1), 83–90.

Rohrbach, L. A., Grana, R., Sussman, S., and Valente, T. W. "Type II Translation Transporting Prevention Interventions From Research to Real-World Settings." *Evaluation and the Health Professions*, 2006, *29*, 3: 302–333.

Rohrbach, L. A., Grana, R., Sussman, S., and Valente, T. W. "Type II Translation Transporting Prevention Interventions From Research to Real-World Settings." *Evaluation and the Health Professions*, 2006, *29*, 3: 302–333.

Rosenstock, I. M. "Why People Use Health Services." *Milbank Memorial Fund Quarterly*, 1966, *44*(3), 94–127.

Rossi, P., Freeman, H., and Lipsey, M. *Evaluation: A Systematic Approach*. (6th ed.) Thousand Oaks, Calif.: Sage, 1999.

Rutman, L. (ed.). *Evaluation Research Methods*. Thousand Oaks, Calif.: Sage, 1997.

Sabatier, P. A. *Theories of the Policy Process*. Boulder, Colo.: Westview Press, 2007.

Sabatier, P. A., and Weible, C. M. "The Advocacy Coalition Framework: Innovations and Clarifications." In P. A. Sabatier (ed.), *Theories of the Policy Process*. (2nd ed.) Boulder, Colo.: Westview Press, 2007.

Salamon, L. M. "Rethinking Public Management: Third Party Government and the Changing Forms of Government Action." *Public Policy*, 1981, *29*, 255–275.

Salmen, L. F. *Beneficiary Assessment: An Approach Described*. Washington, D.C.: World Bank, 2002.

Sanderson, I. "Evaluation in Complex Policy Systems." *Evaluation*, 2000, *6*(4), 433–454.

Scheirer, M. A. "Program Theory and Implementation Theory: Implications for Evaluators." In L. Bickman (ed.), *Using Program Theory in Evaluation*. New Directions for Evaluation, no. 33. San Francisco: Jossey-Bass, 1987.

Scheirer, M. A. (ed.). *A User's Guide to Program Templates: A New Tool for Evaluating Program Content*. New Directions for Evaluation, no. 72, San Francisco: Jossey-Bass, 1996.

Schorr, L. *Common Purpose: Strengthening Families and Neighborhoods to Rebuild America*. New York: Doubleday, 1997.

Scott, A. G., and Sechrest, L. "Strength of Theory and Theory of Strength." *Evaluation and Program Planning*, 1989, *12*, 4: 329–336.

Scriven, M. "Evaluation Perspectives and Procedures." In W. J. Popham (ed.), *Evaluation in Education—Current Perspectives*. Berkeley, Calif.: McCutchan, 1974.

Seel, R. "Introduction to Appreciative Inquiry." 2008. http://www.new-paradigm.co.uk/introduction_to_ai.htm.

Segal, D. "It's Complicated: Making Sense of Complexity." *New York Times*, Apr. 30, 2010. http://www.nytimes.com/2010/05/02/weekinreview/02segal.html.

Senge, P. M. *The Fifth Discipline: The Art and Practice of the Learning Organization.* New York: Doubleday, 1990.

Shadish, W. "Program Micro- and Macro-Theories: A Guide for Social Change." In L. Bickman (ed.), *Using Program Theory in Evaluation.* New Directions for Evaluation, no. 33. San Francisco: Jossey-Bass, 1987.

Shapiro, J. Z. "Contextual Limits on Validity Attainment: An Artificial Science Perspective on Program Evaluation." *Evaluation and Program Planning*, 1989, *12*, 4: 367–374.

Shapiro, I. *Training for Racial Equity and Inclusion.* Washington, D.C.: Aspen Institute, 2002.

Shapiro, I. "Theories of Change." In G. Burgess and H. Burgess (eds.), *Beyond Intractability.* Boulder: Conflict Research Consortium, University of Colorado. Jan. 2005. http://www.beyondintractability.org/essay/ theories_of_change/.

Shaw, G. B. *Everybody's Political What's What.* London: Constable, 1944.

Shaw, I., and Crompton, A. "Theory Like Mist on Spectacles Obscures Vision." *Evaluation*, 2003, (2), 192–204.

Sherman, L. W. *Policing Domestic Violence: Experiments and Dilemmas.* New York: Free Press, 1992.

Sherman, L. W., and Berk, R. A. "The Specific Deterrent Effects of Arrest for Domestic Assault." *American Sociological Review*, 1984, *49*, 261–272.

Sibthorpe, B., Glasgow, N., and Longstaff, D. *Complex Adaptive Systems: A Different Way of Thinking About Health Care Systems. A Brief Synopsis of Selected Literature for Initial Work Program—Stream 1.* Canberra: Australian Primary Healthcare Research Institute, Australian National University, 2004.

Siemens, G. "Connectivism: A Learning Theory for the Digital Age." *International Journal of Instructional Technology and Distance Learning*, 2005, *2*(1), 1–8.

Sitaker, M. "Adapting Logic Models over Time: The Washington State Heart Disease and Stroke Prevention Program Experience." *Preventing Chronic Disease: Public Health Research, Practice, and Policy*, 2008, *5*(2), 1–8.

Skinner, B. F. *Science and Human Behavior.* New York: Macmillan, 1953.

Smith, M. F. *Evaluability Assessment: A Practical Approach.* Norwood, Mass.: Kluwer, 1989.

Smith, M. F. "Evaluability Assessment: Reflections on the Process." *Evaluation and Program Planning*, 1990, *13*(4), 359–364.

Smith, N. L. "Using Path Analysis to Develop and Evaluate Program Theory." In L. Bickman (ed.), *Advances in Program Theory.* New Directions for Evaluation, no. 47. San Francisco: Jossey-Bass, 1990.

Smith, N. L. "Clarifying and Expanding the Application of Program Theory-Driven Evaluations." *Evaluation Practice*, 1994, 15, (1), 83–87.

Snowden, D. J., and Boone, M. "A Leader's Framework for Decision Making." *Harvard Business Review*, Nov. 2007, pp. 69–76.

St. Pierre, R., Layzer, J., Goodson, B., and Bernstein, L. "Report on the National Evaluation of the Comprehensive Child Development Program." 1996. http://www.researchforum.org/project_abstract_166.html.

St. Pierre, R., Layzer, J., Goodson, B., and Bernstein, L. "The Effectiveness of Comprehensive Case Management Interventions: Evidence from the National Evaluation of the Comprehensive Child Development Program." *American Journal of Evaluation*, 1999, *20*(1), 15–34.

St. Pierre, R., and Rossi, P. "Randomize Groups, Not Individuals: A Strategy for Improving Early Childhood Programs." *Evaluation Review*, 2006, *30*, 656–685.

Stacey, R. D. *Strategic Management and Organizational Dynamics.* London: Pitman, 1993.

Stame, N. "Theory-Based Evaluation and Types of Complexity." *Evaluation*, 2004, *10*(1), 58–76.

Stufflebeam, D. L. "The Use and Abuse of Evaluation in Title III." *Theory into Practice*, 1967, *6*, 126–133.

Stufflebeam, D. L. "The CIPP Model for Program Evaluation." In G. F. Madaus, M. Scriven, and D. L. Stufflebeam (eds.), *Evaluation Models: Viewpoints on Educational and Human Services Evaluation.* Boston: Kluwer, 1983.

Stufflebeam, D. (ed.). *Evaluation Models*. New Directions for Evaluation, no. 89. San Francisco: Jossey-Bass, 2001.

Stufflebeam, D. L. "The 21st Century CIPP Model." In M. C. Alkin (ed.), *Evaluation Roots*. Thousand Oaks, Calif.: Sage, 2004.

Stufflebeam, D., and Shinkfield, A. *Evaluation Theory, Models, and Applications*. San Francisco: Jossey-Bass, 2007.

Successworks. *Evaluation Framework for the Stronger Families and Communities Strategy, 2000–2004*. Canberra: Department of Families and Community Services, 2001.

Suchman, E. *Evaluative Research*. New York: Russell Sage Foundation, 1967.

Sutton, S. "Recovery and Relapse: Back to the Drawing Board? A Review of Applications of the Transtheoretical Model to Substance Abuse." *Addiction*, 2001, *96*, 175–186.

Torvatn, H. "Using Program Theory Models in Evaluation of Industrial Modernization Programs: Three Case Studies." *Evaluation and Program Planning*, 1999, *22*(1), 73–82.

Toulemonde, J., Fontaine, C., Laudren, E., and Vincke, P. "Evaluation in Partnership." *Evaluation*, 1998, *4*(2), 171–188.

Truman, C., and Triska, O. H. "Modelling Success: Articulating Program Theory." *Canadian Journal of Program Evaluation*, 2001, *16*(2), 101–112.

Tversky, A., and Kahneman, D. "Judgment Under Uncertainty: Heuristics and Biases." *Science*, 1974, *185*, 1124–1131.

U.K. Department for International Development. "Guidance on Using the Revised Logical Framework." Feb. 2009. http://mande.co.uk/blog/wp-content/uploads/2009/06/logical-framework.pdf.

United Way of America. *Measuring Program Outcomes: A Practical Approach*. Alexandria, Va.: United Way of America, 1996.

University of Wisconsin Extension. Program Development and Evaluation Unit. "Logic Model." 2003.

Unrau, Y. A. "Using Client Exit Interviews to Illuminate Outcomes in Program Logic Models: A Case Example." *Evaluation and Program Planning*, 2001, *24*(4), 353–361.

U.S. General Accounting Office. *Managing for Results: Critical Actions for Measuring Performance*. Washington, D.C: General Accounting Office, 1995.

Ware, J. H. "Investigating Therapies of Potentially Great Benefit: ECMO." *Statistical Science*, 1989, *4*(4), 298–340.

Ware, J. H., and Epstein, M. D. "Comments on 'Extracorporeal Circulation in Neonatal Respiratory Failure: A Prospective Randomized Study' by R. H. Bartlett et al." *Pediatrics*, 1985, *76*, 849–851.

Warrener, D. *Synthesis Paper 3: The Drivers of Change Approach*. London: Overseas Development Institute, 2004.

Weiss, C. H. *Evaluation Research: Methods of Assessing Program Effectiveness*. Upper Saddle River, N.J., Prentice Hall, 1972.

Weiss, C. H. "Nothing as Practical as Good Theory: Exploring Theory-Based Evaluation for Comprehensive Community Initiatives for Children and Families." In J. Connell, A. Kubisch, L. B. Schorr, and C. H. Weiss (eds.), *New Approaches to Evaluating Community Initiatives*. Washington, D.C.: Aspen Institute, 1995.

Weiss, C. H. "Theory-Based Evaluation: Past, Present and Future." In D. Rog and D. Fournier (eds.), *Progress and Future Directions in Evaluation: Perspectives on Theory, Practice and Methods*. New Directions for Evaluation, no. 76. San Francisco: Jossey-Bass, 1997.

Weiss, C. H. *Evaluation: Methods for Studying Programs and Policies*. Upper Saddle River, N.J.: Prentice Hall, 1998.

Weiss, C. H. "Theory-Based Evaluation: Theories of Change for Poverty Reduction Programs." In O. Feinstein and R. Picciotto (eds.), *Evaluation and Poverty Reduction*. Washington, D.C.: Operations Evaluation Department, World Bank, 2000.

West, K. P., and others. "Double Blind, Cluster Randomised Trial of Low Dose Supplementation with Vitamin A or Beta Carotene on Mortality Related to Pregnancy in Nepal." *British Medical Journal*, Feb. 27, 1999, 570.

Westhorp, G. S. "Development of Realist Models and Methods for Evaluation in Small-Scale Community Based Settings." Unpublished doctoral dissertation, Nottingham Trent University, 2008.

Westley, F., Zimmerman, B., and Patton, M. Q. *Getting to Maybe: How the World Is Changed.* New York: Random House, 2006.

What Works Clearinghouse. *Procedures and Standards Handbook Version 2.0.* Washington, D.C.: U.S. Department of Education, Dec. 2008. http://ies .ed.gov/ncee/wwc/references/idocviewer/doc.aspx?docid=19&tocid=1.

White, H. *Theory-Based Impact Evaluation: Principles and Practice.* Washington, D.C.: International Initiative for Impact Evaluation, 2009.

Wholey, J. *Evaluation: Promise and Performance.* Washington, D.C.: Urban Institute Press, 1979.

Wholey, J. *Evaluation and Effective Public Management.* New York: Little, Brown, 1983.

Wholey, J. "Evaluability Assessment: Developing Program Theories." In L. Bickman (ed.), *Using Program Theory in Evaluation.* New Directions for Evaluation, no. 33. San Francisco: Jossey-Bass, 1987.

Wholey, J., Nay, J., Scanlon, J., and Schmidt, R. "Evaluation: When Is It Really Needed?" *Evaluation,* 1975, *2,* 2: 89–93.

Williams, B., and Imam, I. (eds.). *Systems Concepts in Evaluation: An Expert Anthology.* Point Reyes, Calif.: Edgepress, 2007.

Williams, K., and Mattson, S. (eds.). *Qualitative Lessons from a Community-Based Violence Prevention Project with Null Findings.* New Directions for Evaluation, no. 110. San Francisco: Jossey-Bass, 2006.

Williams, R., and Hummelbrunner, R. *Systems Concepts in Action: A Practitioner's Toolkit.* Palo Alto, Calif.: Stanford Press, 2010.

Williams, V. "Designing an Evaluation Framework for the Western Australian Aboriginal Justice Agreement." Paper presented at the 2008 International Conference of the Australasian Evaluation Society, Perth, 2008. http://www.aes.asn.au/conferences/2008/presentations/Vivki%20Williams%201455%20-%201525%20Vague%20Clarity.pdf.

Wilson-Grau, R. "Comments for Julia Coffman on Developing Effective Public Policy Strategies." AEA 365 Blog 22, Mar. 2010. http://aea365 .org/blog/?p=303.

W. K. Kellogg Foundation. Logic Model Development Guide. Battle Creek, Mich.: W. K. Kellogg Foundation, 2004. http://www.wkkf.org/knowledge -center/resources/2010/Logic-Model-Development-Guide.aspx.

World Bank. *Maintaining Momentum to 2015? An Impact Evaluation of Interventions to Improve Maternal and Child Health and Nutrition in Bangladesh.* Washington, D.C.: World Bank, 2005.

Worrall, J. *What Evidence in Evidence-Based Medicine? Causality: Metaphysics and Methods.* London: Centre for Philosophy of Natural and Social Sciences, London School of Economics, 2002.

Yampolskaya, S., Nesman, T. M., Hernandez, M., and Koch, D. "Using Concept Mapping to Develop a Logic Model and Articulate a Program Theory: A Case Example." *American Journal of Evaluation,* 2004, *25*(2), 191–207.

Yates, B. "Applying Diffusion Theory: Adoption of Media Literacy Programs in Schools." Paper presented to the Instructional and Developmental Communication Division, International Communication Association Conference, Washington, D.C., 2001.www .westga.edu/~byates/ applying.htm.

Zahariadis, N. "Ambiguity Time and Multiple Streams." In P. A. Sabatier (ed.), *Theories of the Policy Process: Theoretical Lenses on Public Policy.* Boulder, Colo.: Westview Press, 1999.

Zaman, A. "Causal Relations via Econometrics." Paper presented at the South Asian and Far Eastern Meetings of the Econometrics Society, July 2008, Singapore. http://mpra.ub. uni-muenchen.de/10128/.

Zimmerman, B. "Ralph Stacey's Agreement and Certainty Matrix." Edgeware Aides. 2001. http://www.plexusinstitute.org/edgeware/archive/think/main_aides3.html.

Zimmerman, R., and Tomsho, R. "Medical Editor Turns Activist on Drug Trials." *Wall Street Journal,* May 26, 2005, pp. B1–B2.

INDEX